IMAGES OF TIME

Images of Time

Mind, Science, Reality

George Jaroszkiewicz

The University of Nottingham, UK

OXFORD
UNIVERSITY PRESS

OXFORD
UNIVERSITY PRESS

Great Clarendon Street, Oxford, OX2 6DP,
United Kingdom

Oxford University Press is a department of the University of Oxford.
It furthers the University's objective of excellence in research, scholarship,
and education by publishing worldwide. Oxford is a registered trade mark of
Oxford University Press in the UK and in certain other countries

© George Jaroszkiewicz 2016

The moral rights of the author have been asserted

First Edition published in 2016
Impression: 1

Published in the United States of America by Oxford University Press
198 Madison Avenue, New York, NY 10016, United States of America

British Library Cataloguing in Publication Data
Data available

Library of Congress Control Number: 2015949945

ISBN 978–0–19–871806–2

Printed and bound by
CPI Group (UK) Ltd, Croydon, CR0 4YY

Links to third party websites are provided by Oxford in good faith and
for information only. Oxford disclaims any responsibility for the materials
contained in any third party website referenced in this work.

Preface

This book is about time. Specifically, it is a commentary on all of the different mathematical representations of time in physics that I am familiar with. There is an ultimate purpose in giving this commentary: I have my own ideas about time, how it should be interpreted and how it should be discussed. There are many strange conjectures about time, many unwarranted assertions. Most of these I find dubious to say the least. With the great motto of the Royal Society of London in mind, *nullius in verba* (take no one's word for it), I developed a prescription for classifying such conjectures, such assertions. In Chapter 2 I explain this prescription and use it at various places in the book to show what I think of some of these assertions. My guiding principle in all of this is the avoidance of metaphysics, for I believe that time is a physical phenomenon and should be discussed as such.

The subject of time has fascinated me ever since, as a schoolboy, I discovered Isaac Asimov's greatest novel *The End of Eternity* [Asimov, 1955]. The tremendously original ideas in that novel triggered in me a passionate interest in time which has never gone away and, indeed, has intensified during my research career. Asimov went further than many others who wrote about time travel, incorporating into his story immensely suggestive concepts such as *physiotime* and *the inertia of reality*. Physiotime is Asimov's term for the inevitable time of ageing endured by the Eternals, the human observers who stand outside the normal physical universe and try to change the course of history. Asimov postulated that these observers would find that changes that they had made to the fabric of reality at one point in normal spacetime would ripple out with increasing effect, but then these effects would start to fall off and eventually die away, as if reality had some sort of inertia, or resistance to change.[1]

I found both of these concepts inspirational and thought provoking. The concept of physiotime is associated with *observers* whilst the inertia of reality is associated with the phenomenon of *persistence*. Both of these concepts are intimately involved with the meaning and interpretation of time. It is my long held belief that too many discussions in physics do not address either of these points adequately. I think that the time concept is actually synonymous with the observer concept, which I hope to explain in detail in this book.

Not long after I discovered *The End of Eternity* I came across John Dunne's remarkable book *An Experiment with Time* [Dunne, 1934]. There I discovered that whilst there is always a gulf between our intuitive feelings about time and our ability to model those feelings mathematically, there is some value in at least trying

---

[1] There are echoes of these ideas in the section on 'Spreadsheet time travel' in Chapter 20 of this book.

to develop a scientific theory or model of phenomena. Dunne attempted to do this by taking the concept of an *infinite regress* seriously, but he was doomed to failure: there was no empirical content in the model. By this I mean that there was nothing that could be tested in the laboratory to validate the model.

The point is, I learned from this the value of mathematical modelling in place of verbal speculation. Dunne's infinite regress had at least the merit of requiring some precision in the concepts.

Much later still, I encountered G. J. Whitrow's *The Natural Philosophy of Time* [Whitrow, 1980] which also had a significant effect on me. Whitrow's approach was to give as comprehensive a survey of time as he could at the time. I would like to think that the book I have written is a tribute to his book that adds some modern perspectives.

Time is the greatest of all physical enigmas. It appears to be synonymous with change; but then why did the Ancient Greek philosopher Parmenides of Elea believe that nothing changes? Everyone is affected by change, because everything in the physical universe changes. This appears to be a universal rule. Despite centuries of effort, however, we do not completely understand the physics of change yet. We know some of the rules for describing change, but there are still many unanswered questions.

In one form or another, time and by implication change have been involved in all the great questions about existence and reality that humans have ever asked. From the earliest myths of our primitive ancestors about the creation of the universe, past the polytheistic and monotheistic religions of the classical world, through the religious dogmas of the Dark and Mediaeval ages, past the scientific revolution right up to the present day, humanity's search for an understanding of the universe has involved time in one form or another. When all is said and done, time will probably remain incomprehensible. However, that should not deter us from attempting to discuss time and its perceived properties.

The adjective *perceived* seems advisable here. We do not know what time really is. Is it a thing, a process, a metaphor, or what? Surprisingly, despite not knowing what time is, it seems we can manipulate it to some extent. For example, special relativity says that it is possible to slow down the rate of passage of time of physical objects by increasing their speed relative to us. This phenomenon is known as time dilation. One day, it may be time dilation which allows humans to colonize the furthest corners of the universe, far beyond our solar system or galaxy.

It is remarkable that humans are able to understand some of the properties of time through the power of mathematics. Some parts of this book will appear mathematical to some readers. I would encourage such readers to be patient, particularly with themselves. The subject of time requires contemplation, patience, intuition, and occasionally, some calculation. Sometimes, a quick sketch is all that is required to understand a point.

Humans seem privileged above other animals in having a more developed capacity to appreciate time. This capacity seems linked with our particular form of consciousness. This is not an assertion of human superiority over other species but merely an observation about the world. Human consciousness is a precarious thing. It can be degraded relatively easily and then our sense of time can be distorted. When we are tired or excited, or affected by chemicals, time can seem to flow slower or faster. When we are asleep, time seems not to pass for us at all and our level of consciousness seems to sink to less than that of an inactive woodlouse.

When we are awake, most of us have the ability to sense the passage of time, a strong feeling that time exists as a dynamical phenomenon, that it 'flows'. We don't always have this feeling at the forefront of our thoughts of course, but when circumstances dictate, we become aware of time, through perhaps a sense of urgency about some deadline. However, precious little can be done with this feeling. Most readers will be sophisticated enough to understand that this notion is itself circular. What is this flow measured relative to?

In current Western culture, there are many speculative concepts that have made their way into the popular imagination. Antigravity and time travel are common themes in science fiction, but current technology does not at this time allow such possibilities to be realized. Most likely, if we have the usual well-read background, we will have been seduced by the popular notion of time as the fourth dimension, based on the assertion that time is like space, a dimension. Unfortunately, that is a just a mathematical model, and there are some good reasons for regarding this notion with scepticism. For instance, we cannot travel in time in the way we can travel in space. We have the freedom to change our position is space, but we can do virtually nothing about our position in time. We cannot at this time even be sure that 'time travel' in the ordinary sense of the phrase is a physically meaningful concept.

Time seems to most people to have a flow, a direction that cannot be reversed. 'Nobody can dip his feet twice in the same river', according to Heraclitus. We go through life once and that appears to be it. We can never go back to the past. One of the illusions that constantly conditions our thoughts and beliefs is that places we have lived in are always there, unchanged, so that when we visit them we feel we are returning to the past. As I see it, the best view of reality is that such feelings are illusions based on a phenomenon we shall discuss in this book which we call *persistence*. Places we think we knew in the past and are revisiting are quite different parts of spacetime to those we once lived in, but there may be sufficient similarity between how they are now and our memories that we are deceived into thinking that there have been no changes. So when we move into those new parts of spacetime, we are deceived into believing that we are revisited the past. Nothing could be further from the truth. The old adage 'you can never go back' represents such a fundamental truth that it should perhaps be called the first law of time.

One of the great mysteries about time is that many of the conventional laws of physics do not appear to take this asymmetry in the direction of time into account. When we solve Maxwell's equations for the electromagnetic radiation emitted by an accelerating charge, we find we have to put in by hand the notion that radiation flows outwards if we want to make sense of time and causality. We have to exclude the so-called 'advanced' solutions by fiat. An interesting exception to this idea was once discussed in a famous paper by the American theorists John Archibold Wheeler and Richard Feynman [Wheeler & Feynman, 1945]. However, Feynman subsequently renounced this paper.

An important issue is that contemporary science cannot accommodate or explain why we have the subjective feeling of this temporal flow. Because of their faith in science, this leads many physicists to subscribe to the view that this flow is an illusion. But it must be said that there have been scientists such as Prigogine who disagree, for the very good reason that our sense of time is an empirical fact and not something that can be argued away.

No one currently understands time in the way that they understand, for example, the structure of atoms or the bits and pieces that make up a car. In such cases, scientists formulate their explanations in terms of simpler components such as electrons and nuclei in the case of atoms, and various bits of metal and plastic in the case of cars. We are still experimenting with many speculative mathematical models of what time might be, but it could easily be the case that we are wide of the mark by a long way. They are most probably all wrong or too simplistic. Just let's put ourselves in the place of nineteenth-century physicists trying to modify the continuum fluid equations of hydrodynamics to incorporate the notion that water is really made of atoms. Without any knowledge of quantum mechanics, we would almost certainly find ourselves making very crude and erroneous models of a phenomenon that was vastly more complex than anything we could ever imagine at that time. Likewise, our current ideas about discrete time, for example, are equally speculative and simplistic.

Nevertheless, it is possible to have a handle on time, via mathematical models, and through them to manipulate time in the laboratory to a certain extent, as relativistic time dilation shows.

I give two cautions. First, time is an ideal subject for philosophical and metaphysical speculation. That is definitely not what this book is about. My interest here is solely in asking the question 'what might the best model of time in the physical sciences be?' We shall exclude purely speculative, philosophical lines of enquiry, but not natural philosophical lines of enquiry. For us the difference will be that the former does not generally use mathematics whilst the later invariably does. Moreover, there should be some notion that a theory should have testable predictions. Unfortunately, this last proviso is difficult to insist on with most mathematical models of time, given the current level of technology.

The other caution concerns this use of mathematics itself. Hopefully, this book will be of interest to a diversity of readers, because time after all affects every one of us. I have tried to make this book readable and accessible, but there are some places where some mathematics is required, as in any subject trying to understand the physical universe. For this reason, I include at the back of the book an *Appendix*, in which I review some of the mathematical technologies used in some of my topics.

Throughout this book I make extensive use of acronyms, for which I ask the reader's understanding and forbearance. It is easier to write SUO a hundred times rather than 'system under observation', although I am aware the effect may make the pattern of letters on some pages resemble a message being decoded at Bletchley in 1941. Generally, every acronym used in a given chapter will be defined the first time it occurs in that chapter.

Throughout this book there is frequent mention of *space-time* and *spacetime*. The former refers to the Newtonian paradigm, in which time and space are regarded as separate, whilst the latter refers to the Minkowski paradigm, in which space and time are part of a unified, four-dimensional continuum.

I take this opportunity to acknowledge and thank the various people who have stimulated and influenced me in my thoughts about time. In particular, I mention three of the most important influences: Nicholas Kemmer's lectures on electromagnetic fields introduced me to questions concerning causality and wave propagation, Rufat Mir-Kasimov introduced me to Hartland Snyder's theory of quantized space-time [Snyder, 1947a,b], and Lino Buccheri introduced me to the debate between endophysics and exophysics.

I take now this opportunity to thank all members of my family, past and present, who have helped me in one form or another. In particular, I thank my wife Małgorzata for her endless patience and my daughter Joanna and her husband David, who gave me a neat definition of life.

Finally, I would say that everyone has some vision of what time is. The past may be inaccessible, but memory remains: I am immeasurably grateful to my parents for the many happy memories that they left me.

George Jaroszkiewicz

Contents

1

Introduction

Plan of this book

This book is about theories of time, not about time itself, because no one really knows what time is. There is no consensus as to whether time is a *thing* or a *process*. Time is far too complex and vague a concept to be understood from any single perspective, so it is not surprising that there have been many different views about it throughout history. Some of these views have evolved within particular cultures and are associated with specific religious and mythological belief structures current in those cultures. Other views about time were developed by free thinkers such as the medieval cleric Nicholas of Autrecourt, who challenged the conventional view that time and matter are continuous.[2]

Whatever their origins, however, all of these views share a common feature: they are human perspectives on time. Since we can give no single account of time as it relates to our own species, it stands to reason that we should not expect to have an account of time that is meaningful to other species. Social insects such as soldier ants will behave instinctively as if time had importance only in terms of the survival of their colony, whilst humans view time in terms of their personal survival. If we humans ever encounter other intelligences in our galaxy, there is no certainty that both species will share the same view of time.

This book divides naturally into three themes, each having its particular focus. The first theme, *Concepts of Time*, consists of Chapters 1 to 9 and addresses the diversity of perspectives on time. Whilst none of these can adequately capture the complete essence of what time is, many of them do have some flavour of truth or value, so it is useful to review those with which we are most familiar. Some are scientific, meaning that some aspects can be empirically validated, such as time dilation. Most perspectives on time, however, are metaphysical and have no empirical content *per se*. The general purpose of this book is to guide the reader towards scientific, mathematically based images of time. Our aim in this theme is to alert the reader to the plethora of images of time that have no scientific foundation.

---

[2] He was condemned to burn his writings, which he did in 1347.

Images of Time. First Edition. George Jaroszkiewicz.
© George Jaroszkiewicz 2016. Published in 2016 by Oxford University Press.

The second theme, *Classical Time*, consists of Chapters 10 to 23 and deals with the mathematical structures employed by mathematicians and scientists to describe time before the advent of quantum mechanics (QM). This includes special relativity (SR) and general relativity (GR) which, although developed after Planck's introduction of quanta in 1900, were based on a classical perspective of time. In the third theme, *Quantum Time*, from Chapters 24 to Appendix, we take into account the fundamental shift in perspective that QM forces on us.

The Appendix consists of a chapter reviewing basic mathematical structures, and this is followed by a bibliography, and an index.

What is time?

Historical records such as Zeno's paradoxes reveal that humans have worried about the nature of time for thousands of years, but the mystery remains. To quote St Augustine:

> What then is time? If no one asks me, I know what it is. If I wish to explain it to him who asks, I do not know.   [Augustine, 398]

The study of time is hampered by the difficulty of defining precisely what time means. Is time something intrinsic to the universe and independent of any observer, or is it a way for an observer to organize a description of the observed universe? To illustrate how much even scientists can disagree, consider cosmology. Just over 50 years ago, there were two competing, fundamentally incompatible views of time in cosmology. The Steady State model dared to imagine a universe that had existed indefinitely into the past, whereas the Big Bang model postulates that time had a beginning.

The subject seems impossible. Perhaps we should start slowly and just write down any sort of half-reasonable idea. Let us do that now. Numerous possibilities spring to mind and a reasonably long list can be assembled in minutes. For example, we could speculate that

1. Time is a measure of change in one system compared to another system.
2. Time is a measure of change in the universe.
3. Time is a relationship between different states of the same system.
4. Time is a measure of disorder or entropy.
5. Time is a measure of acquired information.
6. Time is what prevents everything happening at once.
7. Time is a numerical count of a fundamental unit of time known as the *chronon*.

8. Time is a standard of change agreed between observers, serving to correlate individual experiences and providing a more objective temporal reference frame than any one individual's estimate of time.

9. Time is a process.

10. Time is memory.

11. Time is one of the four dimensions of four-dimensional spacetime.

Each of these statements has some merit in some context, but every one contains implicit assumptions which, when identified, make that statement appear unsatisfactory. For instance, to assert that time is a measure of change begs the question: *who or what is doing the measuring, how do they measure it, what precisely is being measured, and what will be done with that information?* If we want to understand better what time is, we really should identify the hidden assumptions we would make were we to adopt any of the above views of time. Chapter 2 gives some guidelines in the identification of these assumptions and on that account is essential preparatory reading before the rest of the book is read.

Paradigms

Time is fundamental to humans. It structures their lives, their plans, and their thoughts. It has always fascinated people. In particular, the beginning of time and the origin of the universe have been the setting for many ancient mythologies. The one that springs immediately to mind is that of the Ancient Greeks. In their creation mythology, *Chronos* was the personification of time, being involved in the creation of the universe (Chronos should not be confused with the Titan *Cronus*, who was the father of the gods in Greek mythology). Where Chronos came from seems to be an obvious question without any answer.

In modern times, interest in time has if anything become an obsession. We are bombarded frequently with speculative films and stories involving time travel, alternative universes, and such like. This is not surprising given our interest in science and science fiction. What helps legitimize such speculation is the fact that GR, Einstein's theory of space and time, a greatly respectable and good theory of large-scale gravitation, does not exclude the exotic possibility of spacetimes containing *closed timelike curves*, or CTCs as they are generally referred to. A timelike curve is a path in spacetime that a physical observer could follow without violating any of the known laws of classical physics.[3] If such a curve were closed, then an observer could in principle travel along such a curve to return to their past, that is, travel in time.

---

[3] What happens along a CTC when quantum mechanics is involved is still the subject of investigation.

We have two aims in this book. First, we want to describe and review various facets of the time concept, particularly focusing on those encountered in various mathematical physical theories about the universe. Second, and more importantly, we want to advance a particular view of time: that all concepts in physics are contextual. It is our assertion that time should not be discussed as an absolute thing or in isolation but only in terms of processes of observation.

Although it seems unscientific, we cannot escape the thought that the concept of time is intimately involved in how *humans* think. Various observations lead us to this view. For instance, animals do not seem at all concerned about the remote past and their interest in the long-term future is generally instinctive and centred on survival strategies: many animals gather and store food during summer in preparation for the coming winter. Certainly, dogs can remember where they buried a bone and can plan strategies to get around obstacles, but they do not hold services of remembrance or enrol onto three-year university degrees. They are very much creatures of the present, the enigmatic 'moment of the now'. Humans, on the other hand, often seem overly preoccupied about the very distant past, such as what really happened on the day Julius Caesar was assassinated. This is undoubtedly directly connected with memory, the transmission in time of quasi persistent patterns in complex molecular structures in our brains. We should not make the mistake of thinking that these patterns are necessarily faithful three-dimensional model images of what we observe visually. As with the compact discs that are used to play back music and films, information can be stored in many ways and it is the decoding and interpretation of it that matters.

In addition to using memory to review the past, humans have the ability to make extraordinarily complex *plans* for the future. Indeed, anyone who does not make plans for their next meal, their careers, or their retirement, is unusual. Plans can be thought of as maps of proposed, possible, or potential futures. As far as we know, no matter how many alternative plans for tomorrow we make, only one of them at most can ever materialize as reality. However, some quantum theorists disagree. We shall discuss that point later on in this book.

Human interest in the past or future is often taken to extremes and turned into entertainment in the form of historically based films and novels depicting events that are known never to have happened and, in the case of science fiction, stories about possible futures based on fictitious laws of physics.

We can be sure that animals other than humans make plans: we have only to witness a pride of lions organizing an attack on a herd of gazelle to see sophisticated planning in operation. We may dismiss such planning as merely instinctive, but we should not forget that we humans are animals too, albeit with more complex mental faculties. We may believe that we have free will, but neurological evidence suggests that humans are much more under the control of their subconscious processes than they ever imagine or care to admit [Halligan & Oakley, 2000].

The development of religion and science required an awareness of time. Both systems of thought are attempts to understand humanity's place in the physical universe, although each system is based on very different standards of logic and evidence. It is difficult to believe that any organism lacking some concept of the past would develop either religion or science. As far as we know, non-human animals do not have religious beliefs or interest in science.

The conventional dictionary meaning of the word *paradigm* defines it as *an example* or *model*, particularly in semantics. In recent usage it has come to mean something deeper and specific, particularly as far as science is concerned. In this book, we shall use this word in this latter sense. It will denote a relatively complete and distinct mental attitude or perspective held by a theorist with associated belief structures concerning some aspect of the universe. When a person thinks according to a specific paradigm, their thoughts and beliefs tend to be channelled into certain directions whilst other directions are excluded because they appear inconsistent with that given paradigm. For example, a person who believes in any particular religion adopts the paradigm provided by that religion and explicitly rejects other religions. Many events and phenomena encountered by that person are then interpreted according to those beliefs. A paradigm is essentially a mechanism for the interpretation of phenomena.

Different paradigms may be mutually inconsistent. For example, the flat Earth paradigm is inconsistent with the spherical Earth paradigm. Sometimes apparently exclusive paradigms are tolerated by individuals despite apparent inconsistencies. This may occur because of personal history, or for convenience.

Science too has its paradigms. For example, scientists may choose to believe in classical mechanics or in quantum mechanics, but they would find it difficult to believe in both simultaneously, although that is not impossible to contemplate. The study of time in particular encounters numerous different and exclusive paradigms, and so it will be necessary for the reader to be ready to switch mental frames of reference whenever necessary.

It is generally accepted that the role of experimental science is to provide objective data, information about the universe that is independent of any observer's belief structures. This is the basis of the scientific motto of the Royal Society of London: '*nullius in verba*', which translates as 'take no one's word for it'. This motto encapsulates a core principle in science. But there is a subtlety here that impinges on all discussions of time and which we should keep in mind: any discussion of any observed process, such as an object falling under gravity, requires us to state who or what is observing it, how they are observing it, and how they are interpreting what they observe. To think otherwise, to believe that things can happen without observers looking at them, is to think classically and metaphysically. The physicist Wheeler went so far as to state the *participatory principle*, which asserts that the only statements that mean anything are our observations. This principle is of critical importance to us in this book: it is our belief that the time concept can be discussed meaningfully only in terms of observers. Essentially, we will emphasize that time is a process and not a 'thing'.

Events in time

There is one feature that is common to all discussions about time and that is its *ordering* property. Before this can be explained, however, the notion of an *event* has to be pinned down. An event is anything that happens over such a brief interval of time and within such a restricted part of space that it can be reasonably modelled as having occurred at a single moment in time and at a single place in space.

The concept of an event is a simplification in general because most physical objects are extended in space and most physical phenomena occur over extended intervals of time. Moreover, any attempt to detect any process involves apparatus which is non-local in space and operates over an extended interval of time. Inevitably, any attempt to pin down a precise time and location for an event will be liable to errors and uncertainties.

Although our definition of an event makes reference to physical objects and phenomena, it could be argued that perhaps that is a mistake. Instead, we could think of events as points in *spacetime* that are independent of whether any physical object (that is, something carrying energy and momentum) was actually associated with them. In this latter view, spacetime is analogous to a piece of amber, with events analogous to insects embedded in it.

This shows that even the simplest notion about time does not come without a context: it will be accompanied by a retinue of hidden assumptions about the nature of reality. For instance, our first notion of event above focused on physical objects (whatever that may mean). It does not necessarily imply that space and time exist by themselves. The second notion, on the other hand, requires us to imagine spacetime as a separate entity (the amber), independent of any matter (the insects) in it.

These two viewpoints have their respective merits and pitfalls. The physicist Ernst Mach was an advocate of the first point of view. For him, space did not exist: he regarded it as a manifestation of the distance relationships between all the physical particles in the universe:

> The physical space I have in mind (which already includes time) is therefore nothing but the dependence of the phenomena on one another. A complete physics that knew of this dependence would have no need of separate concepts of space and time because these would already have been encompassed. [Mach, 1866]

It is frequently said that although Einstein was influenced by Mach, when Einstein constructed GR, he assumed spacetime has its own physical properties, such as a distance or metrical structure, independent of matter. In 1920, in an address delivered at the University of Leyden, Einstein said that space was

> endowed with physical quantities, but that this ether may not be thought of as endowed with the quality characteristic of ponderable media . . . The idea of motion may not be applied to it. [Einstein, 1920]

That this metric structure may be influenced by matter is beside the point here. In SR and GR, spacetime is regarded as a meaningful concept independent of matter.

The point of view that empty space (the *vacuum*) has some independent physical significance is assumed by the formalism of quantum field theory, which views the vacuum as simply a particular state of reality. In this theory, material particles are subordinate to the vacuum, being interpreted as no more than excited states of it.

Temporal ordering and time labels

We need two or more events to define time. Suppose we have observed two events A and B. Then according to our own sense of time we will generally feel entitled to say that one of three possible mutually exclusive temporal relationships exists between these events. We may say that either A is *earlier* than B, or A is *later* than B, or that A and B are *simultaneous*, that is, occur at the same time.

Sooner or later, we will need to write down our observations. This will require us to agree on some notation. Suppose we decide on the following minimal scheme: if event A has occurred before B we will write $A < B$, whilst if A has occurred later than B we write $A > B$. According to this convention, $A < B$ means the same thing as $B > A$. If the events are simultaneous, we write $A = B$. At this point, we have not introduced real numbers at all: the symbols $<$ and $>$ are just notation for *earlier than* and *later than*.

There is one problem with the above notation. The simultaneity of two events does not mean that they are the same event: $A = B$ is misleading notation for simultaneity. To get around this, we introduce the concept of an *ordered set*, discussed in Chapter 7. A well-known set with these properties is $\mathbb{R}$, the set of real numbers. It is the ordered set most commonly used to model time. The procedure is to associate to each event A a unique real number t_A, called the *time of that event*. The notions of later, earlier, and simultaneous between events are then encoded into the various possible ordering relationships between real numbers. For example, if event A is later than event B we write $t_A > t_B$ whilst if A and B are simultaneous we write $t_A = t_B$. This latter equality then does not mean that $A = B$ in the mathematical sense of equality [Howson, 1972].

The contextuality of temporal ordering

We come now to one of the important points we are trying to emphasize in this book: the *contextuality of temporal ordering*. In the previous section we discussed the temporal ordering between two events as if there was something intrinsic about that, as if it was a property of those two events. This is what the Newtonian vision of Absolute Time discussed in Chapter 14 asks us to believe.

That view of time was undermined by SR, because it is possible in SR for two different observers to have opposite views of the temporal ordering between two events. Specifically, if events *A* and *B* are *relatively spacelike*, then there exist some observers for whom *A* appears *before B*, some observers for whom *A* appears *simultaneous with B*, and other observers for whom *A* appears *after B*. The non-uniqueness of temporal ordering for some pairs of events in SR is known as the loss of absolute simultaneity.[4]

Many years after the development of SR, theorists such as Tangherlini realized that the loss of absolute simultaneity in SR could be attributed to the particular *synchronization protocol* employed by Einstein and others in their explanation of the results of the Michelson–Morley experiment, discussed in Chapter 16. In Chapter 18 we discuss Tangherlini's approach, one in which the physics of SR is reproduced without the loss of absolute simultaneity [Tangherlini, 1958].

Arrows of time

The term 'arrow of time' expresses the idea that a particular process carries some sort of ordering or one-way directionality identified with a direction of time:

> Without any mystic appeal to consciousness it is possible to find a direction of time on the four-dimensional map by a study of organization. Let us draw an arrow arbitrarily. If as we follow the arrow we find more and more of the random element in the state of the world, then the arrow is pointing towards the future; if the random element decreases the arrow points towards the past. That is the only distinction known to physics. This follows at once if our fundamental contention is admitted that the introduction of randomness is the only thing which cannot be undone.
> [Eddington, 1929]

There are many arrows of time, each associated with a specific kind of physical or conceptual process. Here is a list of some important arrows of time:

The thermodynamic arrow of time

Entropy is a measure of the relative disorder of a system under observation (SUO). The second law of thermodynamics states that the entropy of an isolated SUO tends to increase with time: this defines the *thermodynamic arrow of time*. An obvious problem is that the concept of an *isolated SUO* is logically inconsistent: any observation requires an interaction between the SUO and the observer. There are at least four interrelated but inequivalent definitions of entropy. This particular arrow of time concept therefore requires a careful statement as to what constitutes an 'isolated system'.

---

[4] There will be some pairs of events in SR such that all standard observers agree on the ordering.

The cosmological arrow of time

By 1927, the Belgian cosmologist Georges Lemaître had applied Einstein's GR to cosmology, proposing that the universe was expanding [Lemaître, 1927]. Two years later, the American observer Edwin Hubble published data supporting this hypothesis [Hubble, 1929]. Since then, the 'Big Bang' model of the universe has been validated empirically on a number of fronts, notably the measurement in 1967 of the temperature of the cosmic background radiation field [Penzias & Wilson, 1967].

Assuming the expansion interpretation of the data is valid, this then defines the *cosmological arrow of time*: future states of the universe are going to be very different to those close to the Big Bang. A notable empirical fact is that all the empirically based arrows of time are consistent with the cosmological arrow of time, in that none of them has been found to switch direction relative to that arrow.

The radiative arrow of time

When we switch on a torch at night, the beam of light spreads out into the space around us. When this light encounters some nearby object, it may be reflected back to us, as in the case of a mirror, or stimulate that object to emit its own characteristic light, which we then observe. That is what a torch is for. The light that we emit is never observed to return to us completely: some or all of it is inevitably lost irreversibly as far as we are concerned. This asymmetry between what goes out and what comes back defines an arrow of time known as the *radiative arrow of time*.

The psychological arrow of time

There are several notable characteristics of the human experience of time. First, humans generally describe time as a continuous process: time seems to 'flow'. This in no way proves that time is continuous, however. After all, humans describe liquids as continuous, but liquids are discontinuous on atomic and molecular scales. The correct interpretation is that *time is continuous as far as our perceptions are concerned*. The perceived continuity of time must be the result of complex processes in the brain that do not have the capacity to resolve any granularity or discreteness in time. The human sense of time is almost certainly an effective view of an immeasurably more complex reality, in much the same way as our sensation of temperature is.

Another characteristic of human perception is the direction of time: we think of the unformed future as lying ahead of us whilst the definite past is behind us. But this is a culturally conditioned perspective that is the other way around in some cultures. For example, the Aymara language of South America associates the future with what is behind us and the past with what is in front of us. A similar perspective is found in the traditional Chinese view of time likened to what a person rowing a boat sees: they have their back to the direction of motion, so that they can see what has just passed by (the past) but cannot see what is coming (the future).

The relationship between memory and the psychological arrow of time has been discussed by Mlodinlow and Brun using a model incorporating the direction of entropy increase [Mlodinow & Brun, 2014]. They showed that in their model, the psychological arrow of time was aligned with the thermodynamic arrow of time.

The biological arrow of time

Life is a process that seems to defy the laws of thermodynamics, in that reproduction appears to prevent the dissipation of a particular species. Although individual organisms age and eventually die in a temporal direction aligned with the expansion of the universe, the paradox is the appearance of younger versions of the species. The resolution of this paradox is resolved by the same principles that allow refrigerators to cool down objects: the laws of thermodynamics do not apply to components of a system but to complete systems. For example, when an object of temperature T_1 is placed in thermal contact with another object of temperature $T_2 < T_1$, the combined system tends to an equilibrium temperature T_3 such that $T_2 < T_3 < T_1$. If we looked only at the second object, it would appear to have increased its temperature.

The evolutionary arrow of time

Recent fossil analysis has supported the hypothesis that human psychology evolved more recently than human physiology [Brown *et al.*, 2012]. It is believed that Stone Age man fashioned shards of stone, known as microliths, into arrow heads. The new evidence suggests that the earliest microliths were fashioned 71,000 years ago, whereas our species, *Homo sapiens*, is between 150,000 and 200,000 years old. This suggests that the human mind evolved into its near modern form relatively late, crossing the threshold beyond which humanity could develop in the form we have today. Moreover, arrowheads suggest an advanced conception of weaponry unique to humans.

The geological arrow of time

The relationship between the evolution of man and the geological processes shaping the planet is complex and involves an interplay between two contrasting images of time, namely the linear (arrow-like) model and the cyclic model. Over certain time scales, the familiar patterns of cyclic time impinge most obviously on humans, such as the daily rising and setting of the Sun, the monthly lunar cycle, and the year long orbital period of the planet around the Sun. In contrast, the timescales for real geological changes, such as the formation of mountain ranges and the drift of continents, takes place over millions of years and in an irreversible way.

This contrast led to different researchers discussing these timescales with contrasting metaphors. In his influential book, *Time's Arrow, Time's Cycle*, the science writer Stephen J. Gould analysed the mythologies invoked by various researchers in their accounts of such phenomena. He wrote

> Deep time is so alien that we can really only comprehend it as a metaphor. And so we do in all our pedagogy. We tout the geologic mile (with human history occupying the last few inches); or the cosmic calendar (with Homo Sapiens appearing but a few moments before "Auld Lang Syne") . . . John McPhee has provided the most striking metaphor of all (in Basin and Range): Consider the earth's history as the old measure of the English yard, the distance from the king's nose to the tip of his outstretched hand. One stroke of a nail file on his middle finger erases human history.
> [Gould, 1987]

Here, '*deep time*' is a reference to the relatively enormous timescale associated with the Earth's geology, measured in units of billion years, compared to human evolutionary scales, which are measured in units of tens of thousands of years for the development of civilization and a few million years for the emergence of our species.

The causal arrow of time

The sequence *cause* followed by *event* was discussed even in Antiquity. Aristotle is famous for listing four types of causes: *material cause, formal cause, efficient cause,* and *final cause*. His view of phenomena and their relationship to time can be described as *teleological,* or driven by purpose in some way. This view is particularly evident in his discussion of mechanics. For instance, Aristotle would say that clouds float above the earth because that is their natural position: if they found themselves at a different level, they would seek to return to their natural place. For Aristotle, therefore, it would be necessary to invoke some agency that had some form of intent to initiate movements. To this end, he invoked the concept of final cause. We shall return to this theme when we discuss our concept of *primary observer* in the next chapter. The Scottish philosopher David Hume was remarkable in his vision of causality, one that we can fully agree with. He linked causality with the way that we as observers interpret events [Hume, 1739].

The weak arrow of time

In particle physics, charge conjugation (C), spatial inversion (P), and time reversal (T) are symmetry operations that can be made on states of SUOs. The CPT theorem asserts that the laws of physics are invariant to (unchanged by) the combined transformations acting on a state. It has been observed empirically that the laws of physics are not completely invariant to the partial operation CP in certain very subtle interactions [Christenson *et al.*, 1964]. Therefore, if the CPT theorem is valid, the laws of physics are not completely invariant to time reversal. This therefore signals another arrow of time, known as the *weak arrow of time*. This exotic corner of physics may yet turn out to be of the greatest significance to our understanding of time and the universe, as it has been speculated to be behind the longstanding mystery of the observed enormous excess of matter over antimatter in our part of the universe.

The quantum arrow of time

There are two contrasting forms of time evolution in QM, known as *unitary* evolution and *non-unitary* evolution respectively. Unitary evolution describes how wavefunctions change in time if no information is being extracted by an observer. This form of evolution is theoretically reversible. Non-unitary evolution arises whenever an observer extracts empirical information: this form of evolution is irreversible because when an observation is made of some quantum outcome in an experiment, the corresponding wavefunction changes. In addition, the observer registers this information in their memory in one way or another and this is an irreversible process, defining the *quantum arrow of time*. Magnetic resonance imaging (MRI) is an increasingly important technique in chemistry, physics, and medicine that exploits both forms of quantum evolution to extract information about the internal structure of objects such as crystals and living tissue.

The mathematical arrow of time

This arrow is associated with the directionality explicit in the standard definition of a function [Howson, 1972], a rule f that maps elements of a set A into a set B. Mathematicians denote this by $f : A \to B$. The definition is not symmetrical between A and B and this asymmetry defines the mathematical arrow of time referred to. Sets and functions are discussed in more detail in Chapter 7.

The Bayesian (probability) arrow of time

According to the Bayesian approach to probability, probability is not an attribute of a state of an SUO alone but conditional on the information held by the observer. As more information is received, the outcome probabilities are readjusted according to a specific formula known as Bayes' theorem. If the observer has a perfect memory, then accumulation of information provides an irreversible process that can serve as an arrow of time.

The spreadsheet arrow of time

As we will discuss in Chapter 20, a spreadsheet program such as EXCEL can be used to model a one-spatial dimensional SUO evolving according to a mathematical algorithm. If successive states of the SUO are described by data in successive columns going to the right with increasing time, then the calculations induce a left–right temporal asymmetry. Any attempt to read data 'in the future' in order to calculate the dynamics at a given instant of time results in the program halting, with the appearance of a schematic arrowed loop indicating an inconsistent flow of data.

The sociological arrow of time

Although progress is never certain, undisturbed human societies generally evolve over time from primitive forms to more civilized and technologically advanced forms, the direction of this evolution matching all the other arrows of time. For instance, archaeologists expect to find that lower layers in a dig contain more primitive pottery than upper layers.

The narrative, diachronic arrow of time

Time affects everything, particularly the individual. When it comes to any given person, there is a conventional view of them that is predicated on a universal arrow, pointing from their birth to their death: a person's life is marked out as a progression, following a path through childhood, education, employment, retirement, and finally senility. That has been described as a *diachronic* perspective, a view of a life as a career line. Not everyone subscribes to that paradigm, however. A radical alternative is to view life in terms of happenings or events that need not be linearly or causally related. This is the so-called *episodic* perspective.

The two perspectives can be reconciled in our view, in much the same way that we can understand how $M31$, the Andromeda galaxy, can be moving towards us, despite the backdrop of the expansion of the universe. That expansion is a large-scale, statistical description of an inexorable outwards motion of all the galaxies, whilst the motion of $M31$ is a random local variation. So in general a person follows a linear path on average through life but can take occasional random detours, none of which will deflect that person from their inevitable destination.

Arrows versus cycles

A recurring theme throughout this book is a clash of temporal imagery. On the one hand time will be described in terms of a directed line, or line with an arrow, whilst on the other hand, phenomena may be best described in terms of cycles or repetitions of a basic pattern. It is not clear which image is correct. For example, although the universe is widely believed to have started at the Big Bang, there is no evidence against the idea that what we are observing is just one of an infinite number of cycles of expansion followed by contraction followed by expansion, and so on. Because this discussion can easily lead to endless metaphysical debate, we formulate in the next chapter a way of assessing any particular discussion of time, so that at least we can say that is it metaphysics (meaning it cannot be proved or disproved) or physical (meaning that we can think of ways of validating the concepts concerned). Given the nature of time, that may well be the best that we can hope for.

2

Observers and time

Introduction

Time is often referred to in conversation in a way that tells us more about the speaker than about time itself: we frequently hear expressions such as '*I have no time right now*', '*the time has flown*', or '*time stood still*'. This may seem reasonable to us because time is part of the human experience of the physical universe and therefore it should be discussed in terms that relate to us. We should however take heed of what Heisenberg wrote in 1930:

> one should particularly remember that the human language permits the construction of sentences which do not involve any consequences and which therefore have no content at all - in spite of the fact that these sentences produce some kind of picture in our imagination; e.g., the statement that besides our world there exists another world, with which any connection is impossible in principle, does not lead to any experimental consequences, but does produce a kind of picture in the mind. Obviously such a statement can neither be proved nor disproved. One should be especially careful in using the words "reality", "actually", etc., since these words very often lead to statements of the type just mentioned. [Heisenberg, 1930]

In this chapter we pay attention to Heisenberg's warning. First we review the concepts of metaphysics, observers, and levels of observation. This will help us focus on physics rather than metaphysics. Then we present a method for analysing statements about time that allows us to decide whether they are physically relevant or metaphysical. Without some such analysis we risk being flooded with a host of unsupported and unverifiable assertions about time, just as Heisenberg warned.

Metaphysics, physics, and validation

We shall use the words *proposition* and *statement* to mean the same thing: a string of words that carries a meaning. A proposition is not always *true* or *false*: sometimes we have no means of validating a given proposition, that is, establishing its truth value. We define a *metaphysical statement* as one that cannot be empirically

Images of Time. First Edition. George Jaroszkiewicz.
© George Jaroszkiewicz 2016. Published in 2016 by Oxford University Press.

validated. An example is the statement 'God exists'. On the other hand, we define a *physical statement* as one that can be empirically validated.

We have no need to make any value judgement about the respective merits of metaphysics and physics: we are interested in the latter and not at all in the former.

Whilst we advise that scientists should avoid metaphysics, there are two circumstances where it is reasonable for them to make metaphysical statements.

Advances in technology

Science marches behind technology, advances of which can turn yesterday's metaphysical speculation into today's good physics. Atomic theory provides a good example of this phenomenon. Dating from Antiquity, the Fourier–Tolman principle of similitude [Tolman, 1914] asserts that the laws of physics are independent of scale. This principle is based on the erroneous belief that matter is continuous, a natural but false assumption used by Aristotle and others to argue for the continuity of time:

> ... but in respect of size there is no minimum; for every line is divided ad infinitum. Hence it is so with time.   [Aristotle, 1930]

In the nineteenth century, speculations by Dalton, Avogadro, Boltzmann, and others on the existence of atoms, and by implication of a fundamental scale, began to undermine the principle of similitude. Although atoms had been conjectured in Antiquity by Leucippus and Democritus, the atomic hypothesis remained on the borderline between physics and metaphysics up to the early years of the twentieth century, until the empirical work of Perrin validated Einstein's theory of Brownian motion. Not everyone accepted the atomic hypothesis even then, notably the physicist Ernst Mach.

Theoretical convenience

Physicists may use metaphysical concepts in support of theory but should not interpret them as more than that. Useful metaphysical concepts are wavefunctions, particles, and even time and space:

> The mathematical machinery of quantum mechanics is a symbolic expression of the laws of atomic measurement, abstracted from the specific properties of individual techniques of measurement. In particular, the space-time manifold that is the background of any quantum-mechanical description is an idealization of the function of a measurement apparatus to define a macroscopic frame of reference.
>
> [Schwinger, 1958]

The danger here is that overuse of such concepts can condition the unwary scientist to believe that such idealizations are actually 'there', that is, that they exist in some physical sense or are real properties of systems under observation

(SUOs). We should keep in mind for instance that the centre of mass of a ring is a useful concept but *has no material existence whatsoever.*

Observers and the physical universe

Metaphysics has no need of the observer concept because, by definition, a metaphysical statement cannot be empirically validated. In contrast, physics needs observers because, by definition, a physical statement is one that can be empirically validated or disproved. According to Wheeler, only observers are significant in science:

> Stronger than the anthropic principle is what I might call the participatory principle. According to it we could not even imagine a universe that did not somewhere and for some stretch of time contain observers because the very building materials of the universe are these acts of observer-participancy. You wouldn't have the stuff out of which to build the universe otherwise. This participatory principle takes for its foundation the absolutely central point of the quantum: No elementary phenomenon is a phenomenon until it is an observed (or registered) phenomenon.
>
> [Wheeler, 1979]

According to Wheeler's participatory principle, the relationship between observers and physical reality is contextual, that is, it depends on the theory and apparatus involved. This is seen clearly in the differences between classical mechanics (CM) and quantum mechanics (QM) and is a principle we adopt throughout this book.

The contextuality of scientific truth

What does it mean to say that a statement is true? Is truth absolute or relative?

In philosophy, there are two opposing views about truth. The *correspondence theory of truth* asserts that the truth or falsity of a statement is determined solely by whether it accurately describes reality. This assumes that there is an objective reality and that it might be described faithfully. The *coherence theory of truth* on the other hand asserts that the truth or falsity of a statement is contextually dependent on other statements. Many confusing statements about time are based on the correspondence theory of truth.

Before we can discuss our main principle, *contextual completeness*, outlined below, we need to distinguish three forms of truth:

Absolute truths

An absolute truth is a proposition asserted to be true under all circumstances or contexts. An example is '*time is continuous*'. Because such statements are by definition metaphysical, physicists should avoid absolute truths as a matter of principle.

Isaac Newton was aware of the dangers of absolute truths, particularly about space and time. In his book *The Principia* Newton defined his concept of Absolute Time as if it were an absolute truth, but immediately qualified it with a statement about his concept of relative time, which he defined in terms of observation of motion [Newton, 1687]. As in so many matters, Newton knew what he was doing. We discuss Absolute Time in Chapter 14.

Contextual truths

A *contextual proposition* is a statement that means something to a given relevant observer. Such a proposition relevant to one observer need not mean anything to any other observer.

In science, observers aim to assign truth values to contextual propositions on the basis of experimental outcomes. A *truth assignment* is the association, by a relevant observer, to a contextual proposition of one element of a set of truth values. Truth value sets are in general binary, that is, they have two elements denoted 0 and 1 respectively. Element 0 will often represent the truth value *false* whilst element 1 represents the truth value *true*, but the opposite convention could be used when required. Truth value sets containing three or more elements can occur: in Scottish law, the set of potential verdicts in a trial is {guilty, innocent, *not proven*}. Mathematicians have extended the classical Aristotelian theory of binary logic to multi-valued logics.

According to this line of thinking, contextual truth values are not intrinsic properties of propositions *per se* but assignments by relevant observers. In other words, this form of truth is contextual.

In this book, we define *contextuality* as the recognition of the circumstances or context under which a truth value is assigned. A contextual statement consists of a proposition or statement, and an associated context. In propositional calculus, contextual statements are known as *material conditionals*: the contexts are known as *antecedents* and the propositions are called *consequents*. A material conditional is usually written in the form $C \Rightarrow A$, interpreted as 'if context C is true then assertion A is true'.

There are several points to keep in mind about contextual statements, because they impinge on our discussion of time in this book:

1. Contextual statements are not automatically true: truth values have to be determined by relevant observers. In science, truth values are validated by experiments.

2. A falsified contextual statement is not metaphysical: the knowledge that a physical theory gives bad predictions can be of great importance to physicists. Metaphysical statements simply have no truth values, so are of no use to physicists.

3. The truth value of a proposition relative to one context carries no implication as to its truth value relative to any other context, even if the difference

between contexts appears to be small, negligible, or insignificant. This principle was recognized early in the history of quantum mechanics, a famous example being the double slit experiment. If any attempt is made to detect the slit 'through which a particle had gone', then the quantum interference pattern hitherto observed is replaced by a classically predicted pattern that shows no interference.

4. Contextual statements should not be interpreted automatically in terms of cause and effect.

Empirical truths

The laws of physics are examples of *empirical truths*, which are contextual truths that have been validated under such a broad range of contexts that they may be regarded FAPP (for all practical purposes [Penrose, 1990]) as absolute. Examples are the laws of conservation of energy and electric charge. The difference between empirical truths and absolute truths is that physicists can test the validity of the former. For instance, Einstein's principle of equivalence, which is based on the assertion that inertial mass and gravitational mass are strictly proportional, has been validated to an extraordinarily fine precision [v. Eötvös, 1890], but in principle could be proved false.

For two thousand years, the proposition '*space is Euclidean*' was regarded as an absolute truth. Following the work of Gauss, Lobachewsky, and Bolyai, it is now regarded as an empirical truth valid only when gravitation can be neglected.

Contextual completeness

A proposition is defined by us as *contextually incomplete* if it says nothing about any relevant observer, that is, if the question '*for whom is this proposition true or untrue?*' remains unanswered. Contextual incompleteness is characteristic of propositions in metaphysics, philosophy, and Aristotelian physics.

The advent first of classical mechanics, then relativity, and finally quantum mechanics forced physicists to the recognition that observers are fundamental to the assignment of truth values in physics. If a proper scientific understanding of time is desired, we should recognize contextually incomplete statements as such and either supply the appropriate contexts or else classify those statements as metaphysical and hence disregard them.

Given that (a) the proper business of physicists is to establish the truth status of physical statements, and (b) such status can only be established via experimentation, and (c) no experiment provides a universal context, we are led to the following principle of observation:

Only contextually complete statements are physically meaningful.

An equivalent statement is that there are no absolutes in physics.[5]

An immediate implication of this principle is that the laws of physics themselves are contextual and not absolute. It is metaphysical to assert otherwise. The correct interpretation of the laws of physics is that they are no more than generally reliable guidelines in the design of new experiments and that there may be contexts in which they fail.

Two examples will illustrate the point:

1. Cantor defined a set as '. . . *a gathering together into a whole of definite, distinct objects of our perception or of our thought—which are called elements of the set.*' [Cantor, 1869]. Without his reference to 'our perception', his statement would be contextually incomplete.

2. Leibniz' principle of the *identity of indiscernibles*, states that '*two objects or entities that have all of their properties in common are the same object*'. There is no reference to any observer, so this statement is contextually incomplete.

Time and contextuality

Contextual information can involve time, particularly in the case of experiments. These frequently involve several temporal stages:

1. a decision stage during which the observer decides on and plans the experiment;

2. a construction stage, during which the necessary apparatus is made;

3. an implementation stage, that is, the actual running of the experiment, with the collection of data;

4. analysis of the data, its interpretation, and its use.

The normal procedure in science is to report on each of these activities separately. Stages (1) and (2) are generally not regarded as appropriate for commentary in final reports of stage (3) or stage (4). Stages (1) and (2) are critical stages that can take years to complete in some cases, such as the Large Hadron Collider. It is a generally accepted metascientific principle that stages (3) and (4) be kept as separate as possible. The reason is that stage (4) depends critically on the observer's theoretical prejudices in favour of some interpretations of the data and against others. It is regarded as essential that this prejudice does not affect stage (3) in any way.

Examples where all components of contextuality are critical to the discussion at hand are Dirac's theory of electron pre-acceleration [Dirac, 1938], the Wheeler–Feynman absorber theory of electrodynamics [Wheeler & Feynman, 1945], and

---

[5] This statement is not a statement *in* physics but *how to do* physics.

the information loss problem in black hole physics [Preskill, 1992]. The rule physicists should follow in such discussions is to establish clearly how all components are dealt with, otherwise such discussions remain metaphysical. For example, if an entangled two-electron state is prepared outside the Schwarzschild radius of a black hole and one of the electrons is allowed in past the Schwarzschild radius, it is not admissible to compare subsequent spin measurements inside and outside of that radius unless the protocol for carrying outcome information out of the black hole has been established (in this case it happens to be impossible because of the event horizon). We should not make the fundamental error of assuming that the use of a spacetime diagram containing events interior and exterior to the Schwarzschild radius entitles us to a godlike exophysical perspective on physical truth values on each side of that radius. Such a diagram is a metaphysical device that is permissible in a theory, provided we do not misinterpret it. Further, in any discussion of evaporating black holes, arguments about unitary state evolution should not be taken at face value if the precise details of what evaporation means for specific observers are not given. The Unruh effect demonstrates this point: space can appear empty according to one observer but full of particles according to another [Unruh, 1976].

The contextuality of the existential quantifier

The terms '*it exists*', '*existence*' and suchlike crop up repeatedly throughout various branches of philosophy. Used without context, these are dangerous words because they have the potential to lull the unsuspecting reader into a false belief that something meaningful has been said. An example is the *mathematical universe hypothesis*, which asserts that 'All structures that exist mathematically also exist physically' [Tegmark, 2014]. This hypothesis is contextually incomplete. It begs questions such as *relative to whom is 'existence' defined?*, *for whom or what is this noncontextual statement true?* and most importantly, *how can this statement be validated, that is, proved to be true or false?*

Not all philosophers fall into the noncontextuality trap. The philosopher Descartes famously made the statement '*Je pense, donc je suis*' [Descartes, 2006], translated into English as '*I think, therefore I am*'. There have been and remain numerous conflicting interpretations of this statement which we do not comment on. We interpret Decartes' statement as an example of a contextual statement: the contextual assertion {*I exist*} is asserted to be true relative to the context {*I think*}. We make two further points about Descartes' dictum:

1. Descartes does not tell us whether he exists or not regardless of his thinking: his statement makes no assertion about non-existence: lack of thought does not imply non-existence. Who would verify that non-existence?

2. Descartes' statement says nothing about whether '*I think*' is true or not yet his complete contextual statement could still be true. It is logically consistent to make a true assertion that $P \Rightarrow Q$, knowing that in actuality both P and Q are false. An example is the statement 'If {*I died yesterday*} then {*I did not just write this statement*}'. Indeed, relative to us, Descartes does not think, since he has been dead a long time.

If we wish to avoid metaphysics, our recommendation is reject any statements based on the word 'exist' whenever appropriate context is lacking.

Primary observers

It is an empirical fact that observers do not have infinite lifetimes: they come and go. Suppose we attempted to discuss the origin of some physical observer A. This would be meaningful only from the perspective of some earlier physical observer B. But this would immediately raise the same question as to the origin of B, and in this way we would be led to an infinite regress.

To avoid such an infinite regress, we should stop somewhere, placing a veto on questions about the origin of chosen *primary observers*, accepting their existence for granted and defining all contexts relative to them only and no further. This means that we may have to scale down our expectations of physics and accept the possibility that we may never have a complete TOE (theory of everything), including an understanding of the origin of the laws of quantum mechanics.

Primary observers will have a sense of time, memory, and purpose, for without any of these attributes they could not be regarded as observers. At any moment of their time they hold in their memories data that they interpret in terms of a hypothesized past relative to that moment. With sufficient data, observers can even attempt to account for their own origins. But that past is a map of the past and should not be confused with it [Korzybski, 1994]. Different primary observers might construct different relative pasts and there is nothing in physics, apart from the need for consistency should they exchange information, to veto that. This means that an observer's past may be as uncertain for them as their future is. It is metaphysics to assert that 'the past' is unique and absolutely fixed.

The concept of primary observer should not be confused with *solipsism*, a rudimentary view of the universe held by a primary observer that they are the only observer in existence. For a solipsist the passage of time is an entirely personal experience and has no meaning otherwise. All perceptions and sensations from the individual's environment are assigned no objective existence unless there is a direct need to do so, such as when a child interacts with a parent. Solipsism is observed by psychologists in young children who believe that their parents do not exist unless they (the parents) can be seen.

Solipsism has a long philosophical lineage dating at least as far back as the Greek philosopher Gorgias in the fourth century BCE. Over the millennia, metaphysicists have elaborated the concept into a number of forms, the most extreme version being known as *metaphysical solipsism*. According to that extreme view, only the solipsist exists and external reality, including other observers, has no existence.

A strict adherence to the principles of contextuality as discussed here leads to the following judgement on solipsism: it is a valid but limited model of the universe provided that all reference to the non-existence of external observers and external reality is explicitly acknowledged as contextual. In other words, a solipsist should question the assertion that only they exist. Solipsism and contextual completeness can be reconciled if a solipsist identifies themselves as a primary observer.

Strict metaphysical solipsism as defined above has no value according to our principles of contextuality because it is contextually incomplete: the assertion that something has no existence is a meaningless assertion without any context for establishing the truth of that assertion.

Solipsism is related to the concept of *superobserver*, discussed in Chapter 6. Solipsism is ubiquitous throughout mathematics and the physical sciences because traditional mathematics and classical physics is explained in a solipsistic style, as if some primary observer was talking to itself. Galileo cleverly bypassed this style by encoding his ideas in physics in the form of conversations between Salviati the supporter of Copernicus, Sagredo the layman, and Simplicio, an Aristotelian [Galileo, 1632].

Experiments

Primary observers perform *experiments*, but what defines an experiment? When is it meaningful to say that the patterns of energy and mass in a particular corner of space and time constitute an experiment?

Our answer is that experiments are as contextual as anything else in physics, being meaningful only relative to their associated observers. Relative to other observers, experiments need have no physical meaning.

An important consideration in any experiment is *functionality*: a nominally four-legged table with one leg sawn off remains a table if it can still support plates. Once it loses that functionality, it is no longer a table. Any corner of the universe, no matter how small or large, constitutes an experiment if an observer can use it as such.

Heisenberg cuts

A *Heisenberg cut* is a hypothetical line dividing two regimes: on one side is the classical regime, requiring a classical mechanical description, whilst on the other side is the quantum regime, requiring quantum mechanics for its description. It is named after Heisenberg, who wrote

The dividing line between the system to be observed and the measuring apparatus is immediately defined by the nature of the problem but it obviously signifies no discontinuity of the physical process. For this reason there must, within certain limits, exist complete freedom in choosing the position of the dividing line.

[Heisenberg, 1952]

According to Heisenberg, such a cut is contextual because it depends on the observer. We would add the following comments.

1) There are no truly exophysical observers, only primary observers sitting inside the physical universe. Each primary observer must have some classical knowledge about themselves relative to that physical universe. For example, experimentalists believe that they are sitting in three-dimensional physical space. In addition, experimentalists will have classical knowledge about the apparatus defining their experiment, such as the relative orientation of the main magnetic field in a Stern–Gerlach experiment.

2) The observer's classical knowledge falls naturally into two parts. The first part we refer to as *relative internal context* and is defined as that classical knowledge that affects quantum outcomes of the experiment. The other part we refer to as *relative external context* and is that observer's knowledge that has no bearing on quantum outcomes [Jaroszkiewicz, 2010].

3) It is a generally unstated but universal assumption in quantum physics that relative external context is irrelevant to quantum outcome probabilities. For example, the time of day is generally not a factor in a double-slit experiment (although it could be in some experiments)[6]; neither is the sort of clothing the observer is wearing, and so on. Peres made the same point when he wrote that quantum state preparation generally does not impress a memory of relative external context on a quantum state [Peres, 1993]. A note of caution should be given here. Whilst it may be true that the colour of an observer's hat may have no impact on the outcomes of a quantum experiment, the time in the history of the universe may be critical. We appear to be living in an era of stability as far as the laws of physics are concerned. Could we even discuss doing a quantum experiment, or the colour of an observer's hat, just after the Big Bang and before the formation of stable atoms?

We propose that a Heisenberg cut should not be thought of as a fundamental transition from classical to quantum regimes, but as a contextual, observer-defined partition between relative external context and relative internal context in a given experiment.

---

[6] The position of the Moon can affect runs of the Large Hadron Collider [Gagnon, 2012].

Generalized propositions

A *generalized proposition* $\mathcal{P}$ is a statement of the form

$$\mathcal{P} \equiv (P, C_I \,|\, O, C_E), \tag{2.1}$$

where P is a proposition, C_I is the *relative internal context*, relative to which P is to be validated, O is the primary observer for whom that validation is meaningful, and C_E is the *relative external context* that defines O. If C_I, O or C_E are absent or not given, then we represent them on the right-hand side of (2.1) by $\emptyset$, the mathematical symbol for the empty set.

Given a generalized proposition $\mathcal{P}$, we assign to it a generalized proposition classification (GPC) defined as $\text{GPC}(\mathcal{P}) \equiv \alpha + 2\beta$, where α is $+1$ if relative internal context is given and zero otherwise, and β is $+1$ if relative external context is given and zero otherwise. This allows us, upon examining any proposition in any theory, paper or book, to establish its GPC and hence some idea of its value to us scientifically.

Typically, a metaphysical generalized proposition is of the form $(P, \emptyset \,|\, \emptyset, \emptyset)$, so such propositions have a GPC of zero. On the other hand, a *contextually complete* generalized proposition typical of quantum mechanics takes the form $(P, A \,|\, \omega, \mathcal{F})$, where A is a declaration of the apparatus that can be used to validate P, ω is some endophysical primary observer, and $\mathcal{F}$ is a declaration of that observer's frame of reference, and so on. Valid quantum propositions have a GPC of 3.

It is convenient to introduce the concept of *validation function*, denoted $\mathbb{V}$. This is a rule that assigns a contextual truth value 1 to a generalized proposition $\mathcal{P}$ that has been found to be contextually true, written $\mathbb{V}\mathcal{P} = 1$, whilst for a generalized proposition that has been found to be contextually false we write $\mathbb{V}\mathcal{P} = 0$. All generalized propositions except those with GPC of zero can be validated, that is, assigned a contextual truth value. This includes propositions in erroneous physics theories, such as the phlogiston theory of heat. Metaphysical generalized propositions have no context in which they can be validated, so Pauli's designation '*not even wrong*' applies to them [Woit, 2006]. Such propositions have no truth values.

Our concern about contextual completeness in any proposition about time has two purposes: it serves both as a warning against unscientific speculation and as an encouragement to speculate in the right way. Speculation is not meaningless *if* it is accompanied by some thought about validation. We have here in mind the long history of the atomic hypothesis. Boltzmann was not wrong to develop his approach to thermodynamics using the atomic hypothesis, which at that time appeared metaphysical. Indeed, Boltzmannn attracted much hostility on that account. But his work helped stimulate great efforts to validate the atomic hypothesis, and he was proved right. It is unfortunate that he killed himself two years before Perrin finally settled the matter empirically.

We started this book by warning the reader that time was a complex and vague concept and that we could not hope to have a single perspective on it. It is reasonable, therefore, to look at as many perspectives as we can on the subject. But we should also listen to Heisenberg's warning given at the start of this chapter. We should at the very least take some care to identify those ideas that are scientifically sound and those that amount to unsupported and unprovable assertions. There are few of the former, but an endless supply of the latter.

3

Subjective images of time

In this chapter we discuss what some of the great thinkers believed and wrote on the subject of time.

The Greeks

There are few more remarkable periods in the history of human thought than the age of the ancient Greek philosophers. Powerful and subtle minds looked at the world around them and by force of intellect developed enduring philosophies of space, time, and matter, philosophies that remain relevant to our discussion of time in this book. No doubt there have been great minds in all cultures speculating on the mysteries of time: whilst Plato and Aristotle walked the streets of Athens, shepherds out on the vast plains of Scythia, masons trimming stones in Egyptian quarries destined for the pyramids, and fishermen on the banks of the Ganges, all of these would have been just as likely to ponder the nature of their existence as anyone else. An essential difference between them and the Greeks is that the latter left written records of their thoughts that continue to be of interest.

Regarded as one of the first Greek philosophers to view the universe from a rational, scientific perspective, Anaximander of Miletus postulated the abstract *Apeiron* as the origin of the universe. Apeiron is analogous to *Chaos*, the formless state of Ancient Greek cosmogony, considered the principle from which everything else appeared.

The philosopher Heraclitus, living in Ephesus on the Ionian coast of Asia Minor around about 500 BCE, held a view of time that accords with experience: '*everything changes*'. Also in the fifth century BCE, Antiphon the Sophist asserted that [Freeman, 1983]

> Time is not a reality, but a concept or a measure.   [Antiphon, *On Truth*]

The views of Heraclitus and Antiphon are compatible with *process time* [Britannica, 2000], which asserts that time is a *process* [Whitehead, 1929] rather than a 'thing' in its own right.

Images of Time. First Edition. George Jaroszkiewicz.
© George Jaroszkiewicz 2016. Published in 2016 by Oxford University Press.

Process time is incompatible with *manifold time* [Britannica, 2000], which asserts that space and time form a geometric entity with physical properties such as a metric, or distance rule. Special relativity (SR) and general relativity (GR) are based on manifold time, so they are incompatible with process time. In particular, SR does not admit absolute simultaneity.

In stark contrast to Heraclitus, Parmenides and his student Zeno of Elea took the opposite view. The single known work of Parmenides consists of fragments of a poem in which he argues that '*Reality is One*', that *change is impossible*, and that *existence is timeless and uniform*.

At first sight these assertions seem obviously incorrect, because it is undeniable that we experience change all around us. However, Parmenides' ideas can be related to manifold time and the contemporary paradigm of space, time, and matter known as the *Block Universe* [Price, 1997]. This paradigm underpins SR and GR and is discussed further in Chapter 8. According to the Block Universe, past, present, and future all coexist in the four-dimensional spacetime continuum and there is no special 'moment of the now'. Physical observers are embedded in spacetime and have an illusory sense of time.

Attributed to Parmenides is the following 'paradox':

> Before an object can move any distance, it must first move through an infinite series of fractions of that distance; but since one can never actually get through an infinite series of steps, no distance can be moved through at all.

This apparent paradox and others like it attributed to Parmenides and his pupil Zeno involve the mathematical concept of a *limit*, which in their day was not understood properly at all. It took mathematicians over two thousand years from the time of Zeno to formulate a mathematically sound definition of a limit. Mathematicians now routinely discuss adding up an infinite number of terms that add up to a finite sum, as we demonstrate as follows.

Two halves make a whole, which we may express mathematically as $1 = \frac{1}{2} + \frac{1}{2}$. Applying this principle to the last term on the right-hand side of this expression, we have $\frac{1}{2} = \frac{1}{4} + \frac{1}{4}$, so the original expression can now be written as

$$1 = \frac{1}{2} + \frac{1}{4} + \frac{1}{4}. \tag{3.1}$$

Replacing the last quarter by two eighths, we now have

$$1 = \frac{1}{2} + \frac{1}{4} + \frac{1}{8} + \frac{1}{8}. \tag{3.2}$$

It is clear we could go on for as long as we wished, expressing a finite number, in this case one, as the sum of an unlimited number of numbers. Mathematicians express this idea in the form of a convergent infinite sum:

$$1 = \sum_{n=1}^{\infty} \frac{1}{2^n}. \tag{3.3}$$

There is a fundamental question here that touches upon the images of time discussed in this book: can we meaningfully discuss an infinite number of operations, such as the sum of an infinite number of terms, as in equation (3.3)? Just who is going to have the time to check that (3.3) is true?

One way to answer this question would be to make a clear distinction between mathematical idealization and physical practicality. In the real world, we could not actually perform an infinite number of summations. Each term in the summation would take a real mathematician a finite interval of time to add to the sum. Even the fastest computer operating with a central processor cycle time of the order of a Planck unit of time (approximately one thousand million million million million million million millionth of a second) could not do it. A more physically realistic representation of the summation in equation (3.3) is

$$1 = \lim_{N \to \infty} \sum_{n=1}^{N} \frac{1}{2^n}, \qquad (3.4)$$

where $\lim_{N \to \infty}$ is an instruction to take the limit as N tends to infinity of a finite expression in N. The advantage of this second notation is that it makes clear that we are dealing with a mathematically defined *process*, rather than an actual infinity of steps.

Limits are needed in order to discuss the infinitesimal calculus, developed independently by Newton and Leibniz. Calculus is necessary for a proper understanding of velocity and acceleration. These concepts are fundamental to the science of mechanics, a knowledge of which underpins modern technology and hence the accurate measurement of time.

Zeno of Elea, a student of Parmenides', is famous for a number of '*paradoxes*' concerning motion. These challenge the listener to draw conclusions that support the philosophy of Parmenides that motion is impossible.

In Antiquity and for many centuries afterwards, Zeno's paradoxes were the centre of much debate. One modern view is that his paradoxes seem paradoxical to the non-mathematically trained listener only because of a limited understanding of $\mathbb{R}$, the continuum of real numbers, time being generally modelled in physics by this set of numbers.

Zeno is known to have formulated about 40 paradoxes but only 10 are currently known. Of these, three concern motion, and therefore time. These are *The Achilles*, *The Racetrack*, and *The Arrow*. We discuss these in turn.

The Achilles

In this paradox, otherwise known as *Achilles and the Tortoise* or *The Hare and the Tortoise*, the runner Achilles is trying to catch up with a tortoise that starts a hundred paces from Achilles and is moving away so as to avoid capture. Achilles can run 10 times as fast as the tortoise can walk, so it is intuitively obvious that Achilles will quickly catch the tortoise. Or will he?

Zeno planted the seeds of doubt into his listeners' minds by the following reasoning why Achilles would *never* be able to catch the tortoise. It takes Achilles 10 seconds to reach the initial position of the tortoise, but during that time, the tortoise has moved a further 10 paces away. Therefore Achilles needs to run another 10 paces to reach that new position. This takes him one more second. But during that extra second, the tortoise has moved another pace. Therefore Achilles still has not caught up with the tortoise.

Continuing this line of argument, Zeno concluded that Achilles would never catch the tortoise, because by the time Achilles reaches the place where the tortoise had been, the tortoise had moved on and was somewhere else.

In the real world, no system can perform an infinite number of tasks. Therefore Zeno's view that Achilles performs an actual infinity of tasks (specifically, each task being to run from his current position to the current position of the tortoise) is physically impossible.

Nevertheless, even if we grant Zeno's analysis legitimacy, his conclusion is incorrect. Equation (3.3) above shows that, expressed as a limit, an infinite sum of terms can converge to a finite sum. Zeno's conclusion has no mathematical validity.

The convergent infinite series is a mathematician's solution to a physicist's problem. There are aspects to this paradox that need to examined from a physicist's perspective. Time is, after all, a physical concept.

A physicist's objection to the mathematician's convergent infinite sum solution is that the original paradox is formulated entirely with a classical mechanical (CM) view of the physics: the positions of Achilles and the tortoise are assumed to 'exist' in an absolute sense at all times. Moreover, their individual positions are imagined to be precise at all times. But matter is never concentrated at a point and objects such as Achilles and the tortoise are extended complexes of matter and energy obeying the quantum laws of physics. According to the principle of contextual completeness discussed in the previous chapter, Zeno should tell us how we could know all the information that he uses to draw his conclusion.

The Racetrack

In this paradox, an athlete can never complete a race because he takes a finite time to cover half of the remaining distance between his current position and the finish. Therefore, he would need to take an infinite number of such finite times, which he cannot actually do.

We raise the same objections to Zeno's analysis of the Racetrack as those we raised for the Achilles: Zeno should tell us how we could know where the athlete was at any instant of time.

The Racetrack is reminiscent of modern particle decay experiments. Quantum systems monitored 'continuously' for decay can exhibit the strange behaviour known as the *quantum Zeno effect*. In brief, too much monitoring can prevent a particle from decaying, a phenomenon first discussed by Turing and investigated experimentally and theoretically many times since. Misra and Sudarshan showed that the questions asked by an observer of a decaying system could influence the dynamical evolution of that system [Misra & Sudarshan, 1977].

The Arrow

In this paradox, an arrow moving through the air instantaneously occupies a certain volume of space. But space does not move. Therefore an arrow cannot move.

This perhaps is the easiest of Zeno's paradoxes to dismiss because it is based on an incomplete understanding of motion. The dynamical state of an object in CM requires a specification of position *and* velocity, but Zeno's reasoning assumes only position is needed.

Aristotle

The polymath Aristotle was much more than a philosopher, writing influential books on an extraordinary range of subjects. Unfortunately, whilst he was good on many topics, science is not on that list. Because of the authority commanded by his writings, his views of mechanics and therefore of time had a severe and negative influence that lasted the best part of two thousand years. On the subject of motion, Aristotle wrote that

> Everything that is in motion must be moved by something.   [Aristotle, 1930]

This was refuted by Newton's first law of motion, also called Galileo's law of inertia.

On the subject of time, Aristotle wrote that

> time is 'number of movement in respect of the before and after', and is continuous since it is an attribute of what is continuous.   [Aristotle, 1930]

Since Aristotle based this image of time on the assumption that matter is continuous, which we now know to be untrue, we have to reject his argument that time is continuous. This does not prove that time is discrete.

The dogmatists

Dogmatism is a statement of a belief as if it were an established fact. In this section we discuss a number of religious dogmatists not because we attach any particular merit to their underlying beliefs but because their images of time continue to influence some important contemporary cultural images of time discussed in the next chapter. If we want to understand religious conflicts in the modern era, we could do no better than to understand how religious authorities long after the foundations of their respective religions interpreted time.

In a book aiming to identify and avoid metaphysical speculation, it may seem strange that we discuss the views of dogmatists. This is to be expected, however, if we consider the constraints dogmatism imposes on its practitioners. By the nature of their approach to truth and the fact that no written source can possibly comment on all aspects of the physical universe, these dogmatists have to devise clever but contextually incomplete arguments to account for concepts such as free will and the creation of the universe.

The fifth-century Christian theologian Augustine of Hippo wrote an influential essay called *Confessions*, in which he explored some concepts of interest to us. Attempting to reconcile free will and the nature of God, Augustine saw God as what we would describe as an exophysical observer having a complete manifold view of time: past, present, and future. We can readily understand the inherent temporal paradox of this image of time: humans are endophysical observers embedded in the time manifold with a limited knowledge of the future. Therefore, whilst Augustine's God knows who is saved and who is not, each individual human has to behave as if their future was undecided and that they had free will.

Augustine speculated on the physical structure of time, writing

> If an instant of time be conceived, which cannot be divided into the smallest particles of moments, that alone is it, which may be called present. Which yet flies with such speed from future to past, as not to be lengthened out with the least stay. For if it be, it is divided into past and future. The present hath no space. [Augustine, 398]

This quotation reveals more about the standards of logic in theology and philosophy than it does about the nature of time. Even if we allow concepts such as time flying *from future to past*[7] Augustine's assertions are contextually incomplete: he does not tell us who has observed time as flying 'with such speed', or by what comparisons has established that its speed is fast.

Some of Augustine's points we can agree with. He discusses for example the relationship between time and memory, writing

> For if times past and to come be, I would know where they be. Which yet if I cannot, yet I know, wherever they be, they are not there as future, or past, but present. For if there also they be future, they are not yet there; if there also they be past, they are no longer there. Wheresoever then is whatsoever is, it is only as present. Although when past facts are related, there are drawn out of the memory, not the things themselves which are past, but words which, conceived by the images of the things, they, in passing, have through the senses left as traces in the mind. Thus my childhood, which now is not, is in time past, which now is not: but now when I recall its image, and tell of it, I behold it in the present, because it is still in my memory.
>
> [Augustine, 398]

---

[7] Note that this is the reverse direction of time normally discussed in physics: time is normally thought of as running from past to future.

On the relationship between past, present, and future, Augustine takes a process time perspective, 'existence' being defined in terms of the observer's current memory. The following is therefore a contextually complete view of past, present, and future, if we assume he is referring to himself as a primary observer. He wrote:

> What now is clear and plain is, that neither things to come nor past are. Nor is it properly said, 'there be three times, past, present, and to come': yet perchance it might be properly said, 'there be three times; a present of things past, a present of things present, and a present of things future.' For these three do exist in some sort, in the soul, but otherwhere do I not see them; present of things past, memory; present of things present, sight; present of things future, expectation.   [Augustine, 398]

Augustine identifies the observer's 'moment of the now', or 'present of the present' in terms of the acquisition of information, specifically *sight*, but despite some of his ideas being remarkably reasonable and well put, he was in the long run constrained by his adherence to a religious perspective.

During the 'Golden age of Islam' traditionally dated from the mid seventh century to mid thirteenth century, there were notable individuals such as Avicenna, Averroes, and Alhazen who discoursed on many fields beyond the purely religious. In that sense they were not dogmatists. Of importance to our discussion here, however, is Al-Ghazali [1058–1111], who left an indelible mark on Islamic culture in his interpretation of causality. In *The Incoherence of the Philosophers* he embraced theological occasionalism, the theory that all causal events and interactions are not products of physical interactions but occur because of the 'will of God'. His ideas on causality have been compared with those of quantum mechanics (QM):

> Initially, it might appear unlikely that there would be any significant similarities between the thoughts of Al-Ghazali (eleventh century CE) and the ideas of quantum theory in the twentieth century. Although separated by culture as well as several centuries, many of the same ideas are incorporated into these two bodies of thought. Important similarities are seen in the role of causality in the natural world, the nature of physical objects, and the extent to which the behavior of objects is predictable
> [Harding, 1993]

Many centuries after Augustine of Hippo, the Italian Christian theologian Thomas Aquinas [1225–74] wrote on the same temporal dichotomy that had worried Augustine: process time versus manifold time, otherwise referred to as the 'A-theory' and the 'B-theory' of time [Brenner, 2010; McTaggart, 1908] respectively.

For Aquinas, God represents the *first cause*, the starting point of all things. Aquinas' dogmatism was criticized by the philosopher Russell [1967] on the mathematical grounds that Aquinas believed that it was impossible for a series to have no first term. Russell's argument is the series of negative integers *ending* at the integer −1.

From our point of view, Aquinas's position at least has the merit of being contextually complete: his primary observer is God. Russell's argument is manifestly contextually incomplete: we have to ask: *Who is talking to whom? Who exactly is doing the checking that the series ends at* −1?, and so on. As with so many metaphysical discussions, it does not take too much effort to see there are hidden assumptions and a contextual incompleteness that makes the arguments inapplicable to the physics of time. Our objection to Russell's glib objections to Aquinas is that a purely mathematical argument cannot be used to prove anything about the physics of time. This does not prove that Aquinas was correct, of course.

Mircea Eliade [1907–86] was a Romanian religious historian whose formulated a novel binary model of time and existence as it relates to societies. The two contrasting components of Eliade's model are commonly referred to as the *sacred*, or religious, and the *profane*, or secular. In the sacred domain, myth and tradition are dominant factors in society: many religions are based on a return to a mythical past. Sacred time is 'circular' in that sense. On the other hand, in the secular domain, the universe develops conventionally and time is linear.

Eliade's views have been criticized by anthropologists and sociologists on the grounds that whilst they may apply to Australian aboriginal beliefs about the 'dream time' and some other cultures, they do not fit in with well-known cosmogenies such as those of the ancient Greeks, who did not yearn for a return to a mythical past. Another criticism is that according to Eliade, linear time is secular and cyclic time is sacred, but this is not the case for Buddhists, Jain, and some Hindu, who regard reincarnation as a natural (i.e. secular) phenomenon.

An influential religious dogmatist-cum-scientist was Pierre Teilhard de Chardin [1881–1955], who attempted to infuse his scientific work in paleontology with a dogmatic theological view of cosmological evolution. Virtually all his cosmological writings are contextually incomplete. However, there are two quotations that attributed to him that we find agreeable:

> The universe as we know it is a joint product of the observer and the observed.

and

> The time has come to realise that an interpretation of the universe-even a positivist one-remains unsatisfying unless it covers the interior as well as the exterior of things; mind as well as matter. The true physics is that which will, one day, achieve the inclusion of man in his wholeness in a coherent picture of the world.
>
> [de Chardin, 1947]

The sceptics

Scepticism is the questioning of given facts and opinions and has proven to be a dangerous practise in some cultures. We only have to think of Giordano Bruno's fate (burning at the stake) and the threat of it to Galileo to appreciate the danger. Whilst scientific scepticism is the foundation of progress in our modern world and is generally regarded as a good thing by society, religious scepticism can appear to challenge religious authority and that can be fatal.

In the fourteenth century, the French theologian Nicholas of Autrecourt [*c* 1299–1369] developed scepticism on the basis that it was right to question beliefs that were not primary, that is, fundamental. He would not have denied the existence of God, for example, but he questioned common assumptions about the structure of space and time. He considered that matter, space, and time were made up of discrete parts, that is, atoms, points, and instants respectively. He went further and attempted to explain the processes of change, that is, regeneration and decay, as due to rearrangements of these discrete constituents.

Nicholas had an empiricist view of knowledge. According to one analysis,

> Nicholas laid it down that the sole sources of certitude are the immediate data of consciousness and the law of non-contradiction . . . and every certain conclusion must be deduced from principles firmly and exclusively based on evidence which only the immediate data of consciousness can supply.   [Weinberg, 1942]

Although to us his scepticism was primarily scientific, it was seen in his day as religious scepticism and so he was forced by the Church authorities to recant and burn his writings.

The Age of Reason

Starting in seventeenth-century Europe, the Age of Reason or *Enlightenment* is characterized by a shift in thinking by intellectuals, replacing the dead hand of tradition with rational debate. Recognizably scientific modes of thinking were applied to great issues, emphasizing the use of rational argument based on principles derived from experience. In the list of intellectuals associated with the Age of Reason are many familiar European names such as Descartes, Spinoza, Locke, and Voltaire. A characteristic of these thinkers is that they did not complete the scientific process and test their conclusions against experimental evidence.

Although he is regarded as the co-founder with Newton of the infinitesimal calculus, we place Leibniz [1646–1716] in this section rather than with Newton, as many of Leibniz' assertions are contextually incomplete and hence metaphysical in nature. Leibniz is noted for several principles that have been much used in metaphysics. These include *the principle of sufficient reason* (PSR), which states that '*nothing is without a cause*', and is regarded as one of the four laws of logic. Our interpretation of this principle is that it is an appeal to contextuality: if something is observed to happen, such as an explosion, we should seek to find what caused that explosion, that is, the context in which it occurred. PSR has been linked to another assertion, dating from Antiquity, that '*nothing comes from nothing*' (*ex nihilo nihil fit*).

According to PSR, all propositions that are regarded as true within a system should be deducible from the axioms of that system. The logician Gödel, however, showed that there are mathematical systems for which statements can be regarded as true but for which there is no validation process within the axioms of that system.

Leibniz considered the laws of mathematics to be derived from certain principles. His *law of identity*, also known as the *identity of indiscernibles*, states that two objects or entities that have *all* of their properties in common are the same object.

As it stands, this statement is contextually incomplete: there is no reference here to any observer. The principle begs the question: *who or what is establishing that the two objects have properties in common?* Indeed, who or what has established that there are two objects in the first place?

This principle on its own is metaphysical, but when contextuality is taken into account it has relevance in QM. Elementary particles such as electrons and photons are classified as belonging to certain types or species, each with its specific characteristics. For example, all electrons have one negative unit of electric charge, a spin angular momentum of $\frac{1}{2}\hbar$, a specific mass in free space measured to have value $9.10938291 \times 10^{-31}$ kilograms, and so on. In particular, electrons are regarded as *indistinguishable*, in the sense that if we knew that we had a physical state in the laboratory consisting of two electrons, we could not say which one of these electrons we were dealing with. In essence, we could not mark, or notch, one of these electrons so as to keep track of its movements. In QM, indistinguishability plays an extraordinary role in maintaining the structure of the universe: the Pauli exclusion principle asserts that no quantum state can exist in which two electrons have identical quantum numbers: this prevents multi-electron atoms from collapsing into so-called degenerate states.

In his *principle of non-contradiction*, also known paradoxically as *the principle of contradiction*, Leibniz asserts that no object can have a property and its negation at once. Again, this statement as it stands is contextually incomplete.

Leibniz's views on time and space are surprisingly modern. He was a relationalist in the sense that to him, space and time should not be thought of as things in which objects exist and move about, but as systems of relations holding between things. This is a precursor of Schwinger's views on time and space, discussed further on in this chapter. According to Leibniz [Leibniz, 1717],

> As for my own opinion, I have said more than once, that I hold Space to be something merely relative, as Time is; that I hold it to be an Order of Coexistences, as Time is an Order of Successions.   [Leibniz's Third Paper to Samuel Clarke]

According to the *Stanford Encyclopaedia of Physics*, Leibniz regarded space and time as continuous, homogeneous, and infinitely divisible.

Of greater relevance to us here is David Hume [1711–76], who based his influential books *A Treatise of Human Nature* [Hume, 1739] and *An Enquiry Concerning Human Understanding* [Hume, 1748] on the philosophical approach known as

empiricism, the guiding principle of which is that the only true source of knowledge is experience. His views on the nature of time and its consequent attributes,
such as causation, are scientific in principle and very much worth reading today.
Hume was an atheist and therefore not influenced by the metaphysics of religion
in the way Descartes, Berkeley, and others were.

In the same way as he discussed purely philosophical themes such as morality
and loyalty, Hume discussed causation in terms of the human observer. He interpreted truth values as holding for the individual, not as absolutes. His approach is
contextually complete, because we can interpret Hume's individual as a primary
observer. Unlike Descartes, Hume did not attempt to relate the individual to any
deeper metaphysical construct such as God.

Hume's discussion of time and space in the *Treatise* is remarkable: of the many
relevant quotes we could take from it, the following one sums up his style of
analysis.

> The capacity of the mind is not infinite; consequently no idea of extension or dur-
> ation consists of an infinite number of parts or inferior ideas, but of a finite number,
> and these simple and indivisible: 'Tis therefore possible for space and time to exist
> conformable to this idea: And if it be possible, 'tis certain they actually do exist con-
> formable to it; since their infinite divisibility is utterly impossible and contradictory.
> [Hume, 1739]

This seems to us to be an argument against the continuity of space and time. If
Hume were alive today, his ideas would surely find resonance with those of the
mathematical physicists Wheeler, Schwinger, and Snyder, discussed later in this
chapter.

Hume postulated a number of principles underpinning his analysis of causation.
His *Copy Principle* states that all aspects of our thoughts come from experience:

> All our simple ideas in their first appearance are deriv'd from simple impressions,
> which are correspondent to them, and which they exactly represent. [Hume, 1739]

Hume's Fork is the segregation of an observer's thoughts and knowledge into
two categories: *relations of ideas* and *matters of fact*. This is in accordance with our
notion of primary observer, a system that obtains information via experiment and
then analyses it in terms of abstract mathematical models.

Hume defined cause in two ways, the shorter of which is

> An object precedent and contiguous to another, and where all the objects resembling
> the former are placed in like relations of precedency and contiguity to those objects
> that resemble the latter.   [Hume, 1739]

We interpret Hume here as saying that what we think of as cause and effect is
the result of how our minds process information. He does not say cause and effect

is intrinsic in the world outside our sensations. Hume's causality therefore is but a convenient mental construct. This is in accordance with QM, which does not discuss causality in the strict deterministic fashion of CM but as a probabilistic attribute of our experimental procedures.

The classical mechanists

In this category we place those rare individuals such as Galileo and Newton who left an enduring mathematical legacy of relevance to us in this book. An important understated aspect of mathematics is its endurance. It withstands changes in philosophical, physical, and cultural paradigms. The mathematics of Euclid, Archimedes, Galileo, and Newton is as valid today as it ever was, even though the particular motivating philosophies of those individuals is of no particular interest to us today except to historians of science.

The first great name that we encounter here is Galileo, generally regarded as the 'Father of Science'. The start of the empirical study of time is often dated to the year 1602, when Galileo discovered that the period of oscillation of a pendulum is independent of the arc of swing, provided this arc is small in some sense. This observation requires minimal equipment: we can readily demonstrate it by constructing two identical pendula side by side and then setting them off performing small oscillations with different individual arcs of swing.[8] We have observed that, whilst the one with the greater arc of swing moves faster, it has to traverse a greater distance during each period and in consequence the two periods are the same to within experimental accuracy.

As with Foucault's pendulum, which hints at the rotation of the Earth relative to the distant stars, Galileo's pendulum experiment could easily have been done in Antiquity. Perhaps it was and perhaps the independence of the period from arc of swing was noted, but no theory that we know of was constructed to account for it. It is likely that cultural conditioning played a decisive hand in this lack of development.

In the Museo Galileo in Florence, it is demonstrated by many fine exhibits dating to before the time of Galileo that technology had been developing at a great pace before he came along. The driving force was *warfare*, not a love of science for its own sake. At that time, Italy was divided into powerful competing city states such as Florence, Rome, Pisa, Lucca, Padua, Genoa, Venice, and Siena. Whereas in previous times a young man from a respectable family could go into the military knowing only how to wield a sword and ride a horse, now he (and it was inevitably a *he*) also had to be educated in logistics, mathematics, and, specifically, ballistics. The cultural scene was settling into a position by Galileo's time for a scientific

---

[8] Two pieces of string each of length two metres, two full bottles of water, and a garage roof truss are all that is needed to perform this experiment: we do not need a clock.

revolution. At the centre of this revolution would be mechanics and the necessity of measuring time accurately.

A detailed mathematical explanation of the physics of Galileo's observations had to wait until Newton gave the general principles of CM in his great book *The Principia* [Newton, 1687]. Newton well understood the need for clarity on the nature of space and time, so relatively soon into his book, Newton took the trouble to present a statement of his concept of time. He discussed two concepts: *Absolute Time* and *Relative Time*. The former time is by definition non-contextual and discussed further in Chapter 14. If he had stopped there, we could accuse Newton of engaging in metaphysics, but he avoided that trap by immediately commenting that Relative Time is based contextually on observation.

When calculations are done using Newtonian mechanics, Absolute Time is usually used as the temporal evolution parameter because that is most convenient. We could use Relative Time if needed. In Chapter 15 we show how Dirac's constraint mechanics can be used to replace Absolute Time by any convenient parameter, such as that given by a less than perfect (i.e. *unequable*, in the language of Newton) clock in our laboratory.

Newton's Absolute Time remained a central feature in physics until the last decade of the nineteenth century, when relativity began to be developed. During that 200-year period, the operational calculus of Newtonian mechanics was refined, first by Lagrange and then by the Irish mathematical physicist William Rowan Hamilton. In Hamilton's approach to CM, now universally known as Hamiltonian mechanics, there is a mathematical construct known as 'the Hamiltonian'. Technically, the Hamiltonian is a *generator of displacements in time*, having this role in both CM and QM. Knowledge of the Hamiltonian for a given SUO allows the theorist to calculate how states of that SUO evolve in time. We shall discuss both the Lagrangian and Hamiltonian formulations of CM later in this book.

Hamilton's views on time are remarkably prescient. In the course of his researches on algebra, he wrote:

> The Author of this paper has been led to the belief, that the Intuition of Time is such a rudiment. This belief involves the three following as components: First, that the notion of Time is connected with existing Algebra; Second, that this notion or intuition of Time may be unfolded into an independent Pure Science; and Third, that the Science of Pure Time, thus unfolded, is co-extensive and identical with Algebra, so far as Algebra itself is a Science. The first component judgement is the result of an induction; the second of a deduction; the third is the joint result of the deductive and inductive processes.   [Hamilton, 1837]

The metaphysicists

Henri Bergson [1859–41] was an influential French philosopher. His views on time have much to recommend them in that they do not endow time with an independent existence. Neither do they consider time to be illusory. He believed that time had a property he referred to as *duration*. Although metaphysical in essence, this concept is useful because Bergson associated duration with creativity and memory, which are manifestations of real, physical brains at work. In other words, Bergson tried to relate time to the observers of time, believing that psychical events could not be related to the quantitative time of physics [Buccheri, 2000].

An important name in process time theory is Alfred Whitehead [1861–1947], a mathematician and philosopher. For Whitehead, reality and time are not defined by objects existing in the spacetime manifold: they are defined by *processes*.

Another important process time philosopher was John Ellis McTaggart [1866–1925], who wrote an influential article [McTaggart, 1908] in which he argued the case for the unreality of time. He classified temporal propositions as one of two types, called A Series and B Series. These are discussed further in Chapter 13.

Ilya Prigogine [1917–2003] was a chemist known for his work on chemical reactions which could evolve in one direction and then appear to reverse their behaviour. Such reactions are observed with the observer's sense of time running in the direction consistent with the external universe. The extent to which reactions can reverse themselves is discussed in Chapter 22.

Prigogine rejected conventional statistical mechanics because he did not believe that irreversibility could be derived from reversible principles of mechanics. He was influenced by the metaphysical speculations of Bergson.

The mathematical physicists

As the twentieth century unfolded, great mathematical physicists constructed complex theories of space, time, and matter. Their technical achievements are discussed elsewhere; what concerns us are their views on what space and time meant to them.

Julian Schwinger [1918–94] shared the Nobel prize with Richard Feynman [1918–88] and Sin-Itiro Tomonaga [1906–79] for the construction of the most accurate scientific theory we have to date: quantum electrodynamics (QED). That was just one of his interests. Single handedly, Schwinger developed *source theory*, a powerful approach to observer–SUO interactions that in principle can be used by an observer to extract all the available information about quantum states of that SUO.

Towards the end of his life Schwinger trod a lonely path, developing his source theory on his own, in an attempt to include the observer in the formulation of quantum field theory. This was something that his more conventionally minded colleagues generally chose to avoid, regarding such discussion as metaphysical, when in fact nothing could be further from the truth.[9]

---

[9] Blind adherence to quantum formalism without any attempt to understand it is scientific complacency and worse than metaphysics.

We saw on page 15 the depth of Schwinger's views about QM and the meaning of time and space. In the same article, he pointed out two fundamental facts about observation [Schwinger, 1958]. First,

> Now one may ask whether there is any other general property of matter in bulk that may be significant in its role as the substance of the instrument of measurement. It is tempting to cite here the evident requirement of permanence-if the measurement device is to fulfil its final function of providing a record of the results obtained-which would appear to limit its atomic constituents to a stable configuration of particles or of antiparticles.

Here he is making the easily overlooked point that *timescales* are crucial to this discussion. All apparatus has a finite lifetime: it is necessary that that lifetime be relatively long compared to the timescales of the phenomena being observed. He went on:

> And, from another direction, all measurement involves thermodynamically irreversible changes in the measurement instrument, which could hardly be detected over the vastly more drastic annihilation processes of matter and antimatter.

Here he is referring to the fundamental point about observation: it is an irreversible process.

In every age, there are great thinkers whose thoughts lie fallow for decades, until those thoughts spring to life when the time is right. One such individual was Hartland Snyder [1913–62], who dared to think of the space-time manifold itself as something observable in a quantum sense. By this we mean that in QM, observed numbers are eigenvalues of observables. Snyder gave a model where classical time and space coordinates were replaced by respective quantum operators. The set of eigenvalues would correspond in some way to the picture given by Schwinger of the space-time manifold as an effective model of something going on 'underneath the radar'. Snyder's vision can be related to current theories of loop quantum gravity and spin networks. His vision of anticommuting geometry continues to stimulate the imagination of theorists. We shall discuss his quantum spacetime model in Chapter 27.

Another notable individual was Peiro Caldirola [1914–84], who explored the idea that time was discrete rather than continuous. In a series of papers he demonstrated that discrete time could be a consistent basis for classical and quantum theories, going so far as deriving an explanation for the big mass difference between the electron and the muon [Benza & Caldirola, 1981]. In his work, there is a fundamental unit of time, commonly referred to as a *chronon*, with a value of about 10^{-24} seconds, approximately the time for light to cross a proton. His pioneering approach to discrete time continues to this day [Farias & Recami, 2010].

John Archibold Wheeler [1911–2008] exerted an enormous influence on twentieth-century physics. He initiated the program known as *geometroynamics* that aimed to explain all of physics in terms of curved spacetime geometry. The name *black hole* is attributed to him. In the speculative programme of quantum gravity, he developed with Bryce DeWitt the equation for the quantum state of the universe generally referred to as the *Wheeler–DeWitt equation*. This equation asserts that the quantum state of the universe does not change, reminiscent of Parmenides' view that nothing changes. This equation is discussed in Chapters 8 and 15. Our specific interest here is Wheeler's view on the universe and its relationship to observation, his *participatory principle*, quoted in Chapter 2.

In the context of QM, the physicist W. L. Bragg [1890–1971] said '*Everything in the future is a wave, everything in the past is a particle*', a sentiment that encapsulates the difference between the quantum information held by an observer *before* a quantum outcome is observed and the classical information held *after* it is observed. This inspired the quantum probabilist V. Belavkin [1946–2012] to develop *eventum mechanics*, in which quantum theory is used to predict the future, relative to a specific moment, that is the current present. The past is defined by all the certain knowledge that the observer has built up of observations:

> We show that the quantum stochastic unitary dynamics Langevin model for continuous in time measurements provides an exact formulation of the Heisenberg uncertainty error-disturbance principle. Moreover, as it was shown in the 80's, this Markov model induces all stochastic linear and non-linear equations of the phenomenological 'quantum trajectories' such as quantum state diffusion and spontaneous localization by a simple quantum filtering method. Here we prove that the quantum Langevin equation is equivalent to a Dirac type boundary-value problem for the second-quantized input 'offer waves from future' in one extra dimension, and to a reduction of the algebra of the consistent histories of past events to an Abelian subalgebra for the 'trajectories of the output particles'. This result supports the wave-particle duality in the form of the thesis of Eventum Mechanics that everything in the future is constituted by quantized waves, everything in the past by trajectories of the recorded particles. We demonstrate how this time arrow can be derived from the principle of quantum causality for nondemolition continuous in time measurements.   [Belavkin, 2003]

The above quote from [Belavkin, 2003] refers to *offer waves from the future*. This is a reference to the remarkable ideas of John G. Cramer, who developed the Transactional Interpretation of Quantum Mechanics. In his daring vision of quantum processes, Cramer discusses the 'flow' of information in time and how it can be related to observers and the Born outcome probability rule, something that has defied explanation ever since Born postulated it in 1926 [Born, 1926]. A very readable account of the Transactional Interpretation is given in the recent book by R. Kastner [Kastner, 2013].

4

Cultural images of time

Cosmogony and theogony

Cosmology is the study of the universe as a whole, its constitution, its past and its future as observed by us in the current present. *Cosmogony* is that part of cosmology that aims to study and explain if possible the origins of the universe. Any discussion of time inevitably involves cosmogony because the two concepts are inextricably linked.

Of necessity, cosmogony is a highly speculative metaphysical discipline frequently lending itself to outlandish imagery. It is the subject of the many creation mythologies to be found throughout history in all cultures, religions, and philosophies. That branch of cosmogony that deals with the origin and descent of gods in ancient cultures is known as theogony. The potency of theogony should not be underestimated as it strongly conditions the way cultures view their status in the world and how they are allowed to treat their enemies. What a culture believes about its origins invariably dictates where it wants to go.

In this chapter we discuss some of the views about time held by cultures throughout recorded history. These views are invariably linked to that culture's historical origins (as distinct to what that culture actually believes): when one culture appears within the context of an older culture, the newcomer usually incorporates elements of existing cosmogony and theogony into its own version of 'the truth'. So it was with Hinduism and Buddhism, the Greeks and the Romans, and so on. In comparative studies of Christianity and Mithraism, it has been suggested at various times that one modelled its theogony on that of the other, but which was the borrower and which was the lender remains the subject of debate [Yamauchi, 1974].

Non-scientific cosmogonies are generally inconsistent or circular when examined in detail, invariably describing the creation of the world in conventional temporal terms. For example, in the Biblical account of Genesis, the creation of the world, God took six days to create the universe and then rested on the seventh. The obvious question then is, was God itself subject to time? The best that theologians can come up with is to explain Genesis as a metaphor, which is no explanation at all.

Images of Time. First Edition. George Jaroszkiewicz.
© George Jaroszkiewicz 2016. Published in 2016 by Oxford University Press.

A frequently occurring theme in mythologies is the juxtaposition of two incompatible temporal images. One variant is of God or gods operating in their own version of process time (presumably), looking over a fixed spacetime manifold in which humans exist. These humans believe that they have free will but in fact they do not from the perspective of the all-seeing God or gods. A variant incompatibility is found in Hinduism, where there is a linear time-based process of divine creation conjoined with a human/animal cyclic time process.

Another frequently occurring feature is the appearance of a *temporal zero*, some unique event in a culture's history from which years can be counted. For example, a peasant being told that '*In the third year of King John's reign* . . . ' would understand perfectly the immediate context of the event being described, whereas being told '*in the year* 1202 . . . ' requires a more sophisticated knowledge of history that a typical uneducated peasant would not relate to. For this reason, many cultures would mark their years from the accession of a pharaoh, king, emperor, or consul.

There is one surprising fact about theogonies that is hard to ignore: they are contextually complete. Religious truth values are invariably related to some God or gods, clearly defined in any given culture, often via a sacred book, acting as a primary observer. Notably, there is usually no attempt at deception: religious believers make a point of saying that they cannot 'prove' anything about their concept of primary observer; it is all a matter of faith.

In contrast, some modern paradigms such as the Multiverse [Deutsch, 2004] refer to no primary observers whatsoever and cannot be validated, yet claim to be scientific.

To distinguish between the absolute primary observer/s (sitting *outside* spacetime) referred to in various theogonies, and the scientifically defined primary observers (sitting *inside* spacetime) that proper physics can deal with only, we need to refine our definition of contextual completeness. If an absolute godlike primary observer is referred to, we shall call that *exophysical* contextual completeness, whereas if a physical primary observer is implied, we shall call that *endophysical* contextual completeness. Exophysical contextual completeness does not lead to a generalized propositional classification (GPC) of 3 because there can be no meaningful statement given of relative external context in this situation. At best, the existence of a book relative to which contextual truth values can be validated leads to a GPC of 1. That does not make any contextual 'truth' in such cases scientifically valid.

Stone Age calendars

The *Stone Age* is a period of human prehistory ranging approximately from about 3.4 million years ago [McPherron *et al.*, 2010] to about 10 – 8,000 years ago, characterized by the use of stone tools. During this age humans evolved from using primitive stone tools to creating sophisticated art and practising ritual burial.

The end of the period is conveniently marked by the development of metalworking, but this took several thousand years in different parts of the world.

The Stone Age is frequently divided into three periods known as the Palaeolithic Age, the Mesolithic Age, and the Neolithic Age. There is no generally agreed demarcation line between these periods, different parts of the world appearing to make transitions at different real times. Paleolithic humans hunted in tundra and lived in caves, Mesolithic humans fashioned tools and hunted in forests as well, whilst Neolithic humans started agriculture and settlements.

The development of mortuary ritual has been traced as far back as 100,000 years, a development linked to the evolution of the brain structures known as the temporal and frontal lobes [Joseph, 2011]. The temporal lobe plays a fundamental role in the formation of long-term memory, without which a sense of the past could not be meaningfully developed. The frontal lobe is involved in the future: it helps the brain process information about current actions and predict future consequences, as well as providing a decision mechanism for choosing the best alternative, according to whatever conditioning the individual has. The transition from the Paleolithic to the Mesolithic has been seen as a change in metaphysical thought [Otte, 2009]: Paleolithic man was the victim of nature whereas Mesolithic man could attempt to control it. The Neolithic period is marked by the development of agriculture and its consequents, such as settlements.

The development of temporal consciousness must have led to a sense of a universe beyond the immediate environment, there being credible evidence for carved Paleolithic lunar calendars and abstract cave art as far back as 30,000 years ago [Pásztor, 2011]. According to one astro-paleontologist,

> Although it is impossible to date these discoveries with precision, it can be concluded that spiritual consciousness first began to evolve over 100,000 years ago, and this gave birth to the first heavenly cosmologies over 20,000 years ago.
>
> [Joseph, 2011]

The Mesopotamians

As the Stone Age came to an end, hunter–gatherer and nomadic herder lifestyles were gradually supplanted by ones based on agriculture. This led naturally to the rise of settlements localized around crop-growing regions along with produce surplus and trade. This led to the development of the first cities. The Sumerian civilization that developed in Mesopotamia is commonly acknowledged to be the first on the planet in this respect.

The Sumerians and Babylonians

The date for the foundation of the oldest known Sumerian city, Eridu, is *c*5400 BCE. Historians believe that the Sumerians invented the system of time that we use to this day, one based on the number 60. This number has the mathematical

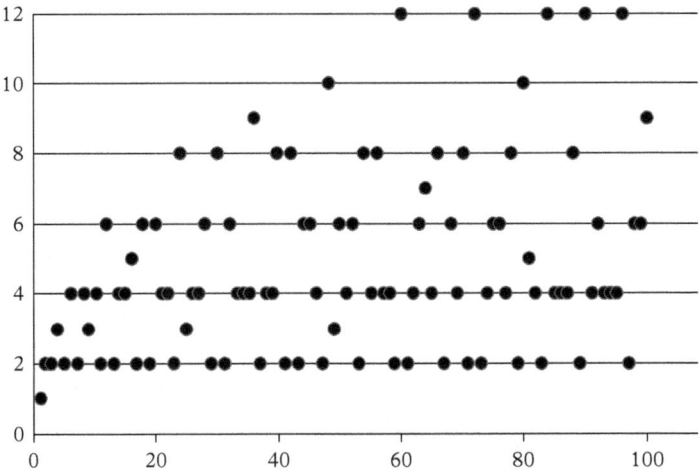

Fig. 4.1 *Graph of integers versus the number of their factors.*

property of being *highly composite*, meaning that it has relatively many factors given its size. It is the smallest number divisible by 12 different integers (counting one and itself) and it is the smallest number that is divisible by the integers 1, 2, 3, 4, 5, and 6. Figure 4.1 is a graph of the first 100 integers against the number of their factors.

The Sumerians divided hours into 60 minutes and applied the same system to geometry, dividing one complete rotation into six multiples of 60 degrees, with each degree being divided into 60 minutes.

Every year at the start of their New Year festival Akitu, the ancient Babylonians re-enacted their creation myth, the *Enuma Elish*. Recovered from clay tablets found at the ruined Library of Ashurbanipal at Nineveh and dated as far back as the sixteenth or eighteenth century BCE, this theogony myth recounts the rise of the chief Babylonian god Marduk and the creation of humans.

The Zoroastrians

Zoroastrianism is one of the oldest extant religions. There are two time concepts in Zoroastrianism: one involves a measurable time with a beginning and an end whilst the other is infinite time, without beginning or end.

Mithraicism

The central figure in Mithraicism is Mithras, a being standing outside of space and time [Ulansey, 1991], that is, an exophysical being existing in the space beyond the Cosmos.

The Middle East

The Ancient Egyptians

Time figured in two distinct ways in Ancient Egypt. On the religious level, the pyramids were constructed to protect the pharaoh's body after his death. In life the pharaoh represented the god Horus, who would raise the Sun every day. In death, he represented Osiris, who set the Sun each day. The pyramid protected the body of the previous pharaoh to ensure a cosmic balance. The pyramids had to be massive to ensure this process would continue.

On a more mundane level, the Ancient Egyptians developed the temporal structures we use today, dividing the day into hours and minutes and constructing elaborate calendars. These were fundamental to Egyptian civilization, as its agriculture depended critically on the annual flooding of the Nile.

The Egyptian calendar had three components. The lunar calendar was for religious purposes and based on 12 lunar months. A 13-month calender was used to intercalate with the rising of Sirius. The civil or administrative calendar had 12 months of 30 days, with an additional five so-called *epagonal days* added each year, to give the required 365. This third calendar was used to match the other two calendars.

To keep track of time, the Egyptians used sundials, but they are credited with the invention of *clepsydra*, or water clocks.

There were two ways ancient peoples defined hours. *Temporal hours* divided the period of sunlight into 12 equal hours, but the problem with this is that such hours change with the seasons, being shorter in winter than in summer. Around 127 BCE, Hipparchus of Niceae, who was based in Alexandria, proposed dividing the day into 24 equal periods known as *equinoctial hours*, and these would have been much more useful to Egyptian astronomers than temporal hours.

Judaism

According to Jewish sources, the image of time as an entity or continuum (manifold time) was absent in ancient Judaism; reality and change were viewed as processes [Stern, 2007]. There are many major events in Jewish history that retain enormous cultural significance but are not used as temporal zeros (i.e. significant events used as starting points in a historical record). These include the Exodus from Egypt dating to around 1300 BCE, the destruction of the First Temple of Solomon by the Babylonians in 587 BCE, the destruction of the Second Temple by the Romans in 70 CE, and the more recent Holocaust. If any event were to be used as a temporal zero, it would probably be the creation of the world, calculated as 5751 BCE.

Christianity

This religion is emphatically based on a linear image of time: God made the world in six days, Christ was born about 4 BCE and the day will come when the world ends. Bishop Ussher took this image so seriously that he analysed the Bible and concluded that:

> In the beginning God created the heaven and the earth. This beginning of time, according to our chronology, happened at the start of the evening preceding the 23rd day of October in the year of the Julian calendar, 710 (i.e., 4004 BCE)
> [Ussher, 1658]

Islam

Like the Christians, Islamic theologians adopted the idea that time is linear. They went further in emphasizing its irreversibility and the value of not wasting it. Therefore, life should be structured around the most important activities. In Islam the most important activity is religious duty, so this leads to the regular, five-times-a-day prayer observation characteristic of that religion.

Islam developed an image of time that reverberates to this day. In the eleventh century CE, the Islamic polymath Avicenna [980–1037] doubted the existence of physical time, suggesting that time exists only in the mind due to memory and expectation. This is in accordance with our discussion earlier of the function of the temporal and frontal lobes.

The enduring legacy of the very influential Islamic scholar Algazel [1058–1111], discussed in Chapter 3, is the view in Islam that causality is synonymous with the will of Allah.

The Europeans

The Julian Calendar

The Romans used 753 BCE, the traditional year of the Founding of Rome, as a temporal zero, but that was a convenient fiction. Unlike the Greeks, who started their hours at sunrise, the Romans started their day of 24 hours at midnight. However, despite the availability of regular water based clocks known as clepsydra, from the middle of the second century BCE, the Roman *hora*, or hour, was variable: daylight was divided into 12 hora, as was night. Accurate timekeeping was important only in special situations such as law courts, where the amount of time a speaker could have was regulated.

The Roman preoccupation with time has left us an enduring legacy of Roman calendars. Before the reforms of Julius Caesar, the Roman calendar year was drifting away from the solar year, requiring the insertion of an intercalary month to close the gap. Caesar addressed the problem on the advice of astronomers, resulting in the Julian calendar, consisting of 12 months, each with its own fixed number of days apart from February. There is a useful rule based on the knuckles of the human hand for remembering the number of days in each month. Starting with January on the outer knuckle of the left hand, we associate February with the space between the next knuckle, March is associated with the next second knuckle in, and so on. July is associated with the innermost (fourth) knuckle of the left and then August is associated with the innermost knuckle of the right hand, and so on. For each knuckle month we associate 31 days, whilst for each space month we have 30, apart from February, which has 28 normally and 29 on a leap year.

The Gregorian calendar

Despite stabilizing the Roman calendar for many centuries, the Julian reforms did not fully succeed. Eventually there was a perceptible disagreement between the Julian calendar and the solar year. Pope Gregory reformed the calendar in 1582 to the form used in the West to this day. The Julian Calendar is still used in some parts of Europe, notably by various branches of the Orthodox Christian Church, celebrating Christmas on 25 December, Julian calendar, which corresponds at this time to 7 January, Gregorian calendar.

Medieval Europe

The development of a linear view of time in Western Europe has been attributed to the appearance of church clocks. In the early fourteenth century, big mechanical clocks began to appear in Italian city church towers. These clocks were driven by gravity rather than the flow of water. Over the centuries, various improvements were made, such as the use of springs to power the movement and the pendulum as a regulator.

Before the appearance of clocks, Western Church authorities divided the day into 12 daylight hours and 12 night-time hours. As time-keeping became more accurate, the ringing of church bells synchronized with ever more reliable and accurate clocks would have impacted on the ordinary person's perception of time. In particular, labourers and their employers would have had a better basis to define and monitor the hours of paid employment.

Time was generally poorly synchronized until the development of railways, which showed up differences in local time standards across the country. Greenwich Mean Time was set up as the legal standard in Britain in 1880, driven by the needs of the railways.

Eastern philosophies

Hinduism

Cyclic time is a well-known concept in Hinduism, linked to a belief in *reincarnation* and the law of cause and effect known as *karma*. Solipsism too has been referred to in Hindu philosophy, in the *Brihadaranyaka Upanishad*, dated to the first century BCE.

Buddhism

The Buddhist view of the physical universe is a remarkably modern one, in which permanence and stability are the exception, not the rule. Indeed, the Buddhist view goes further than the prevailing Block Universe model of General Relativity in asserting that time is not a geometric quantity but a process [Goleman, 1996]:

> In Buddhism, the concept of linear time, of time as a kind of container, is not accepted. Time itself, I think, is something quite weak- it depends on some physical basis, some specific thing. Apart from that thing, it is difficult to pinpoint- to see time. Time is understood or conceived only in relation to a phenomenon or a process. [Interview with H.H. the Dalai Lama]

Like the Hindus, Buddhists believe in reincarnation, viewing *cyclic time*, that is, *reincarnation*, with apprehension, that is cyclic time is *profane*.

A central concept in Buddhism is *dependent origination*, the assertion that all things occur because of explicable cause. This is the motivation for the practice of meditation: understanding the cause of a problem, particularly with the way a person's mind thinks about it, leads eventually to the coming to terms with it.

Australian Aboriginal concepts of time

Dream time is a commonly misunderstood term used to encapsulate Australian Aboriginal beliefs about time and space. The Aboriginal image of time is based on a perception of an individual's place and role in their society that is different to modern Western thinking. Dream time is a view of reality that encompasses the beginning of life, the influence of ancestors, and of death, a view that transcends the time and space of ordinary experience. Aborigines think of it in terms of an 'all-at-once' image of reality rather than a 'one-thing-after-another' image of time that Western thinking holds to. Dream time can be experienced through altered states of consciousness induced by tribal mythology and rituals.

The Central Americans

The Maya

The Maya are famous for their sophisticated calendar and their system of recording time, a system still in use in parts of Central America. The Mayan system is based on mathematical relationships between a multitude of cycles of astronomical, religious, and cultural significance. These cycles were principally measured in days but some of the longer cycles ran to many thousands of years.

Mayan mathematics was *vigesimal*, that is, based on base 20, in contrast to our standard base 10. As we saw with the Babylonians above, the choice of base in any number system is an important factor in the flexibility and power of that system. Interleaving with their base 20 were some important factors outside of the Maya's influence, such as the time the Earth takes to orbit the Sun. The solar year is approximately 365.25 days, which cannot be described exactly by any normal cycle. The Maya attempted to model the solar year approximately using integer-based cycles, forcing them to use some unusual prime numbers that in the long run gave rise to periods of extraordinary long duration.

Some of the cycles in their calendar are the following:

1. The *Tzolkin cycle* T lasted 260 days, which means we may write $T = 2^2 \times 5 \times 13$.

2. The *Haab cycle* H was related to the solar year and lasted 360 days, so we may write $H \equiv 360 = 5 \times 73$. Clearly, H is a poor approximation to the solar year of 365.25... days. Haab was defined as 18 Mayan 'months' of 20 days duration each, plus five 'nameless days', or *Wayeb*.

3. The *Calendar Round cycle* C consisted of 52 Haab cycles. This number is determined as the shortest time that contained an integral number a of Tzolkin cycles and an integral number b of Haab cycles. The calculation requires us to solve the Diophantine equation

$$a \times T = b \times H, \tag{4.1}$$

which has an infinite number of solutions. The Calendar Round cycle involves the smallest non-trivial values, which are readily found to be $a = 73$ and $b = 52$.

We see here the (perhaps undue) dependency in the Mayan theory of time cycles on what is effectively a random number, that is the length of the solar year in days. The Haab cycle is very close to the solar year. Suppose that the solar year was precisely 367 days rather than 365.25 and that the Haab cycle was then taken to be 367. Keeping the Tzolkin value $T = 260$ then the modified Calendar Round cycle would require us to solve the diophantine equation

$$c \times 260 = d \times 367 \tag{4.2}$$

for integers c and d. However, 367 is a prime number, so we deduce that c must be a multiple of 367, leading to the conclusion that $c = 367$, $d = 260$ is the required solution. This means that the Calendar Round cycle jumps from lasting 52 'Old Haab' cycles to lasting 260 'New Haab' cycles, that is, just over five times longer.

4. The Venus cycle: as keen astronomers, the Maya were aware of the rising and setting of the planet Venus and associated a Venus cycle with it consisting of 584 days.

The Mayan calendar has been described as a 'Long Count', starting on a specific day, the date the Maya associated with the creation of the world and identified as 11 August, 3114 BCE in the proleptic Gregorian calendar and 6 September 3113 in the Julian calendar. Starting with the *k'in*, or one standard day, the Mayan calendar worked with a sequence of ever greater spans of time. After the k'in came the *winal*, which was the 20 day Mayan 'month', followed by the *tun*, a Long Count 'year' of 18 winal, and so on. At the other end came the *Alautun*, equivalent to a staggering 23,040,000,000 days, or 63,081,429 solar years.

A remarkable point about the Mayan's scale of time is that the Alautan dwarfs Bishop Ussher's scale of time (approximately 6,000 years) by a factor of about ten thousand.

The Aztecs

The Aztecs had a philosophy of time in which every day had a religious significance. For them, time went in cycles of repeated destruction and recreation of the world. The Aztec calendar system was less precise than that of the Maya but it shared a number of features common to Central America. There was a 365-day agricultural cycle called the *year count* and a 260-day ritual cycle called the *day count*. These two cycles together formed a 52-year cycle called the *calendar round*.

The Incas

The Incas regarded space and time as a single concept named *pacha*, translated as 'world' or whole cosmos in their language of Quechua, but including also a temporal context, such as a specific moment in time.

The Far East

The Chinese

In common with all cultures, the traditional Chinese view of time impacts on daily life. Time is valued, so punctuality is important. However, time is spent in thought and mental preparation, reflecting on actions to be taken.

The Japanese

In Japanese traditional culture, time is not measured according to its duration but to the quality of what is done. This is exemplified by the famous Tea Ceremony, a ritual that requires respect for the proper order of the ceremony, courtesy, and tradition. On the other hand, the modern Japanese worker will be meticulous in their work rate and attention to time schedules.

5

Literary images of time

Introduction

The twentieth century is remarkable for many things, one of which is the extra-ordinary diversification of literature into many genres and formats. Of interest to us in this chapter are those novels and films that discuss time travel. Many such novels and films are stories where time travel is just a literary plot device with little or no thought given to consistency or to physical implications. In many stories, extraordinary events happen because of 'magic' or super-physical powers granted to individuals. The Greek myths contained such elements. Another common plot device in many time-travel stories is to introduce 'alien technology'. This plot device allows a writer to maintain a pseudo-scientific facade allowing any sort of unphysical possibility into a story on the logically false basis that 'because we cannot prove the non-existence of aliens and such technology, we are entitled to assume that they exist'.

There are benefits and dangers in reading time-travel fiction and watching films based on time travel.

Benefits of science fiction

Properly conducted science is hard: real theoretical and empirical advances can take centuries. During the long periods between paradigm shifts, scientists tend to be conditioned into modes of thought that become fixed in the prevailing para-digm and ultimately require revision. A good example is the advent of special relativity in the early years of the twentieth century. Before 1905 many theorists had an ingrained space-time view of Absolute Time and Space but after Einstein's landmark paper [Einstein, 1905b] and Minkowski's famous talk on the geomet-rization of spacetime [Minkowski, 1908], scientists' view of time was in terms of spacetime.

An obvious benefit of science fiction is to provoke its audience out of condi-tioned modes of thinking and into novel directions. In particular, a good film or novel can inspire young people to use their imaginations to think about the past and the future as not being fixed but both possibly subject to alteration. The value

Images of Time. First Edition. George Jaroszkiewicz.

here is that when they become better trained as scientists, these young people can use the rigours of their scientific disciplines to explore the validity of the science-fiction concepts they were inspired by as children. Many scientists will acknowledge the tremendous excitement they felt watching *Dr Who*, the archetypical time traveller, first screened on British Television in 1963, still running and immensely popular around the world. Many scientists would have been stimulated to go into science because of such programmes.

A particular feature of science fiction is that it can appeal to all classes of audience: Isaac Asimov's novel *End of Eternity* can be just as thought provoking to a seasoned mathematical physicist as it was to this author when, as a 16-year-old school boy, he read it just before a high school physics examination. In order to achieve a broad appeal, a cardinal rule in science fiction is that mathematics should be avoided at all costs. It is then down to the skill of the author to present their concepts in words rather than symbols.

Dangers of science fiction

The downside to science fiction is that unwary readers can forget that second word, *fiction*, and start to believe that inconsistent or scientifically baseless assertions in novels and films represent actual physical reality. This leads to a phenomenon readily identified in online science forums and debates: many contributors, whether scientifically trained or not, discuss time travel by *assertion*, as if the mere statement of a nonsensical or impossible concept has to be given as much weight as those for which there is at least a modicum of evidence.

One of our aims in this book is to give some guidance, via the principle of contextual completeness, in sifting out those concepts that are unverifiable and those that could make sense. For instance, and with the greatest respect to both authors, if we were journal editors and had to review Leibniz' article on monads, we would classify it as poor science fiction [Leibniz, 1714], whilst Hume's ideas on causality would be acclaimed as contextually complete and giving some important scientific insights [Hume, 1739].

Time travel novels invariably have to introduce some ad hoc technology such as a time machine, or postulate some extraordinary mental or biological power, in order to justify protagonists moving around in time. There is no known scientific foundation for any such postulates, apart from one spectacular and stimulating concept: the possibility of *closed timelike curves* (CTCs) encountered in some general relativistic spacetimes, such as that of Gödel [Gödel, 1949]. CTCs and Gödel's spacetime are discussed in Chapter 20.

Literary varieties

In this section we discuss some temporal concepts that have received some useful literary attention.

Time as the fourth dimension

The writer H. G. Wells [1866–1946] is famous for pioneering science fiction novels. He wrote the most famous time-travel novel of them all, *The Time Machine* [Wells, 1895]. In that story, the hero first explains that time is just like another dimension of space. Physical objects do not exist only for infinitesimally short instants of time, but have duration, that is, extension in time. Since there are three conventional spatial dimensions (left–right, forwards–backwards, and up–down), time is reasonably regarded as the *fourth* dimension.

In Wells' vision, time and space are to be thought of as unified aspects of a single geometrical entity, a four-dimensional continuum referred to in this book as *spacetime*. What is remarkable is that this image of time preceded Minkowski's geometrization of Einstein's special relativity by over a decade [Minkowski, 1908].

In the *Time Machine*, Wells gives a reasonably convincing explanation for our usual inability to travel back and forth in time. He considers the difference between the vertical spatial dimension (up–down) and the two horizontal spatial dimensions (left–right and forwards–backwards) that we are used to at any place on the surface of the Earth. We have freedom to move back and forth in any horizontal direction, but gravity always forces us downwards in the vertical dimension. It was only with the advent of air balloons and airships that humans could defy gravity and travel at will up as well as down. In his story, the hero has constructed an analogous device, *the* Time Machine of the title, that liberates the time traveller from the forces that push them normally always in the same direction along the temporal dimension.

In order for his story to have some credibility, Wells has to introduce a plot device in the form of the time machine itself, a vehicle in which the hero sits and is transported back and forth in time. Wells gives no real details (how could he?), writing

> . . . we all followed him, puzzled but incredulous, and how there in the laboratory we beheld a larger edition of the little mechanism which we had seen vanish from before our eyes. Parts were of nickel, parts of ivory, parts had certainly been filed or sawn out of rock crystal. The thing was generally complete, but the twisted crystalline bars lay unfinished upon the bench beside some sheets of drawings, and I took one up for a better look at it. Quartz it seemed to be.   [Wells, 1895]

Time as a parameter

Raymond King Cummings [1887–1957] was a science fiction writer of the 1920s, regarded as a founding father of the literary genre known as *Pulp Science Fiction*. Such literature is characterized by a clear disregard for the laws of physics, typical stories being centred on conflicts between good and evil found in comparable genres such as Westerns, but dressed up in a pseudo-scientific cloth. However, on occasion, some interesting observations about the physical world can be found buried deep in otherwise completely forgettable and sometimes embarrassingly bad stories. In what is regarded as his best story, *The Girl in the Golden Atom*, Cummings has one of his characters make the useful and memorable observation:

Time . . . is what keeps everything from happening at once.   [Cummings, 1922]

From our point of view, this is another way of saying that time is one of the parameters used to label different physical states.

The inertia of reality

Isaac Asimov [1920–92] is widely regarded as one of the finest science-fiction writers of the twentieth century. He wrote the *Foundation* series of novels, based on the rise and fall of a human empire in the galaxy. Linked to these stories is one particular novel based on time travel, *The End of Eternity* [Asimov, 1955]. In that novel, Asimov discusses some thought-provoking concepts to do with time travel:

1. There are two flows of time: the usual time of *Reality*, equivalent to the time of the normal universe that we are living in, and *physiotime*, the time of *Eternity*, the realm of existence of the 'Eternals'. In this two-component image of time, we may identify Asimov's *Reality* with manifold time and his *physiotime* with process time.

2. The Eternals, humans living in Eternity, stand outside of the normal universe and have an exophysical view of it. They are able to see past and future in the 'current reality' of the normal universe. Their aim is to manipulate events in the current reality in order to improve life for the average person living in it.

3. The Eternals view Reality in a Block Universe form, that is, they are able to see Reality throughout a specific range of normal time. This starts in the twenty-third century (as we would know it), when Eternity was built, and goes 'upwhen' many billions of years. Power for the maintenance of Eternity outside the fabric of the normal universe is from 'Nova Sol', that is, the energy generated by the conjectured final supernova explosion of our Sun in several billion years' time.

4. An Eternal can enter any point of Reality at will. From the perspective of normal individuals embedded in Reality, such an Eternal would appear to materialize from thin air.

5. An Eternal can change critical events in the current Reality. Ordinary causality within Reality then spreads out waves of change from a given critical event into the future of the current Reality (as seen by Eternals back in Eternity). Assuming the Eternals see the laws of Reality operating according to standard physics, such as special relativity, the reality changes done

by the 'Technicians' of Eternity would all be located inside and on a re-
tarded lightcone in the spacetime of Reality, with its vertex centred on the
critical event. Lightcones are discussed in Chapter 17.

6. Asimov did not assume that induced Reality changes increased in an unbounded way. He developed the concept of *the inertia of Reality*: when a
critical change is made by an Eternal in Reality, waves of change spread
ever outwards, but get reduced in magnitude the further the Eternals look
in the Block Universe view that they have.

Asimov was a trained research scientist and academic, and his discussion of
time gives a strong impression that he thought carefully about the logical structure
of the *End of Eternity*.

Time and irreversibility

In Tom Stoppard's play *Arcadia* [Stoppard, 1993], the action revolves around
the precocious Thomasina, a 13-year-old girl who writes about, and discusses
with her tutor Septimus Hodge, the irreversibility of time and chaos, years before
these concepts were developed by mathematicians and scientists. An unusual and
effective dramatic technique is to interleave action from the years 1809–12 and
the present.

Stoppard's play impacts on audiences in several ways. First, it brings fundamental concepts of time to the attention of the non-scientist in the audience in a
way that they can relate to and discuss themselves. Process time is, after all, experienced by everyone. The second impact is to show that thinking about time
may be unusual but can be a reasonable activity for non-professional scientists,
given enough imagination.

Hyperspace

One of the predictions of relativity is that there is a maximum speed, the speed of
light, so that no physical object, effect or signal can travel faster than that speed.
This is not a significant factor on small distance scales such as over the surface
of a planet: light can travel seven times around the world in one second, which
means that the ordinary human observer can disregard this limit under normal
circumstances.

A significant problem arises for science fiction writers when they have to deal
with communication between well-separated planets in different solar systems or
even in different parts of our galaxy. This is because conversations would stretch
over many years between planets when their separations are measured in light-
years (approximately 6 million million miles). This is a clearly inconvenient plot
factor in a novel, so many authors, including Asimov, have simply ignored the
problem by postulating a short cut referred to as *hyperspace*. A space ship with

hyperspace travel capability, usually referred to as a *warp drive or hyperdrive*, enters hyperspace near the ship's point of origin and emerges instantly (as far as the ship's occupants are concerned) many light year away, close to its destination planet.

A variant approach is taken by Frank Herbert in his *Dune* novels: the Guild Navigators can 'fold space' when under the influence of the spice Melange, and this enables space ships to pass instantly from one part of the galaxy to another. Such a concept can be mathematically described as a wormhole in Einstein's theory of general relativity.

Alternate universes

A common theme in science fiction is the existence of parallel universes, where different versions of reality play themselves out. In films such as the *Back to the Future* trilogy, the hero Marty has to change events in his past in order to return to the parallel universe that he started off in.

In this particular case, photographs are posited as having a semi-permanent record of alternate realities, and it is only by having one of these that Marty knows that reality is different to what it should be. Without this convenient plot device, Marty would have no incentive to change the future.

Alternate futures

In the acclaimed *Star Trek* episode, 'City on the Edge of Forever', Captain Kirk and Mr Spock have to decide whether the social worker Edith Keeler should die or not. Depending on their answer, they will either prevent her from dying and consequently changing the future that they want to restore and return to, or she will die and the future reverts to what it 'had' been before they were stranded in the past. Behind this conundrum is a machine-like being, the Guardian of Forever, that plays the role of an exophysical observer, relative to which existence and time are defined and can be seen to change.

Information from the past

The television police series *Crime Traveller* uses the plot device of a time machine that enables a detective to travel back in time to solve crimes. There is nothing in the laws of physics that forbids the uncovering of information from the past: archaeologists do it all the time. The use of a time machine in this series just adds some interesting novelties to police procedures. Any attempt by the detective to change the past is always thwarted in some way: a murder cannot be prevented but the identity of the murderer can be discovered. In one episode, the hero detective goes back in time and places a winning bet in a lottery, knowing its outcome, but the attempt fails due to one oversight or another. This is the weakest element in these stories, because information is contextual: marks on a piece of paper may be a winning lottery number to a man but meaningless to a dog. That the universe would know the difference and conspire to erase information from the future depending on its context seems not credible to say the least.

Temporal consistency

In the fourth *Star Trek* film '*Star Trek IV: The Voyage Home*' (the one with the whales), whilst back on the Earth of their past, Dr McCoy asks the Engineer Scotty if giving information from the future to the past about transparent aluminium might not cause a temporal paradox. Scotty replies that they don't know for sure that such a temporal circle of information flow was not responsible for transparent aluminium existing in the future in the first place. That answer satisfies McCoy, as it is logically consistent. The moral here is that time travel need not cause inconsistencies, a point made by Novikov [Novikov, 1998].

6

Objective images of time

Introduction

Any discussion about time will be influenced by numerous factors such as our familial and national cultures, our personal experiences, the people we regularly talk to, our mental outlook, and our senses. Even the language we think in will play a role, for in some languages there is not the same richness of tense, that is, expression of past, present, and future, as in others. All of these factors play a role in how our brains are conditioned to interpret the information that we receive from our immediate environment. In this chapter we review those factors that are dictated by the laws of physics, starting with light.

Optics

Humans are primarily visual creatures: our eyes are superb detectors of the optical frequency regime of the electromagnetic radiation spectrum, commonly referred to as light. Vision provides the brain with sufficient information to create the illusion that the world around us is a three-dimensional geometrical space in which objects exist and move about in a continuous way. This illusion comes at a cost: we become conditioned to believe this geometrical model *is* reality, whereas the model may be no closer to reality than a map is to the country it represents. Therefore if we wish to understand time we should keep the following scientific facts about light in mind.

The optical spectrum is limited

As a wave process, light is characterized by its wavelength λ or equivalently by its frequency ν. These are related by the rule $\lambda \nu = c$, where c is a universal constant known as the *speed of light*. That part of the light spectrum that we can detect with our eyes is in the frequency regime that runs from red to blue colours. This regime is a relatively miniscule part of the entire frequency spectrum of light: in principle, electromagnetic radiation (for that is what light is) can contain frequency

Images of Time. First Edition. George Jaroszkiewicz.
© George Jaroszkiewicz 2016. Published in 2016 by Oxford University Press.

components ranging from near zero to near infinity. When scientists analyse information from other frequency regimes such as x-rays or infra-red light, the universe looks very different.

Animals such as bats and dogs build complex and successful mental images of their environment based on non-visual information: bats use sound waves and dogs have a powerful sense of smell. Likewise, in the absence of visible light or a breakdown of vision, humans have developed alternative mental images of their environment. In medicine, for instance, humans exploit non-visual techniques such as x-rays, thermal and magnetic imaging, and ultrasound to give radically different images of the human body compared to visual images. Generally, these technologies are translated into visual images that are then seen by the human eye as if they were optical images.

The speed of light is relatively great

A crucial fact underpinning the illusions created by vision is that c is enormous relative to speeds ordinarily encountered by humans: light travels over a hundred million times faster than a car moving at a hundred kilometres an hour. This is why humans have developed a sense of time which deceives us into believing that we see events as they happen: in ordinary human terms the speed of light is effectively infinite. This contributes to the conditioning that things are just happening 'out there' and that we can know all about them right there and then as they happen. This underpins the notion of simultaneity that classical mechanics (CM) has built into its foundations.

Thunderstorms remind us, however, that the speed of transition of signals is a crucial factor in our mental images: there is usually a delay of several seconds between us first seeing a lighting strike and then hearing the accompanying thunder. Light travels close to a million times faster than sound in Earth's atmosphere.

If light took several seconds to reach us from nearby objects, our conditioned view of time would probably have evolved to be very different to the one we have now. We experience something like this when we are in a dense fog: our perceptions then become dominated by sounds and touch rather than vision. Under such circumstances, when delays become significant, our sense of time can appear distorted and we no longer have the immediacy of vision.

On rare occasions, we may be reminded that the speed of light is relatively enormous compared to the speed of sound. The author once saw a farmer driving a fence-post into the ground with a large hammer, the farmer being on the other side of a valley. The sound of the hammer blows was heard halfway between the sight of the hammer hitting the post, creating an illusion that the sound was being generated by the hammer hitting the air above the farmer's head.

Sunlight is strong

Sunlight is relatively bright on the surface of the Earth, even though our planet is about 150 million kilometres from the Sun. During daylight hours, the world around us is flooded with sunlight, so much so that our eyes receive what appears to be a continuous visual information stream, a veritable flood of light. This flood can be so intense that it poses dangers to our eyes if we observe it for too long and contributes greatly to our sense of time. When we see things in bright daylight, there is so much optical information that our brains interpret it in terms of objects moving continuously in time and space.

Quantum physics gives us a different perspective. In 1900, the physicist Max Planck was trying to understand the observed frequency spectrum of light being emitted from enclosed containers, the so-called 'black body radiation'. He found that the continuous flood model of light did not stand up to critical examination. Instead, a better fit was a model where atomic detectors in the walls of the containers could absorb light only in discrete lumps called quanta [Planck, 1900]. Five years later, Einstein suggested that light itself consists of particles carrying energy in discrete lumps [Einstein, 1905a]. The name *photon* for these particles was not coined until 1926 and is attributed to the American physical chemist Gilbert Lewis [Lewis, 1926].

There is evidence that the human retina can respond to individual photons [Hecht & Pirenne, 1942], but transmission of information from the retina to the brain is filtered so that small numbers of photons do not count. Otherwise the world would look chaotic and very different to what we normally think we see.

At night, when sunlight is absent, or when there is a fog that disperses sunlight, our visual interpretation of the world around us can make us believe that time has slowed or even stopped.

Normal objects do not shine of their own accord

All objects are subject to the laws of thermodynamics, which means that under normal circumstances, they have a temperature and emit electromagnetic radiation characteristic of that temperature. However, most objects do not emit enough light during the night to stimulate our vision. For that we require sources of light such as torches to stimulate objects to emit light. This is also true during the day, for the temperature during the day is not greatly different to that at night in absolute terms. It is easily forgotten that the magnificent vistas of mountains, prairies, oceans, and clouds that we perceive during the day are created by sunlight: when that goes, so do the illusions.

In addition, it is easily overlooked that what we see is generally only skin deep: we see light from relatively thin surface layers of atoms and molecules of the objects we are looking at.

Light momentum is negligible

Consider an ordinary object such as a tennis ball moving from a server S to their opponent O. Such objects carry energy and momentum from S to O according to the laws of classical mechanics (CM). Newton's third law of motion tells us that S will experience a back-reaction on their racket as they serve whilst O will feel the momentum of the ball as their racket hits it in reply. Visible light appears not to obey these rules: it seems to carry no momentum. Contrary to Newton's third law of motion, there is no noticeable back reaction on objects that we are looking at and our eyes do not feel any pressure as we look. This apparent lack of 'back reaction' contributes to the impression that vision is non-destructive to both the source of light and to our optical systems.

Again, nothing could be further from the empirical facts. Light does carry momentum and it can be highly destructive. Even optical frequency light can degrade surfaces such as photographs and works of art. The dangers of strong sunlight are now well established: a severe suntan is not so much a fashion statement as a signal of skin damage.

Optical transparency

A critical physical factor that allows the brain to create the space-time illusion is that air is transparent to visible light. When we are in a fog or diving in a murky pool of water, that illusion is destroyed and the universe around us suddenly becomes a very different place.

In early universe cosmology, the reverse is held to have happened: space became transparent to light only around 377 thousand years after the Big Bang, at the point where hydrogen and helium atoms could form. These are electrically neutral and therefore do not hinder the passage of light in the same way as charged particles. Before this so-called recombination time, the universe is best described in terms of a succession of phases or epochs, depending on the dominant form of interaction at each stage.

Other factors

The laws of physics impinge on our model of time and space in other ways besides the visual. Here are some points to think about.

Timescales

The timescales on which humans operate are crucial to the illusions created by our brains. Our brains can process complex visual information in about 150 thousandths of a second [Thorpe *et al.*, 1996]. When we look at structures that persist for say many hours or even weeks, such as icebergs, then we are conditioned to believe that those objects are relatively permanent. It then becomes natural to objectivize such structures, assigning them names, positions, and other properties as if they were indeed real objects. We would never assign names to individual waves in the sea or gusts of wind, as they disappear too quickly in relative terms to mean anything to us in context. Storms and hurricanes can last for days on Earth and are commonly objectivized and named, but tornadoes and sunspots, which do not last long, are not generally given names. A notable example in astronomy of objectivization is the Great Red Spot on Jupiter, a storm that is known to be many centuries old and has therefore been objectivized as an object in its own right.

Thermodynamics

On a warm summer's day, when we look around us and see a tranquil scene, it is too easy to forget that this image is but an illusion, masking a microscopic world teeming with movement on atomic scales, a seething mass of atoms and molecules in dynamical thermodynamic equilibrium. We do not see the electromagnetic energy pouring from the Sun constantly onto the trees, houses, and mountains in the distance. We do see some of that radiation when those objects reflect or re-emit the light that falls on them: it is that feeble light that induces out brain to create the images of stability that we perceive. If the Sun stopped for some reason, our perception of the universe would change dramatically.

Observational context

Owners of pets will know that some animals do not appear to recognize images in mirrors or on television screens. As for humans, lawyers know that witnesses who saw the same event may give honest but contradictory accounts of what happened. All this suggest that the mental images of the world around them created by an observer depend as much on that observer as on the properties of the SUOs being observed. It may be unwise therefore to believe in an objective image of time based on human experience. If an observer such as a dog can interpret visual information so differently to a human, then the human view of time may be irrelevant to them.

Observer independence

The flood of information transmitted by sunlight is so pervasive and cost free that the illusion is created that different observers can see the same objects to the same extent. Again, nothing is further from the facts. This illusion necessarily breaks down when the light intensity falls to the regime where individual photons are being registered, because a photon cannot be absorbed by more than one detector, by definition. It is at this point that the laws of quantum mechanics have to be used instead of the classical laws of physics.

This particular illusion has had a critical influence on all theories of time and space. In general relativity, it leads to the concept of *general covariance*, the principle that the laws of physics are independent of observers. Provided a discussion is restricted to classical mechanical principles, which are based on a godlike exophysical primary observer with total knowledge, this concept is reasonable. In the quantum domain it can be manifestly incorrect and misleading if applied carelessly. For instance, it is physically meaningless to say that a photon of frequency ν

in one frame has a different frequency ν' in another frame that is moving relative to the first frame. The fundamental fact is that a photon can be detected only in one detector, not in more than one.[10]

This brings us to the metaphysical concept of counterfactual definiteness. A counterfactual is a contextual proposition based on a context that is known not to have happened. In classical mechanics, counterfactuals are in order, whereas in quantum mechanics they have to be treated with caution. Suppose an electron is prepared in a specific quantum state. Then we can observe its position or we can choose to observe its momentum. But we cannot observe its position and momentum simultaneously. It is a fundamental logical error to assert that just because we *could have* observed some property of an SUO, that property had to '*be there*' all the time.

None of this means that we cannot be sure of anything. In QM, we get around counterfactual uncertainty by statistics: we can perform many runs, or repetitions, of a basic preparation protocol to create a conceptual object called a state. Then we perform various quantum outcome experiments so as to determine the average properties of the state, such as the initial momentum of the incoming particles in a scattering experiment. Then we have some confidence that the same preparation protocol will give us the same state, to within the uncertainties that arise, and then we can allow the state to evolve further into the experiment proper.

It is for this reason that the realist position that quantum wavefunctions have an objective reality cannot be supported. Not only are quantum states contextual, the temporal architecture involved in setting them up is simply not linear. Schrödinger unitary evolution is much more like a police report of a crowd disturbance than an account of an actual event.

Time in the laboratory

In contemporary physics, time is generally partnered with three-dimensional *physical space*, the three-dimensional space of position, to form a four-dimensional geometrical structure, the spacetime continuum. We discuss the mathematics of continua in Chapter 7.

In non-relativistic mechanics, the distinction between time and space is maintained, so in that context we shall refer to the space-time continuum as *space-time*. In contrast, in special and general relativity, time and space are merged into a single structure that we shall refer to as *spacetime*, that is, without a hyphen. The model of the spacetime continuum as a four-dimensional *manifold* (discussed in the Appendix) is referred to by philosophers as *manifold time,* leading to the *Block Universe* paradigm, discussed in Chapter 8. Points in the spacetime manifold are referred to as *events*.

---

[10] Our personal view is that we should not say that a photon has hit a detector: rather, all we should say is that a detector has been triggered. The former statement is metaphysical, the latter is factual.

Manifold time is contextually incomplete in two ways. First, it says nothing about any primary observer, about who or what is 'looking in'. In other words, it does not address the question *for whom is the existence of the spacetime manifold relevant or even defined?* We should keep in mind Schwinger's view quoted in Chapter 3 that space and time are abstractions derived from the context of our current apparatus and human modes of perception, a view reinforced by Meschini in his doctoral thesis [Meschini, 2008].

Second, in addition to a lack of reference to any primary observer, manifold time says nothing about how events are related to the processes of information extraction. This question is generally ignored by theorists because of a widespread belief that physics is all about the properties of SUOs. It is our contention that this is wrong: quantum physics is all about how observers interact with SUOs. Therefore we need to focus on the information extraction question (IEQ): *how do we actually extract information during an experiment?*

The IEQ is resolved empirically in particle physics experiments by the use of three-dimensional arrays of detectors. These come in a number of forms such as the Wilson cloud chamber, the Glaser bubble chamber, the spark chamber, and the streamer chamber. Their common feature is that they attempt to detect signals at a sufficiently large number of points in spacetime to allow an effective picture to be built up of what might be going on in the idealized spacetime manifold.

The IEQ remains a difficult and as yet unresolved problem and theoretically matters could not be in poorer shape. It suffices to see how many different opinions there are about quantum wavefunctions to appreciate this point. The *intrinsic approach* explicitly avoids the IEQ by discussing quantum wavefunctions as if they had objective identities rather than being contextual theoretical devices. This was the conceptual error, in our opinion, that Schrödinger made in 1926 when he published a remarkable series of papers on wave mechanics [Schrödinger, 1926a,b,c,d,e,f,g]. He originally interpreted the wavefunction of an electron state as *the* particle. Although better interpretations of QM have been developed over the decades, there remains a core of theorists who assign an objective reality to quantum wavefunctions [Bohm, 1952].

Corresponding to the intrinsic approach in quantum mechanics is the *principle of general covariance*, which asserts that the laws of physics take the same form in all suitable frames of reference. A corollary to the intrinsic approach is that it does not matter which frame of reference or coordinate patch is used to describe components of tensors (the intrinsic objects of interest in GR), and therefore it is permissible to use any convenient set of coordinates.

As it was classically motivated, the principle of general covariance takes no account of the IEQ. It is our view that the reason *quantum gravity* (the programme of quantizing gravity) has made no significant progress over several decades is because the IEQ is ignored twice: once on the gravitational side and once on the quantum side. Used carefully, the principle of general covariance has been successful when it is restricted to relative external context. Examples are the laws of black hole thermodynamics as formulated by Bekenstein [Bekenstein, 1973] and

Hawking [Hawking, 1976], and explored by Unruh [Unruh, 1976]. In such cases, quantum information extraction is discussed in relatively localized experimental contexts with laboratories embedded in a classical relativistic background.

The problem with current quantum gravity is that no analysis of the primary observer is made usually. Our view is that physical primary observers have to look in two unrelated directions: they look *inwards* when describing quantum processes of SUOs and there they can use the rules of quantum mechanics; they also have to look *outwards* to the idealized spacetime that they believe they occupy and there they use the classical rules of general relativity. The two sets of information are about different things: quantum information coming from experiments is about the observer–apparatus–SUO relationship, whilst classical information about the background spacetime in which these experiments are embedded is more about the observer–environment relationship. Trying to equate the two sorts of information is an example of what is called a *category error*. It makes no sense physically and has proven mathematically near impossible to apply a quantum description outwards. Attempts to do this ultimately lead to the metaphysics of the Multiverse [Deutsch, 2001].

Laboratory frames and observer choruses

From a properly analysed contextual physical point of view, the principle of general covariance makes no practical sense: every experiment is necessarily associated with equipment, apparatus, observers, and their laboratory. Moreover, quantum principles emphasize that observation is *not* cost free: every piece of information extracted by an observer must involve some change in the state of the object being observed. Wavefunction collapse is not a blemish on an otherwise beautiful subject, as some purists maintain, but an essential part of the information extraction process. Therefore, a careful approach to space and time physics requires us to specify our laboratory frames of reference very carefully.

This was understood by Einstein to apply even in the realm of classical GR. His *principle of equivalence* states that the laws of physics relative to a localized, freely falling, non-rotating laboratory are the laws of Special Relativity. Such a laboratory is hardly identifiable with an arbitrary coordinate frame.

Whenever we refer to an abstract mathematical frame of reference, we shall refer to it as a *coordinate patch*. When such a patch is identifiable with the time and space coordinates in some localized physical laboratory, such as those referred to in the above equivalence principle, we shall refer to it as a *lab frame*. The main characteristic of a lab frame is that one of its coordinates, usually denoted by t (for time), is associated with the set of clocks synchronized in that laboratory. In this context, such a parameter plays the ambivalent role of a localized Absolute Time.

In contrast to arbitrary coordinate patches, lab frames require careful specification and come with some limitations. A significant fact is that laboratory frames cannot cover the whole of GR spacetimes as a rule, that is, they are not global.

This makes the particle concept in quantum field theory somewhat problematical [Colosi & Rovelli, 2009]. There are several reasons for this. First, observers and their laboratories are *inside* the spacetimes they discuss, that is, laboratories are endophysical by nature. Second, laboratories and their observers do not exist forever: they are created and destroyed in the course of time, something that is hard to factor into current theory generally. Third, laboratories and their observers are invariably of finite spatial extent, that is they are generally extended in space but never infinitely so. Finally, the spacetime may in principle contain *closed timelike curves* (CTCs), or loops in time. These are discussed in Chapter 20. Whilst there is no evidence for CTCs, the possibility exists theoretically in classical GR [Gödel, 1949] and under such circumstances the notion of an observer needs to be radically overhauled. In particular, CTCs and quantum mechanics seem fundamentally incompatible.

In the laboratory description of both space-time and spacetime, idealized endophysical observers are modelled as *worldlines* threading their way through the continuum. Worldlines are discussed in Chapter 14. Such observers have memories and carry clocks that are used to record the times of significant events occurring along their worldlines. Observer clocks are usually assumed to operate continuously in time, leading to the use of real numbers as a mathematical model of time, discussed in Chapter 7. However, it is possible to model the passage of such clocks by discrete time, discussed in Chapter 23.

Laboratories have non-zero spatial extent whilst worldlines have the spatial properties of points. Therefore, we have to introduce the concept of *observer chorus*. This is the theoretical analogue of a cloud or bubble chamber in experimental particle physics: a continuum of idealized observers each located on its own worldline, threading its way through the region of spacetime concerned and primed to record the times of phenomena occurring at or close to their position. Worldlines of different members of the chorus never intersect, so an event is observed by only one member of the chorus. This effectively gives that event a unique spatial position relative to the other members of the chorus and a unique time as recorded by the observer concerned.

Chorus protocol

The setting up of an observer chorus requires a *protocol*, that is, a careful definition and preparation of the elements of the chorus *before* any information is collected. A protocol will involve a number of requirements.

Frame geometry

A specification of the spacetime geometry of the flux of worldlines in a chorus is fundamental to the physics being reported. For example, the worldline flux of an inertial frame chorus will be very different to that of a laboratory undergoing acceleration. This can have a significant impact on the nature of the observations, as the Unruh effect demonstrates [Unruh, 1976]. According to that effect, a particle detector accelerating in apparently empty spacetime is predicted to detect a thermal spectrum of particles.

Standardization

All members of an observer chorus should have an agreed standard of measurement. If this is not done carefully, disasters can occur. A notable example is the failed Mars Climate Orbiter disaster in 1999. This occurred because information being processed on Earth was assumed to be in Imperial units (pounds, etc.) whilst the same information being processed by the orbiter itself was assumed to be in Standard International units (kilograms, etc.).

Synchronization

This is the procedure by which all the individual clocks in a chorus are set to zero or some suitable initial time. Assuming the clocks carried by the members of the chorus are all standardized, synchronization then gives an empirical definition of *simultaneity*, relative to that chorus and the protocol employed. We shall see in later chapters that synchronization can have a profound affect on the description of physics. In particular, the synchronization protocol assumed in Special Relativity leads to the loss of absolute simultaneity: two events that appear simultaneous relative to one inertial frame chorus need not be simultaneous relative to another inertial frame chorus.

Once synchronization has been established for a given observer chorus, *hyperplanes of simultaneity* can be identified. A hyperplane or hypersurface of simultaneity is a collection of events observed by a chorus that have the same time coordinate relative to that chorus, as shown in Figure 6.1. A fundamental criterion for such a surface is that any two points on it cannot causally affect each other. In principle, this could include pairs of events that lie within the same *lightcone*,[11] as long as there is adequate shielding preventing dynamical interaction between the events. For this reason, the concept of hyperplane of simultaneity has been generalized to that of *stage* [Jaroszkiewicz, 2008a].

The synchronization process itself is generally a non-trivial process, usually based on signals sent back and forth between an origin and the other members of a chorus. In Newtonian space-time physics, it is assumed that all chorus observer clocks can be synchronized with an idealized fundamental time parameter called Absolute Time. The advent of Special Relativity made it clear that not only is synchronization a non-trivial task but it might not even be consistent in certain situations, such as the *spinning disc* discussed in Chapter 19.

---

[11] Lightcones are discussed in Chapter 17.

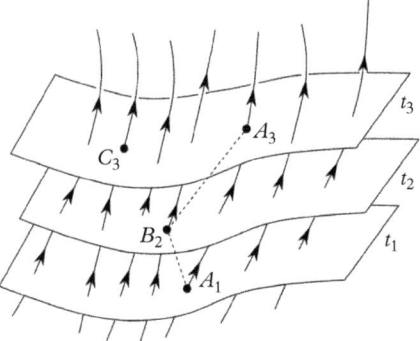

Fig. 6.1 *Hyperplanes of simultaneity labelled by a time parameter t, defined for a chorus of observers. Chorus members synchronize clocks by signalling with light or other means, shown by dashed lines between A and B. By definition, events A_3 and C_3 cannot causally affect each other.*

Superobservers

Humans are singular objects, in that their visual perceptions lead to the creation of an illusion in the brain that they, the observer, is the centre of their environment, an illusion that taken to extremes leads to the solipsistic outlook discussed in Chapter 2. The illusion has to be maintained in the face of the fact that light signals from different places and times can arrive at the retina at the same time. The brain is capable of unscrambling that information because of two facts: (i) because of the structure of the eye, distant object appear smaller in angular terms than nearby objects of the same real size; and (ii) the speed of light is so great compared to most speeds encountered by humans that temporal discrepancies can be ignored normally.

A superobserver is a special member of a laboratory frame chorus, receiving information from all the other member of the chorus, usually via light signals. In principle, there will be retardation effects due to the relative distances of other members of the chorus and to the finite speed of light. In classical physics, these factors contribute to time *delays* in signal reception, according to the observer's own clock, never advancement. The situation is not so clear cut once quantum principles are used, simply because certainty that information has been transmitted cannot be established by the superobserver until observational protocol (i.e., context) has been established and implemented. Then and only then can a classical picture of events in the superobserver's imagined spacetime be constructed with any degree of physical consistency.

7
Mathematical images of time

Introduction

The human sense of time is qualitative: we have the feeling that time 'flows' continuously. In contrast, real events such as birth, death, the daily rising and setting of the Sun, the annual cycle of seasons, the monthly phases of the Moon, and other memorable phenomena, all of these punctuate that continuity. Memory and the needs of society to record and predict such events undoubtedly led humans to develop a quantitative and discrete description of time using counting numbers (integers). We still count years, days, hours, and minutes in those terms.

Integers are of fundamental importance to mathematicians. They are regarded as the basis of mathematics by constructivists and intuitionists, those mathematicians who seek a 'natural' basis to mathematics. A fundamental constraint on all mathematicians regardless of their persuasion is that every aspect of mathematics is based on the taking of discrete, countable steps: no theorem is proved 'continuously' but step by step, line by line, even though mathematicians think they themselves are operating in continuous time. When Cantor developed his controversial theory of transfinite numbers, he was constrained to take countable numbers of steps, using discrete lists (ordered sets) of numbers in his 'diagonal' method; when Alan Turing [1912–54] began to theorize about and then build his 'machines', they were designed on discrete principles that eventually led to the modern computer [Hodges, 2014].[12]

Discreteness is also the basis of the physical world despite our subjective feeling to the contrary. Quantum phenomena underpin physical reality and are predicated on discreteness. When any experiment is analysed realistically, in terms of what observers actually do in the laboratory rather than what is *imagined* they do, information, the basis of everything to do with time, always comes in discrete amounts. It took humans a very long time to come to that perspective however: the atomic hypothesis was finally settled only just over a hundred years ago.

The language of integers penetrates all aspects of human existence, even to the extent of our mental processes: a *thought* is a discretization, an objectivization of

---

[12] Before the Second World War started, Turing was seen making metal cogs in his Cambridge College rooms: cogs with teeth are the embodiment of discreteness [Hodges, 2014].

Images of Time. First Edition. George Jaroszkiewicz.
© George Jaroszkiewicz 2016. Published in 2016 by Oxford University Press.

unimaginably complex processes going on inside our heads. A *decision* is a definite, recognized end to a set of thoughts. We count events, thoughts, and decisions with integers: thoughts without conclusion lead to *indecision*.

We saw in Chapter 4 that there is evidence for Paleolithic calendars dating as far back as 30,000 years ago [Pásztor, 2011]. This proves that Stone Age humans were dealing with integers even if they did not appreciate that fact. As humans became more sophisticated, they began to fill in the mathematical gaps between the integers: halves, thirds, and so on, until finally they invented[13] the continuum of real numbers, denoted by $\mathbb{R}$. Currently, the dominant mathematical model of time in science is based on $\mathbb{R}$, which has all the properties required to model a continuum. We shall discuss continua in more detail further on in this chapter. Other mathematical representations of time have been invoked besides the reals: we shall mention some of them also.

Some reasonable requirements

Regardless of which mathematical representation we choose, certain properties of time are regarded as fundamental and common to all mathematical images of time. These are listed below.

Primary observers

Mathematicians routinely make truth statements relative to sets of axioms and postulates stated in contextually incomplete form, no primary observer being required in pure mathematics. However, in Chapter 2 we argued that in physics, all relevant statements should be contextually complete, meaning that we should always ask *for whom is this or that assertion meaningful?* Therefore, whenever mathematicians apply their mathematics to the physical universe, they should identify the primary observers involved.

Two notable examples come to mind where the contextual incompleteness of mathematics created difficulties for physicists, one involving the structure of space and the other the structure of time:

1. For over two thousand years, geometers assumed that the postulates and axioms of Euclidean geometry were absolute, until Gauss, Bolyai, and Lobachevsky independently showed that Euclid's Fifth Postulate is independent of the other axioms and postulates. Assuming homogeneity and isotropy,[14] a primary observer embedded in three-dimensional physical

---

[13] Or discovered, according to the Platonists.
[14] A homogeneous medium has the same physical properties everywhere, whilst there is no preferred direction at any point in an isotropic medium.

space could in principle measure angles and distances and hence determine empirically which one of three mutually exclusive spaces they were in. These correspond to spherical, Euclidean, and hyperbolic geometries respectively.

2. Newton's Absolute Time [Newton, 1687] makes sense if we assume that he was describing the universe as seen by some exophysical primary observer (God) standing outside Absolute Space and Time with total information about everything in it. Classical physicists became as conditioned to think in terms of Absolute Time as mathematicians had become conditioned to believe that Euclidean geometry was absolute. The advent of relativity changed this conditioning.

Localization

Observers discuss systems under observation (SUOs) in terms of relatively localized points in space and time, conceptual objects called *events* located in a conceptual arena called *space-time* in classical mechanics and *spacetime* in special relativity (SR) and general relativity (GR).

Temporal ordering

There is a sense of temporal *order* in the universe: observers classify events as earlier, simultaneous, or later, relative to some chosen reference event.

Temporal architecture

This is the verbal description of the various relationships between events, observers, and SUOs and depends on the mathematical model used. For example, there is an obvious difference between linear time and cyclic time. Another important architectural difference involves the *space-time* and *spacetime* concepts discussed in Chapter 5. Space-time has the traditional architecture used by all human observers sitting in any laboratory, measuring laboratory time with clocks and locating events in three-dimensional space using rulers. On the other hand, spacetime need not be split into two such components in this way: it has an architecture of a four-dimensional continuum independent of any observer, a model introduced by Minkowski in 1908 and reviewed in Chapter 19. We discuss temporal architecture further in Chapter 13.

Sets

In order to use any mathematical representation of time that encodes the above points, we need to understand the primary mathematical concept of a *set*. Virtually the whole of mathematics is based on the *set* concept, discussed in more detail in the Appendix. We gave Cantor's definition of a set in Chapter 2, noting that we classified it as contextually complete.

In its day, Cantor's work was the subject of fierce criticism from leading mathematicians such as Kronecker and Poincaré. Their objections touch upon a fundamental question that affects all aspects of knowledge: *how does human thought reflect reality*? This was a question discussed in Antiquity by Plato in his theory of *forms*: he asserted that non-material concepts (forms or ideas) are 'more real' than the world of our perceptions. The point about Cantor's context-ually complete definition of a set is that it is not physics: from his perspective it was enough to invent a concept with no requirement to validate its existence. Constructivists such as Kronecker disagreed vehemently and rejected those math-ematical concepts of Cantor's that could not be realized by specific examples. In a sense, the constructivists were applying the scientific principle of *nullius in verba* to mathematics. This debate continues to this day, with constructivists such as Bishop [Bishop, 1977] denying the validity of the hyperreal number concept used by Robinson in the construction of non-standard analysis [Robinson, 1966].

The set concept has proven impossible to define to every mathematician's satis-faction. Because it is so useful, it is best to treat it in the same way as our concept of primary observer: a set is an intuitive, primary conceptual structure that is used to discuss other concepts, with a veto on any further discussion about its structure or meaning.

Mathematicians frequently encounter sets with so many elements that some convention has to be used to label their elements. For example, points of spacetime are labelled by coordinates relative to some chosen frame of reference, such as an inertial frame in special relativity.

Physicists often discuss events as if they were physical objects but we should keep in mind Schwinger's assertion quoted in Chapter 2 that spacetime is but an abstraction designed to model the action of apparatus. All sets used by physicists are just models of conceptual entities and need not correspond to anything with an objective existence [Carter, 1998]. Whilst some of these sets appear realistic, such as position in CM, other sets, such as Hilbert spaces in quantum mechanics, are mathematical abstractions that are never regarded as physical in any sense.

Sets can consist of a finite number of elements, a countably infinite number of elements, or even an uncountably infinite number of elements. Here we have invoked the elementary concept of *counting*, as in 'one, two, three . . . ', which we assume at this point is understood by the reader. An example of a set with a finite number of elements is $\{47, A\}$, which has just two elements: the num-ber 47 and the letter A. An example of a set with a countable infinity of elements is $\mathbb{N}^+ \equiv \{1, 2, 3, \dots\}$, the set of positive integers. An example of a set with an uncountable infinite number of elements is $\mathbb{R}$, the set of real numbers.

These examples are of interest to us because time is generally modelled by real numbers, but we shall see presently that there are alternatives.

The concept of a set with no elements in it whatsoever has proven so useful that it has been given its own name, *the empty set* and its own mathematical symbol, $\emptyset$.

We used this concept in Chapter 2 to denote an absence of contextual information. The empty set concept can be useful in physics, such as labelling states of non-existence of apparatus in a laboratory [Jaroszkiewicz, 2010]. Here 'non-existence' is defined contextually, relative to some primary observer who can ascertain empirically at any given time whether a particular piece of equipment exists or does not exist in their laboratory as far as that observer is concerned.

Set theory has a number of auxiliary concepts such as *subset*, *union*, and so on, [Howson, 1972], that have proven useful in a number of disciplines such as probability theory and topology. The concept of a *function* is an important extension of set theory dealing with relationships between different sets and is discussed below.

Temporal ordering and ordered sets

Time is synonymous with *temporal ordering*: if an observer observes two events, one event will be judged by that observer to occur earlier than the other or else they are deemed to be simultaneous, according to that observer's sense of time. This ordering does not require any specific mathematical skill in principle: it is something everyone can do naturally. For example, spectators make such an ordering in their minds when they see the leading runner in a race go past the winning post: that runner has come in first and the others will come in later.

The question of a mathematical representation of this ordering only arises when a record of a temporal ordering has to be made, a record that will be used *later* for some purpose. Memory can serve to keep records of temporal ordering, but only up to a limited point. The problems with memory are that it degenerates with time, is individual to an observer, can be faulty, and cannot be transmitted absolutely faithfully into the minds of other observers. Therefore, some artificial device has to be invented to keep track of temporal records.

To this end, temporal ordering is recorded mathematically by the use of *ordered sets*. Such a set is a non-empty set with a *binary relation* $\leqslant$ between any two of its elements. Given any two elements a, b of the set we have three possibilities: $a < b$, $a = b$, or $b < a$. If $a < b$ we say a is *earlier than b*, if $a = b$ we say a and b are *simultaneous*, and if $b < a$ we say that a is *later* than b.

Functions and the mathematical arrow of time

A *dynamical process* is something that happens in time, starting with an initial state or configuration and ending with a final state or configuration. The mathematical concept of a *function* has the same architecture and so functions can be used to represent dynamical processes.

Functions involve relations between sets. To define a function, we first need to understand the auxiliary concept of a *Cartesian product*. Given two sets A and B, their Cartesian product $A \times B$ is the set of all possible ordered pairs (a, b), where a is an element of A and b is an element of B. There is an implicit assumption here that some primary observer has defined this ordering and can keep track of it. The standard notation for Cartesian products does this tracking automatically for us by the rule that elements from set A are on the *left* whilst those of B are on the *right* whenever the element (a, b) is written.

Example

If $A \equiv \{2, 5, 7\}$ and $B \equiv \{u, v\}$ then

$$A \times B = \{(2, u), (2, v), (5, u), (5, v), (7, u), (7, v)\}. \tag{7.1}$$

The Cartesian product $A \times B$ is not the same as $B \times A$, that is, the ordering matters. The Cartesian product concept is readily generalized to three or more sets.

Functions are defined by mathematicians using *three* sets, A, B, and C, where C is a subset of the Cartesian product $A \times B$. A function f is the ordered triple $f \equiv (C, A, B)$ such that for each element a in A, there is exactly one *image element $f(a)$ of a in B* [Howson, 1972]. C is the set of all ordered pairs $(a, f(a))$, but is not itself a Cartesian product. A schematic representation of a function is shown in Figure 7.1. There may be elements in $A \times B$ that are not in C, there may be elements in B which are not the images of any elements of A, and the possibility that different elements of A have the same image in B is not ruled out.

The image $f(a)$ of a is called the *value of the function* at a. Mathematicians generally insist on functions being *single-valued*, that is, there is only *one* image $f(a)$ of

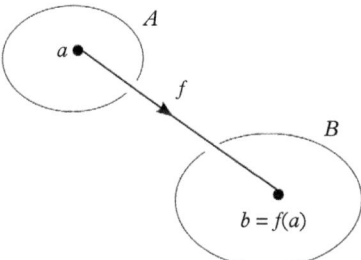

Fig. 7.1 *Schematic of a function f. A is called the domain and B is the range of the function.*

every element *a* in *A*, although the concept of *multi-valued function* is occasionally encountered. This touches upon a fundamental point in the mathematical modelling of time: given a definite past and present, classical mechanics assumes that the future is well defined and unique. We discuss this point further in the section on *prediction* below.

The set *A* is called the *domain* (*of definition*) of the function and *B* is called the *range*. The set of values of a function is the *image* of *A* under *f* and often written $f(A)$. If $f(A)$ is a proper subset, which means that there is at least one element of *B* not in $f(A)$, then $f(A)$ is not the same as *B*: *range* and *image* are not synonymous in general.

We are now at the point where we can define the *mathematical arrow of time*. With reference to the above figure, we see that there is a natural sense of *direction* in the definition of a function $f : A \rightarrow B$: we go *from* the domain of definition *A* to the range *B*, and not the other way round.

Before we can discuss prediction and retrodiction, we need several additional concepts.

1. A *surjection* from a set *A* to a set *B* is a function such that $f(A) = B$, that is, the image of the domain equals the range of the function. Mathematicians say *f* maps *A* onto *B*.

2. An *injection* from a set *A* to a set *B* is a *one-to-one* function *f* from *A* to *B*, which means that if $f(a_1) = f(a_2)$ then $a_1 = a_2$. Injections are fundamental to conventional assumptions about time: given the present, we assume the past is well defined. Humans do not like to imagine that the past is not unique. A representation of a non-injective function *f* is shown in Figure 7.2(a) whilst an injective function *g* is shown in Figure 7.2(b).

3. A *bijection* from a set *A* to a set *B* is a surjective injection *f* such that $f(A) = B$, that is the image equals the range, and for every element *b* in *B*, there is precisely one and only one element *a* in *A* such that $b = f(a)$.

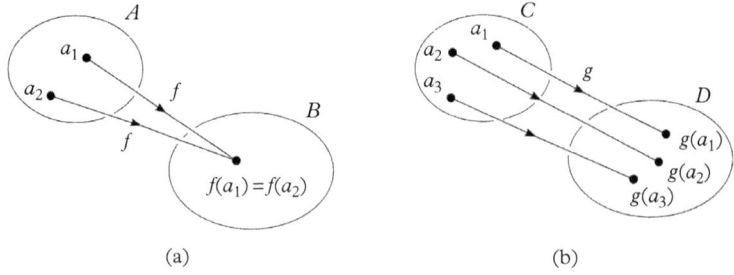

(a)   (b)

Fig. 7.2 *(a) A non-injective function f; (b) an injective function g.*

If f is a bijection from A to B then there automatically exists a bijection f^{-1} from B to A called the *inverse function*. Many-to-one functions and bijections play a fundamental role in both classical and quantum mechanics: they can be used to discuss prediction, retrodiction, reversibility, and irreversibility.

Prediction

The Cartesian product concept allows us to give a mathematical representation of the human image of time. Consider the concepts of *past* and *future*. If anything distinguishes human perception above that of other animals it is surely our ability to imagine alternative futures and pasts. Suppose we denote the current present by P and two possible futures as the Cartesian products (P, F_1) and (P, F_2). If strict determinism holds, then at least one of these Cartesian products must be unphysical, relative to P.

Prediction is the art and science of applying some evolution function to an observer's initial data in order to determine a final state of some SUO, out of a range of possibilities or potentialities, in the observer's future. There are two types of prediction: those that are unambiguous and those that are imprecise in some way. *Deterministic mechanics* is mechanics that allows the first kind of prediction: an evolution function maps an initial state into a unique final predicted state. On the other hand, *stochastic mechanics* takes an initial state and gives us only a probability distribution over a set of possible final states, only one of which will occur.

Newtonian classical mechanics (CM) is deterministic, which led to the following famous statement by Pierre-Simon de Laplace [1749–1827]:

> We ought then to regard the present state of the universe as the effect of its anterior state and as the cause of the one which is to follow. Given for one instant an intelligence which could comprehend all the forces by which nature is animated and the respective situation of the beings who compose it an intelligence sufficiently vast to submit these data to analysis it would embrace in the same formula the movements of the greatest bodies of the universe and those of the lightest atom; for it, nothing would be uncertain and the future, as the past, would be present to its eyes.
>
> [de Laplace, 1812]

There is a caveat that cannot be ignored: the 'intelligence' referred to by Laplace is known as *Laplace's demon*, a god-like exophysical primary observer, all-seeing and capable of infinite precision in its calculations. Human, however, are endophysical observers with limited resources. Laplace noted that fact, continuing the above paragraph with the observation that: '*The human mind offers, in the perfection which it has been able to give to astronomy, a feeble idea of this intelligence.*'

Laplace's demon has been criticized from several perspectives: (i) the universe runs on irreversible mechanical principles, not according to reversible Newtonian mechanics; (ii) the universe runs according to the principles of quantum mechanics, which gives probabilities of random outcomes; (iii) there is a limit to computational power in the universe [Minkel, 2002] and (iv) small changes or uncertainties in initial conditions can lead to catastrophically different final states if the dynamics is chaotic. Chaos is discussed in Chapter 9.

A deterministic classical mechanical system can be modelled by an evolution function $f : A_{\text{initial}} \rightarrow B_{\text{final}}$ that takes us from the space A_{initial} of initial states to the space B_{final} of final states.

Retrodiction

Retrodiction is the black art of finding an answer to the question: given a final state b_{final} in B_{final}, from which initial state a_{initial} in A_{initial} did the system evolve from? Such a question is of prime concern to cosmologists, palaeontologists, and archaeologists.

In order to even consider this question in terms of the evolution function, we first have to assume that b_{final} is actually in the image set $f(A_{\text{initial}})$ and not in its complement relative to B_{final}. Furthermore, if the assumed evolution function f is not an injection, then the answer to the retrodiction question is not unique, which would mean that the assertion that the past is unique could not be validated in terms of that assumed evolution function.

It has always to be kept in mind that our ideas about the uniqueness of the past are generally based on metaphysical assumptions. The best we could do would be to make reference to some assumed mathematical *model* of time, the most popular one being based on some form of injective function from past to present. In that case, if the evolution function is a bijection, we could say that the retrodiction problem in CM does have a solution (i.e. exists in the mathematical sense even if we could not actually compute it).

Cardinality

To understand the assertion that time is a continuum, we need to understand continuous sets: therefore we need to develop some further set theoretic concepts.

Two sets A and B are *equinumerous*, written $A \sim B$, if there exists a bijection from A to B. This means every element of A can be paired with one and only one element of B and vice-versa. This leads to the concept of *cardinality*, a statement of how 'big' a given set is.

For any set A, we assign to it a *cardinal number* called the *cardinality* of A and denoted $\#A$. The cardinality of a set is a measure of how many elements there are in that set. Cardinal numbers come in two varieties: finite cardinal numbers are equivalent to ordinary whole numbers, that is the non-negative integers, whilst infinite cardinal numbers are altogether stranger beasts, the subject of much mathematical debate in the nineteenth century.

The finite cardinals

To construct the set of finite cardinals, we start with the empty set $\emptyset$. Given that the empty set is defined to be a set with no elements at all, we *define* the cardinality of the empty set to be the cardinal number 0, corresponding to the integer *zero*, that is $\#\emptyset \equiv 0$.

Now define the set $\emptyset^{[1]} \equiv \{\emptyset\}$. This is *not* the empty set: it contains one element, which happens to be the empty set. Therefore we define $\#\emptyset^{[1]} \equiv 1$, corresponding to the familiar integer *one*.

Next, defining the set $\emptyset^{[2]} \equiv \{\emptyset, \emptyset^{[1]}\}$, we write $\#\emptyset^{[2]} \equiv 2$.

Next, defining the set $\emptyset^{[3]} \equiv \{\emptyset, \emptyset^{[1]}, \emptyset^{[2]}\}$, we write $\#\emptyset^{[3]} \equiv 3$.

Continuing this process, the general extension is $\emptyset^{[n]} \equiv \{\emptyset, \emptyset^{[1]}, \dots \emptyset^{[n-1]}\}$, with $\#\emptyset^{[n]} = n$.

A set A has finite cardinality n if A is equinumerous with $\emptyset^{[n]}$, that is

$$\#A = n \Leftrightarrow A \sim \emptyset^{[n]}, \tag{7.2}$$

where the 'if and only if' symbol $\Leftrightarrow$ is the mathematician's notation for 'the right-hand side is always true if the left-hand side is true, and vice-versa'.

The natural numbers

The collection of all sets $\emptyset$, $\emptyset^{[1]}, \dots$ defines $\mathbb{N}$, the *natural*, cardinal, or *counting numbers*, written $\mathbb{N} \equiv \{0, 1, 2, \dots\}$. Our definition includes zero. The set of non-zero natural numbers, or positive integers, is denoted $\mathbb{N}^+$.

The set $\mathbb{N}$ itself does not have finite cardinality, so has infinite cardinality. A set that is equinumerous to $\mathbb{N}$ is called *countable*. By definition, the cardinality of such a set is called *aleph null* and denoted $\aleph_0$.

It was found convenient by mathematicians to define $\mathbb{Z}$, the *integers*, an extension of the natural numbers to include negative integers, that is, $\mathbb{Z} \equiv \{\dots, -2, -1, 0, 1, 2, \dots\}$. This set is countable, that is, $\mathbb{Z} \sim \mathbb{N}$, despite the fact that it appears to have twice as many elements as $\mathbb{N}$. Moreover, $\mathbb{Z}$ is a group under addition (groups are discussed in the Appendix) whilst $\mathbb{N}$ is not: the additive inverse of a positive integer is a negative integer, which is not an element of $\mathbb{N}$.

Linear continua

Does time have holes in it?

This seems a strange question, but one that has to be asked of any mathematical image of time. What motivates this question is the fact that the ordering property of events that we discussed in Chapter 1 is insufficient to model time as we feel it should be modelled.

To understand the problem, consider the set $\mathbb{Z}$ of integers discussed in the previous section. These are an ordered set: if p and q are any two integers then either $p < q$, $p = q$, or else $p > q$.

Mathematicians went further and defined $\mathbb{Q}$, the set of *rational numbers*, defined by

$$\mathbb{Q} \equiv \left\{ \frac{p}{q} : p, q \text{ are any integers, with } q \neq 0 \right\}. \tag{7.3}$$

This too is an ordered set and it is countable, a fact that is relatively easy to show. However, $\mathbb{Q}$ has a property that $\mathbb{Z}$ does not have: it is *dense*. This means the following. Suppose r_1 and r_2 are two rational numbers such that $r_1 < r_2$. Then it is always possible to find a third rational number r_3 such that $r_1 < r_3 < r_2$. In other words, we can always fill the hole between any two different rational numbers with other rational numbers. In fact, we can always find a countable infinity of rational numbers between any two different rational numbers. On the contrary, if p is an integer, then $p < q \equiv p + 1$ but there is no integer r such that $p < r < q$.

Greek mathematicians eventually discovered that the set of rationals also has 'holes' in it. By this is meant that there are real numbers that are not rationals, such as $\sqrt{2}$[15]. Equivalently, we could say that the set of all rationals does not include some other real numbers, called the irrationals. These 'holes' in $\mathbb{Q}$ have to be filled in order to give us the real numbers set $\mathbb{R}$, and it is that set that is commonly used to model time.

The existence of the irrationals makes the rigorous definition of a continuum harder than it appears intuitively. Mathematicians often refer to a continuum as a *linear continuum* to reflect this fact. To understand the technical details of linear continua we need to define a few more terms.

Definition *A* partially ordered set *or* poset S *is a set with a binary relation denoted by $\leqslant$, such that*

  i) *for every element x in S, $x \leqslant x$, (reflexivity);*
  ii) *if x and y are elements of S such that $x \leqslant y$ and $y \leqslant x$, then $x = y$, (antisymmetry);*
  iii) *if x, y, and z are elements of S such that $x \leqslant y$ and $y \leqslant z$ then $x \leqslant z$, (transitivity).*

There may be elements u, v of a poset for which the binary relation $\leqslant$ is not defined. This has significant physical application in relativity. Minkowski spacetime $\mathcal{M}^4$, the four-dimensional spacetime of SR, has a Lorentzian metrical lightcone structure that creates this possibility. This is discussed in more detail

---

[15] In the mythology of mathematics, the student who discovered the irrationality of $\sqrt{2}$ was murdered by the other Pythagoreans, because this discovery did not fit into their quasi-religious belief that the universe ran on rational lines.

in Chapter 17. We can pick pairs of different events U, V in $\mathcal{M}^4$ such that in some inertial frames, the times t_U, t_V assigned to them respectively satisfy $t_U \leqslant t_V$, and such that in other inertial frames, we have $t'_V \leqslant t'_U$. Such pairs of events will be called relatively spacelike pairs.

To define a linear continuum, we need to eliminate the possibility of relatively spacelike relationships. We do this by introducing an extra condition, which turns a poset into a totally ordered set:

Definition *A* linearly ordered *or* totally ordered set *S is a poset with the additional property added to the three above:*

iv) *for any two elements x, y of S, then $x \leqslant y$ or $y \leqslant x$, a property known as* totality.

The totality property of a totally ordered set essentially places a veto on finding a spacelike pair in S. An additional binary relation can be defined for each totally ordered set:

Definition *A* strict total order *on a totally ordered set is a binary relation* < *such that $x < y$ if and only if $x \leqslant y$ and $x \neq y$.*

One last step remains. Before we define a linear continuum, we need to define the concept of least upper bound:

Definition *Let S be a subset of a poset X. Then an element b of X is an* upper bound *for S if for every element x of S, $x \leqslant b$.*

and

Definition *Let S be a subset of a poset X. If there exists an element b_0 of X such that $b_0 \leqslant b$ for every upper bound of S, then b_0 is the* least upper bound *or* supremum *for S.*

We note that the supremum, if it exists, is unique.
We now have the structures needed to define a linear continuum:

Definition *a* linear continuum *is a non-empty totally ordered set S such that*

1. *S has the least upper bound property;*
2. *Given any two different elements x and y of S such that $x < y$, then there always exists another, distinct element z in S such that $x < z$ and $z < y$. We write $x < z < y$.*

The archetypical linear continuum is $\mathbb{R}$, the set of real numbers, which is generally used to model continuous time.

Further concepts

To begin to understand some exotic aspects of the mathematics of time, we need to get to grips with even more mathematical technology. There is no simple way around this.

Linearly ordered groups

Even the simplest number system, the integers, has some fundamental properties that were recognized by mathematicians as important. One of these is the concept of a *group*, defined in the Appendix. Groups *per se* do not naturally have the ordering property of an ordered set, but we can add on such a property.

Definition *A linearly left-ordered group G is a group with an extra property, denoted $\leqslant$, such that*

1. *given any two elements a, b of G then $a \leqslant b$ or else $b \leqslant a$. We do not need to think of this relation as 'is less than or equals';*
2. *if $a \leqslant b$, then for any element c in G we have the group multiplication rule $ca \leqslant cb$.*

The set $\mathbb{R}$ of real numbers is a linearly ordered group under addition, because if x and y are any real numbers such that $x \leqslant y$, then we can always write $z + x \leqslant z + y$ for any real number z. Here the relation $\leqslant$ does have the standard meaning 'is less than or equals'.

We can just as easily define linearly right-ordered groups, that is, $a \leqslant b \Leftrightarrow ac \leqslant bc$ for all elements c of the group.

A linearly bi-ordered group is one which is linearly right-ordered and linearly left-ordered.

If a and b are two elements of a linearly ordered group such that $a \leqslant b$ and $a \neq b$, then we write $a < b$.

Infinitesimals and Archimedean groups

At this point we come to some of the deepest concepts in mathematics: numbers that are immeasurably small and immeasurably large in some sense. This topic remains controversial in modern mathematics, a battleground between the constructivists and intuitionists on one side and the followers of Cantor (infinities) and Robinson (infinitesimals) on the other.[16] When Newton and Leibniz independently developed the differential calculus, the cornerstone of mechanics and the study of time, they had to employ powerful intuition to navigate the mathematical subtleties of limits and infinitesimals.

---

[16] By all accounts, Cantor was fiercely hostile to infinitesimals.

For any element g of a linearly ordered group G, we define

$$ng \equiv \underbrace{g + g + \ldots + g}_{n \text{ terms}} \tag{7.4}$$

for any positive integer n. This assumes the group is abelian (group multiplication commutative) and so we use the addition symbol + to denote the group 'product'.

Given any two elements x, y of a linearly ordered group G such that $x < y$, then x is *infinitesimal with respect to y* if for every finite integer n we have $nx < y$. Alternatively, we say that y is *infinite with respect to x*. Essentially, an infinitesimal such as x cannot be scaled up to be 'bigger' than y.

A linearly ordered group is *Archimedean* if there is no pair of elements such that one element is infinitesimal with respect to the other.

These concepts are mind boggling when their implications are understood. Does any of this make sense physically? The reader has to decide for themselves. Our advice is not to focus on the symbols but to reflect on what is meant, just as Newton and Leibniz had to when they engaged with the concepts of infinity and the infinitesimals underpinning the whole of humanity's thoughts about time, space, and motion.

Cyclic time

If we believe in reincarnation, time travel, or a series of Big Bang expansions followed by Big Crunch contractions, we may have to drop the notion that time is linear and use periodic functions. The big danger here is that such discussions can easily be contextually incomplete: we should always ask who is monitoring cyclic time and establishing that it is it indeed cyclic. Our view is that this would require an exophysical observer based on a linear time perceiving an SUO with periodic behaviour.

Signature and multi-dimensional time

The advent of special relativity (SR) led to the idea that time and space were part of *spacetime*, a four-dimensional continuum with a novel distance structure related to Pythagoras's theorem in standard plane geometry. Recall that in the geometry of the Euclidean plane, a right-angled triangle with sides of length x, y, and z satisfies Pythagoras' 'theorem', $z^2 = x^2 + y^2$, if z is the length of the hypotenuse, the longest side. In spacetime, suppose we have two events with relative coordinates x, y, z, and t, as measured by observers in some inertial frame. Then the SR 'distance' s, corresponding to the hypotenuse in Pythagoras' theorem, is given by Minkowski's 'line-element' [Minkowski, 1908]

$$s^2 = c^2 t^2 - x^2 - y^2 - z^2, \tag{7.5}$$

where c is the speed of light. An excellent account of Minkowski's ideas about this concept is given in [Petkov, 2012]. The importance here to us is the *signature*, the pattern of pluses and minuses in (7.5). In our convention, we have the pattern $(+,-,-,-)$. We shall discuss the physical implication of having different signatures, such as $(+,+,+,+)$ or $(+,+,-,-)$ in Chapter 12. In the latter case, the spacetime appears to have *two* time dimensions, something that plays havoc with all the standard notions of ordering and causality that the one-dimensional image of time based on $\mathbb{R}$ embodies. Signature is a fundamental concept in the study of time and is discussed further in the Appendix.

The complex numbers and imaginary time

Eventually, even the real numbers proved insufficient to satisfy the needs of mathematicians, so they extended the set $\mathbb{R}$ to the set $\mathbb{C}$, the complex numbers. In fact, mathematicians have gone beyond $\mathbb{C}$ to bigger, more complicated sets such as the quaternions (discovered by Hamilton) and the octonians. It turns out that quantum physics requires $\mathbb{C}$ but not the quaternions or the octonians directly. No one knows why $\mathbb{C}$ is necessary or apparently sufficient in this respect, but as we shall discuss in Chapters 24 and 26, the use of complex numbers in QM seems related to the processes of observation. These have a natural ordering: first we prepare a state of an SUO and then we observe quantum outcomes. If we want to reverse this process in some way, such as in experiments into time reversal symmetry, as discussed in Chapter 26, then the complex conjugation properties of the complex numbers come into play. The real numbers are inadequate in this context.

Given that the relative external context discussed in Chapter 2 is generally modelled by the real numbers used in CM, and that complex numbers are used in the QM formalism associated with relative internal context, we suggest that the transition from $\mathbb{R}$ to $\mathbb{C}$ in the modelling of time is precisely what is called a Heisenberg cut, that is, the interface between CM and QM.

Discrete time

There can be no direct proof that time is either continuous or discontinuous. The nature of time is contextual, that is, how it is looked at. There are advantages and disadvantages in any mathematical image of time.

Continuous time has had a very successful history by any measure of the word success. However, concerns can be raised at its continued exclusive use in fundamental theory. We need only look at the concept of temperature to see the dangers of keeping to a traditional intuitive perspective in physics. The concept of temperature is a remarkably useful one in many fields but it is now understood as an effective concept that can break down, such as for SUOs not in thermal equilibrium. The continuum based notions of classical thermodynamics have been replaced successfully by quantum statistical physics, which is based on discrete principles.

With this in mind, a number of theorists have explored the notion that time is discrete, either viewing discrete time as a numerical approximation approach to continuous time [Bender & Strong, 1985] or else as an intrinsic concept in its own right [Caldirola, 1978]. A review is given in [Jaroszkiewicz, 2014].

The mathematics of discrete time is particularly rich in phenomena, richer than for continuous time. The reason is that there is a hierarchy of assumptions that leads to traditional Newtonian mechanics: with each step up in the hierarchy, there is a new constraint on what the mathematics can do for us, so it does less. Let us look at this hierarchy from the ground and work our way up.

In the first instance, we may model time simply as a finite index set, such as $\{a, b, c\}$. Then different events can be given different times, or indices, such as events E_a, E_b, E_c. No ordering need be assumed, the temporal index serving merely to implement Cumming's view that '*time is what keeps everything from happening at once*' [Cummings, 1922].

The next step would be to impose an ordering on the index set, such as $a < b < c$. Then E_b is later than E_a and earlier than E_c. A *first-order* discrete time mechanics would be a rule that allowed an observer to use a knowledge of E_a to determine what E_b should be, and then use E_b to determine what E_c should be. Essentially, we would have $E_b = U_{ba}(E_a)$, where U_{ba} is some evolution function that takes us from the set of possible events at time a to the set of possible events at time b. This process could then continue, in the form $E_c = U_{cb}(E_b)$, where U_{cb} need not have any relationship with U_{ba}: they are in principle completely different functions, with the domain of U_{cb} being the range of U_{ba}. Discrete time mechanics was presented in this way by Maeda and collaborators [Ikeda & Maeda, 1978; Maeda, 1980, 1981].

We may bypass time b in this approach by writing $E_c = U_{cb} \circ U_{ba}(E_a)$, where the symbol $\circ$ denotes the *composition* of two functions. In general, we would have to ensure that domains and ranges of functions in a composition make it meaningful. This is usually not an issue because domains and ranges in applications to space-times invariably assume that all domains and ranges are copies of the same domain. This need not be the case however, a prospect that arises when processes of observation are looked at carefully [Jaroszkiewicz, 2010].

A *second-order* discrete time mechanics would require a knowledge of E_a and, independently, a knowledge of E_b in order to work out E_c, that is, a rule of the form $E_c = U_{cba}(E_b, E_a)$. When Newtonian continuous time mechanics is discretized in time in some chosen way, it turns out that the resulting discrete time mechanics usually takes on this second-order form.

There is no natural way of temporal discretization, however, for the very good reason that discretization represents a step *back down* the mathematical hierarchy, with a consequent increase in generality.

If we have a discrete-time mechanics based on integer time, then we can go up the hierarchy and move to (say) rational time, that is, take the time index parameter to be a rational number. Now we have a dense set of states to play with, but there would still be the 'holes' where the irrationals were. Moreover, we would not have a linear continuum. The next step in the hierarchy would be to include the irrationals, so that finally our time would be a continuum. Even so, the dynamics we were using could be different to standard Newtonian mechanics. We could have continuous time but dynamical variables that were not differentiable functions of time, as in the case of Brownian motion, or even discontinuous functions of time. Newtonian mechanics generally supposes that dynamical variables are at least twice-differentiable functions of time except possibly at a finite number of times where impulses occur.

Fuzzy time

The real world is not like the clear, well-defined mental world of the mathematician. Sets in the real world have many facades, many blurred boundaries, and whether any particular object should be counted in this or that set is often problematical. For example, we normally classify people as belonging to either the set of living individuals or to the set of dead individuals. But when does a person actually die? Is there a sharp distinction?

Recognizing the practical difficulties of defining real world sets, mathematicians developed an approach called *fuzzy sets* that has the agenda of taking into account such issues. The approach in fuzzy set theory is to assign to an element a *membership degree*, that can range from say zero to unity. If an element has membership degree zero then it is not in a given set for sure, whilst if it has value one then it is in that set for sure. Values of membership degree between zero and one reflect the degree of ambiguity in deciding whether the element is in the set or not.

The notion of fuzzy set has been extended to time. In the fuzzy time-series approach, for instance, we consider a collection of fuzzy sets $\{F_n : n = 0, 1, 2, \dots\}$ indexed by a discrete time parameter, with various rules for predicting F_N from a knowledge of some or all of the fuzzy sets in the set $\{F_n : n = 0, 1, 2, \dots, N-1\}$. We note that the concept of fuzzy set implies the existence of an observer, because inanimate objects cannot have any quality of indecision about them: indecision is an attribute originating with an observer.

Causal sets

Causal set theory lays out logical relationships between elements of a set that convey patterns of ordering: this theory has been used to discuss cosmology [Bombelli & Sorkin, 1987; Ridout & Sorkin, 2000].

A causal set C is a set with a partial ordering property $\prec$ such that (i) if $x \prec y$ and $y \prec z$ then $x \prec z$ and (ii) if $x \prec y$ and $y \prec x$ then $x = y$. This is the same pattern of relationships that we discussed above with temporal ordering. The point is that causal set theory develops concepts related to causality, growth, familial relationships, and time that can be identified in the real world. For instance, a chain is a linearly ordered subset of a causal set C, that is every two elements of a chain are related by ≺, so this is like a family, with ancestor and successor elements in the chain.

Discrete general covariance is the notion that labels carry no physical meaning: classical causal sets are regarded as having an intrinsic structure independent of observation.

Time is discussed in causal set theory in two ways. External time (exo-time) is the view from outside the causal set and is the time sense in which an exophysical observer can see the causal set as 'growing' in the forwards direction of time, rather like Asimov's notion of the Eternals in *End of Eternity* looking over the Block Universe of Reality and seeing Reality changes patterned into that structure. In conventional causal set theory exo-time is supposed to have no physical meaning, so it is a metaphysical construct.

On the other hand, endo-time, or internal time, impinges on each element of a causal set in that it is born (generated) to the future (further along the causal chain) of all currently existing elements (from the perspective of the exo-observer). Therefore, no element can arise causally in the past of an existing element from which it was generated. Consistency then requires the *irreflexive convention*, the assertion that no element of a causal set can precede itself.

Causal sets are related to *spreadsheet mechanics*, discussed in [Jaroszkiewicz, 2014] and in Chapter 20. In that form of mechanics, a spreadsheet program such as Excel is used to pattern dynamics in a cellular mechanical simulation, with time running horizontally and the current past being fixed to the left of the current time. The irreflexive convention is encoded in the Excel software: if any attempt is made to read information in advance of the current 'time', the program halts because it has decided that an inconsistency will occur.

Many important aspects of physics and time can be discussed in terms of causal sets, including the generation of lightcone–like structures from what are localized rules of dynamics.

p-adic time

We discussed above the 'holes' in the rationals that were filled in with the irrational numbers. It turns out that the p-adic numbers give another way to fill in these holes, to 'complete' the rationals. There is a theorem called Ostrowski's theorem that says essentially that we can 'complete' the rationals in only two ways: either with the irrationals or with a choice of p-adics.

p-adic numbers come with a choice of prime number p, such as 3, 17, and so forth. Given such a p, we first define the p-adic digits $a_0 \equiv 0$, $a_1 = 1, \ldots, a_{p-1} \equiv p - 1$. Then a p-adic integer is a string of the form $\ldots a_{i_3} a_{i_2} a_{i_1}$ whilst a p-adic number is a string of the form $\ldots a_{i_3} a_{i_2} a_{i_1} \cdot a_{j_1} a_{j_2} a_{j_3} \ldots$

One potential difficulty with the p-adics is that, unlike the rationals and the reals, they do not form what is called an *ordered field*. Therefore, if we wish to keep a sense of ordering, we will have difficulty in using p-adics as a model for time, classically. However, we come once again to the Heisenberg cut discussed above. Complex numbers can be used in quantum mechanics even though they do not form an ordered field, because they are used as auxiliary technology in the calculations that lead to real-world predictions, and those are always in the form of real numbers. Complex numbers disappear as we calculate quantum mechanical outcome probabilities. p-adics have been used to model quantum mechanical path integrals, with the time parameter being treated as a p-adic number [Dragovich & Rakić, 2010]. That is consistent, because path integrals by definition are summations over potential configurations that are not themselves observable. So even if p-adic time is used in such path integrals, we are not committed to produce predictions in the form of p-adic numbers.

8

Illusionary images of time

The Block Universe

Perhaps the biggest problem physicists have in trying to understand time is the following paradox: the most successful theories of space and time in physics are special relativity (SR) and general relativity (GR), but these make no reference to the 'here and now', the ever fleeting, illusive instant of process time known as the *present*. Moreover, those theories do not differentiate between past and future on the fundamental level.[17] All references to the present or to any asymmetry between past and future have to be put in by hand.

This paradox is encountered in every branch of physics: Maxwell's equations do not distinguish between past and future, SR has abolished absolute simultaneity, unitary evolution in Schrödinger mechanics is incompatible with wavefunction collapse, white holes are as good solutions to Einstein's field equations in GR as black holes and, most disturbing, time itself has disappeared from the formalism of quantum cosmology. We shall discuss all of these issues in depth in later chapters.

There are two important exceptions: thermodynamics, the laws of which involve one of the arrows of time discussed in Chapter 1; and the Born rule in quantum mechanics (QM), which involves the inherently time asymmetric concept of probability.

If we were asked to account for the above paradox, we would point the finger at two contributory factors: the central roles of *geometry* and *symmetry* in modern physics. Consider geometry. As a branch of mathematics, geometry is quite reasonably discussed without reference to any primary observer: points, lines, and planes just 'exist' as part of the axioms. As for symmetry, this manifests itself in geometry in many places. For instance, a line has a length that does not depend on the direction in which it is measured: the distance from point A to point B on a line is the same as the distance from B to A.

Whenever physics is discussed in geometrical terms, both the timelessness and symmetry aspects of geometry have an impact: timelessness is seen in the

---

[17] We define the fundamental level in this context as the level at which the theorist writes down the basic equations of a given theory.

Images of Time. First Edition. George Jaroszkiewicz.
© George Jaroszkiewicz 2016. Published in 2016 by Oxford University Press.

spacetime (no hyphen) paradigm because that has abolished the moment of the 'now', whilst symmetry is seen in the lightcone structure of those spacetimes: without contextual information we cannot say which one of the two branches of a lightcone goes into the future and which one goes into the past.[18]

According to the principle of contextual completeness discussed in Chapter 2, problems should be expected whenever contextually incomplete mathematics is used in physics. Most modern theories are based on a geometrical model of space and time known as the *Block Universe* [Price, 1997]. This model presupposes that time is a dimension like space and that together these dimensions form a four-dimensional continuum known as *spacetime*. This is a basic assumption in SR and GR.

The Block Universe model is neither true nor false: it is just a useful way of sorting our memories of the past and our plans for the future in the form of a four-dimensional map, with one of the dimensions representing time. But we should keep in mind Korzybski's dictum that memory is a map of the past and should not be confused with it [Korzybski, 1994]. *Memory* is not an object but an ongoing process occurring at an observer's present, allowing us to classify events that may have happened in the relative past. However, there are many questions raised by this process, such as: *'how can any observer know whether a given memory is a faithful record of what happened?'*, *'how can an observer be sure that a memory is not a false account created long after the event in question?'*, and *'if two different observers hold different memories of the same purported event, how can we establish who is correct, assuming such a concept is valid?'*

The assumption that there is a single, observer-independent Block Universe is contextually incomplete, with a generalized proposition classification (GPC) of zero, on a par with metaphysics. A more reasonable view is that a block universe (note the small letters) is a convenient contextual model used by a primary observer to organize their data (which is always limited), with an architecture that depends critically on the current information held by that observer. For example, a primary observer's block universe could have a start time and an end time. Any block universe model consistent with the Big Bang would have a start time, for instance.

A classical block universe represents a point particle not as a point in three-dimensional space but as a *worldline*, a line in four-dimensional spacetime. A worldline weaves its way through spacetime and represents the entire past and future of the particle. On such a worldline there is no special moment of the 'now', no particular instant of simultaneity.

An analogy can be drawn with a long box of strands of raw spaghetti. The box that contains the spaghetti is analogous to spacetime, regarded as the container of worldlines, whilst the individual strands of spaghetti are the worldlines of different particles.

---

[18] Past and future are defined here relative to the expansion of the universe.

Although the Block Universe is relatively limited in the scope of its explanations, we can deduce from it several important clues about how we arise as conscious beings. The Block Universe encompasses the principles of SR and GR, including the metrical (geometrical) and lightcone structure of the spacetime concerned. Lightcones are of critical value in relativity as they help us identify potential causal relations between pairs of events in spacetime, and are discussed in detail in Chapter 17.

The Block Universe model also allows a discussion of endophysics, the idea that individual primary observers are located around worldlines embedded in space-time. According to this perspective, observers are inside the universe and receive information from their environment in a specific way dictated by the structure of lightcones. Now it is undeniable that humans have a consciousness when they are awake. It is a phenomenon arising from processes in an observer's brain in a relatively localized way that can be identified to some extent with a point on the observer's worldline. Only events situated inside or on the instantaneous *past light-cone* with vertex at a given point on the observer's worldline can send signals that converge on the brain at that given point on that worldline. Therefore, the observer's consciousness, which includes current memories and opinions, is different at different points along their worldline, even though the whole spacetime structure, including the observer's worldline, 'is there' all along in the block universe. Our subjective feeling that time is passing probably arises via some process of comparison going on in our brains, comparing what we are currently observing and our memories of what we have just observed.

There is a deep mystery here. Whilst we may think we have explained how consciousness arises along an observer's worldline, we have no explanation as to why we feel the particular moment of the 'now', the present right now being 20:08 in the evening of the 10 April 2015. Oops, it is gone. It is now 20:09.

That is the central mystery of time.

Persistence and transtemporal identity

There is probably nothing in the universe more potent yet more ephemeral than human thought: one flash of insight and the thought is gone, unless we can memorize it. Memory has its limitations, unfortunately. It is not transferable to other individuals except by elaborate invented means such as language. As civilizations developed, they enhanced the oral transmission of knowledge by the development of writing and other means, such as memorial statuary. We may interpret writing as a mechanism for extending the persistence of thought.

A similar problem arises in computer science. When a computer is switched off, the last operating state of the computer is lost, unless some mechanism exists to back it up. In computer science, the ability of a computational state to outlive the processes that created it is known as *persistence*.

The human view of reality is predicated on persistence: the visual information that we receive from our environment is processed on its way into the brain and interpreted in terms of images templated on already familiar patterns. When the brain judges there to be a sufficient correlation between the images constructed from the incoming information and the memories of earlier images, then the brain is deceived into assuming that the environment is relatively persistent. This process is sophisticated enough in humans to give an impression that the person is moving against a fixed background, with the direction of gravity going up–down and the local surface of the Earth generally being given a horizontal assignment. It is possible to undermine this process by not fixing visual attention on the distant horizon or background, so that suddenly a person can have the illusion that they are the centre of the world and the rest of the universe is spinning around them.

One of the greatest illusions that the brain plays on itself is to make it believe that the world around it never changes. In reality, everything changes, nothing lasts forever. In particular, when we visit old haunts, such as schools and houses where we once lived, we should keep in mind the truth expounded by Thomas Wolfe:

> You can't go back home to your family, back home to your childhood, back home to romantic love, back home to a young man's dreams of glory and of fame, back home to exile, to escape to Europe and some foreign land, back home to lyricism, to singing just for singing's sake, back home to aestheticism, to one's youthful idea of "the artist" and the all-sufficiency of "art" and "beauty" and "love", back home to the ivory tower, back home to places in the country, to the cottage in Bermuda, away from all the strife and conflict of the world, back home to the father you have lost and have been looking for, back home to someone who can help you, save you, ease the burden for you, back home to the old forms and systems of things which once seemed everlasting but which are changing all the time–back home to the escapes of Time and Memory.   [Wolfe, 1940]

Related to persistence is the concept of *transtemporal identity*. An object has a transtemporal identity [Stuckey, 1999] if it can exist or endure over observable periods of time, relative to a given observer. Unfortunately, this concept has been generally discussed by philosophers and metaphysicists without much regard to the contextuality involved. As a consequence, it continues to provoke undue and fruitless debate. For example, it has been used to suppose that time and space are necessarily continuous:

> A necessary condition of transtemporal identity would seem to be spatio-temporal continuity. In other words, we can say that *a* is identical with *b* only if there is a continuous trajectory connecting them, or, in other words, that *b* and *a* should both be 'individual stages' in the succession of such stages which corresponds to the 'career' of a single persisting individual.   [French & Krause, 2002]

There is no reference in this statement to any observer who would establish 'spatio-temporal continuity', so it is contextually incomplete.

The question of transtemporal identity was discussed in Antiquity. In the conundrum known as *The Ship of Theseus*, the question is asked: *if every time a ship returns to harbour some of its planks are replaced, does it remain the same ship when all of its original planks have been replaced, or is it a new ship?*

Our answer to this is that in common with all such questions, the contextuality of the terms used must be made more precise and understood. For instance, what do we mean by *a ship?* If we said that it was a certain collection of planks, ropes, sails, and so on, then suppose we removed one plank and asked the same question. If the answer was that it was still the same ship, then we would go on removing bits and pieces until a point came when it could no longer be reasonably regarded as a ship. What would determine that point would not be any absolute definition of a ship, but a functional one: if a collection of planks, ropes, sails, can function as a ship as far as an observer is concerned, then it is a ship.

When various such 'paradoxes' are analysed, there are usually hidden assumptions and it is because of these that the paradox arises. In the case of the Ship of Theseus, we should state clearly whether the ship is originally defined to be a very specific, particular collection of planks, or just a functional object used for transporting goods. If the former definition is used, then replacing just one plank would mean that we had a different ship, by definition. On the other hand, if the latter definition is used, then as long as the ship remained functional, replacing any number of planks would not matter.

The problem of time in cosmology

We live in a universe in which there appears to be a well-defined direction of time. Evidence for this is the second law of thermodynamics and the expansion of the universe.

However, in classical general relativity, there is no obvious way of defining a unique sense of time. Indeed, in various approaches to cosmology, the notion of time disappears altogether, leading to the so-called *problem of time*. This is exemplified by the *Wheeler–DeWitt equation*. This equation arises in quantum cosmology as follows.

1. In standard non-relativistic quantum mechanics, the Schrödinger–Dirac equation

$$i\hbar\frac{d}{dt}\,|\Psi,t\rangle = \hat{H}(t)\,|\Psi,t\rangle \qquad (8.1)$$

describes the evolution of a pure state vector $|\Psi,t\rangle$ for a system under observation (SUO), where t is the laboratory time as measured by an observer external to that SUO and $\hat{H}(t)$ is the Hamiltonian operator.

2. In the approach to quantum cosmology taken by Wheeler and DeWitt [DeWitt, 1967], the SUO is taken to be the universe as described by Einstein's theory of general relativity (GR), with metrical degrees of freedom assumed. The chosen dynamics has certain assumed symmetries, particularly diffeomorphism invariance.[19]

3. Because of these symmetries and before canonical quantization, the classical analogue theory requires the application of Dirac's constraint mechanics, discussed in Chapter 15.

4. It turns out the relevant Hamiltonian H_U for the universe is a so-called *primary constraint*.

5. Such a (primary) constraint has two roles classically:

    (a) it vanishes on the so-called *surface of constraints*, written $H_U \approx 0$ in the notation of Dirac [Dirac, 1964], and

    (b) it is a generator of so-called *gauge transformations*, which are transformations of the dynamical degrees of freedom that do not affect any physical observables.

6. Quantization for such a constrained system then leads to the requirement that physical states $|\Psi_U\rangle$ of the universe must satisfy the constraint equation

$$\hat{H}_U |\Psi_u\rangle = 0, \tag{8.2}$$

where $\hat{H}_U$ is the quantum operator corresponding to the classical Hamiltonian H_U. This is the Wheeler–DeWitt equation.

7. But a (naive) view of quantum cosmology would be that the quantum state for the universe should satisfy an equation formally equivalent to the Schrödinger–Dirac equation (24.1), that is,

$$i\hbar \frac{d}{d\tau} |\Psi_U, \tau\rangle = \hat{H}_U |\Psi_U, \tau\rangle, \tag{8.3}$$

where τ is the relevant temporal evolution parameter.

8. We conclude that physical states of the universe must be timeless,

$$i\hbar \frac{d}{d\tau} |\Psi_U, \tau\rangle = 0. \tag{8.4}$$

This is the so-called *problem of time*.

Although much has been made of this result, particularly by Barbour [Barbour, 1999], the problem here is that the discussion has a GPC of zero. Indeed, the

---

[19] Also known as *general covariance*, this is the invariance of intrinsic properties of a spacetime to smooth reparametrizations of coordinates.

assumption that the universe is a quantum SUO seems profoundly wrong, an example of an extrapolation of a good theory (GR) to well beyond its domain of validity. The subject of quantum cosmology has been criticized by Finke and Leshke [Fink & Leschke, 2000], who make the point that quantum theory is a theory of *observation*, not a theory of *things*. By definition, the universe contains everything and runs just once, so there cannot be any observer standing outside and determining expectation values via multiple reruns of the universe.

9

Causal images of time

Introduction

The concept of time is used by humans to make sense of their relationship with the universe. On its own however, time is too vague to be of much use in philosophy or science. Inevitably, it is supplemented by auxiliary concepts such as *causality*, *determinism*, *fatalism*, *teleology*, *finalism*, *chaos*, *reversibility* and *irreversibility*, *probability*, and more. Each of these concepts represents some particular aspect of time considered important in various contexts. In this chapter we discuss some of these aspects in some detail. Irreversibility is discussed in Chapter 22.

Causality

Causality is hard to define precisely. It is used as a basis for judgements in law, the allocation of responsibility, the logic of planning, and for dynamical effects in science. Classical mechanics (CM) in the form given by Newton is predicated on the principle of causation, which asserts that '*nothing happens without a reason*'. Leibniz enshrined this idea into the *principle of sufficient reason* [Kronz, 1997]:

> All truths have a reason why they are rather than are not.   [Leibniz]

This is an echo of a principle attributed to Leucippus in the fifth century BCE, who said that [Gregory, 2013]

> Nothing happens at random but everything for a reason and by necessity.
> [Leucippus]

Causality begs several questions, such as: '*who has established that something has happened?*' and '*to what extent is the reason for that happening independent of the observer?*' To understand causality properly, therefore, several pieces of a complex conceptual framework have to be in place: we need the concepts of *primary*

Images of Time. First Edition. George Jaroszkiewicz.

observer, *event*, *temporal ordering*, and contextual completeness. These have been discussed in earlier chapters.

Suppose a primary observer O has observed many instances of some event E, each at a specific time, as measured by O's clock. Suppose further that O has noted that, prior to each instance of E occurring, an instance of another event C had always occurred at a definite time before that instance of E. Then O might conjecture that an event E could not occur without an earlier occurrence of an event C. O might even conclude that each event C was the *cause* of a subsequent event E, which would then be interpreted as the *effect of C*.

The problem is, C and E events could be the mutual effects of some other cause A, as in Figure 9.1(b). In such a case all we could say for sure was that C and E were *perfectly correlated*: *every* time we observed an event of type C, we could be sure that E had already occurred, or would occur, even if we did not attempt to observe it.

However, when we look in detail at such scenarios, matters can be more complex than that. Take events happening on macroscopic scales, such as battles. Each is unique, but it is often convenient to ignore what are in context regarded as unimportant differences and invoke a causal explanation for repeated patterns: the Roman Army invariably defeated the Gauls because of superior training.

Locke believed causality is a category (concept of understanding) used to classify experience [Locke, 1690]:

> In the notice that our senses take of the constant vicissitudes of things we cannot but observe that several particulars, both qualities and substances begin to exist and that they receive existence from due application and operation of some other being. From this we get cause and effect. [Locke]

The philosopher John Stuart Mill pointed out a fundamental feature of causality [Mill, 1882]:

> If the whole prior state of the entire universe could again recur, it would again be followed by the present state. [Mill]

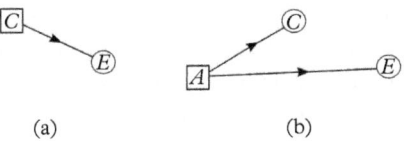

(a)   (b)

Fig. 9.1 *In (a), event C appears to be the cause of event E. In (b), both events are seen to be caused by event A and are positively correlated.*

In logic and mathematics, the concept of *implication* is related to causality in a particular way. Consider two propositions, P and Q. If it is certain that Q is always true whenever P is known to be true, we write $P \Rightarrow Q$. In words, we say 'P implies Q'. But can we say that P *caused* Q?

A moment's thought gives the answer *no*: it could be that P and Q were themselves the inevitable consequences of some other event R, so that the truth values of P and Q are correlated, but not causally dependent on each other.

Reichenbach devised a notation to express his notion of causality, known as the *mark method* [Reichenbach, 1958]. We shall translate his discussion into our terms. Suppose $\mathcal{E}_1$ and $\mathcal{E}_2$ are two contextually complete generalized propositions such that $\mathbb{V}\mathcal{E}_1 = \mathbb{V}\mathcal{E}_2 = 1$, where $\mathbb{V}$ is the validation function introduced in Chapter 2. If $\mathcal{E}_1$ is the *cause* of $\mathcal{E}_2$ we write $\mathbb{V}(\mathcal{E}_1 \rightarrow \mathcal{E}_2) = 1$ otherwise we write $\mathbb{V}(\mathcal{E}_1 \rightarrow \mathcal{E}_2) = 0$.

Reichenbach considers what might happen if $\mathcal{E}_1$ changed, perhaps by a 'small' amount, to a modified proposition $\mathcal{E}_1^*$ and $\mathcal{E}_2$ changed to $\mathcal{E}_2^*$. Then according to Reichenbach, $\mathbb{V}(\mathcal{E}_1 \rightarrow \mathcal{E}_2) = 1$ is consistent with $\mathbb{V}(\mathcal{E}_1^* \rightarrow \mathcal{E}_2^*) = 1$ or even with $\mathbb{V}(\mathcal{E}_1 \rightarrow \mathcal{E}_2^*) = 1$, but never with $\mathbb{V}(\mathcal{E}_1^* \rightarrow \mathcal{E}_2) = 1$.

There are two points here that should be commented on. First, the possible consistency of $\mathbb{V}(\mathcal{E}_1 \rightarrow \mathcal{E}_2) = 1$ and $\mathbb{V}(\mathcal{E}_1 \rightarrow \mathcal{E}_2^*) = 1$ allows for the possibility that $\mathcal{E}_1$ could be the cause of more than one potential or actual outcome: this can make sense both in classical and quantum physics. Second, the inconsistency of $\mathbb{V}(\mathcal{E}_1 \rightarrow \mathcal{E}_2) = 1$ and $\mathbb{V}(\mathcal{E}_1^* \rightarrow \mathcal{E}_2) = 1$ means that Reichenbach regards causes as unique to their effects. This suggests that Reichenbach's image of time is that the past is unique, because causes are always supposed to precede their effects.

Blame and responsibility

Humans like to believe that they live in a universe based on logic and sense, so that if something good or bad happens then *there must be an underlying reason for it*. Such a belief helps us come to terms with a universe where events seem beyond our control. It is a challenge to our dignity and intelligence to realize that perhaps we are not significant in this universe, that perhaps what appears important to us is contextual, based on our ideologies and conditioning, and outside of that has no intrinsic significance to the rest of the universe. In this context, religion has an important role. A belief in a religion based on a god or gods that causes events to happen eliminates the need to account for those events: random events can be attributed to the working of divine beings. Religions also comfort us by placing humans relatively high up on the scale of importance, generally just below the divine being or beings.

There are dangers in such thinking, for if we attribute responsibility for events to superior beings then we are absolved from responsibility for those events. We may be led to a belief in fatalism, the view that nothing that we can do now alters the future.

Believers in fatalism are frequently inconsistent. A person who believes that a divine being has decided the course of future events is often ready to condemn others for their actions in the past, at which point causality is inverted and becomes *blame*. When something happens that we interpret as bad for us, it seems natural to blame another person for that, because there is no comfort in blaming an invisible god or gods: it is usually much more satisfying to punish a real individual for their crimes, real or imagined.

Our conditioning in this respect is powerful: it is very uncomfortable to contemplate the possibility that those who commit the most hideous of crimes should not be held responsible for their actions, perhaps because they were ill. The fact is, human understanding of causality, on the macroscopic scales defined by human social interaction, is virtually non-existent. Take the case of a driver who goes too fast around a bend and kills a child on a bicycle. To what extent did the driver cause that death? Certainly, if they had not driven too fast, the child might have lived. But, equally, the child's parents could have decided to prevent the child cycling on dangerous roads in the first place.

Perhaps our ignorance of causality is just as well. If the day comes when human actions are fully understood and controlled, then perhaps we will no longer be humans but machines, running on pre-programmed, deterministic lines with a total absence of blame or responsibility. Whilst this may seem like Utopia, the problem is that such pre-programming is not intrinsic to the universe and we could find our society running along very unpleasant lines, such as the imposition of mandatory termination of life at age 60 [Asimov, 1950].

Determinism and known unknowns

Determinism is based on the contextually incomplete assertion that what is done at one time will have definite consequences in the future. That is not the same thing as saying that if we do something now then we can predict what the consequence of that action will be. This inability to calculate the consequences of our actions has nothing to do with any imagined randomness that can upset otherwise predetermined paths. The information we have about the future may be limited, or may be such that any attempt to acquire more information will destabilize our predictions.

There is an analogy here with the solution of equations in mathematics: we can prove that there are five solutions (roots) to a quintic equation, but there is no formula that can give them to us in general.

We should therefore distinguish between two sorts of determinism. A type 1 deterministic system is one such that, given initial conditions at time $t = 0$, we can fully predict the state of the system at any subsequent time $t > 0$. On the other hand, a type 2 deterministic system is one such that we know that a unique state exists at any subsequent time but we have no means of predicting it.

Joseph-Louis Lagrange [1736–1813] refined Newtonian mechanics. His equations of motion are equivalent to those of Newton but based on a different temporal architecture. Newton's equations of motion are essentially predicated on process time: a system under observation (SUO) is set in motion at an initial time under conditions prevailing at *that* time: these are usually the initial positions and velocities of all the particles constituting that SUO. The future state of the system is then determined solely by the laws of mechanics.

Lagrange went further and used a manifold image of time. He developed the *principle of virtual work*, which was eventually refined as an action principle based on the Calculus of Variations and Hamilton's principle. These are discussed in Chapter 14. Suffice it at this stage to say that such principles are teleological in flavour, since the application of variational principles requires us to decide what the final configuration of the SUO should be, *before* we work out how the SUO could get there.

In QM, this approach takes on a bizarre flavour. Now we cannot decided what the final configuration should be: there may be many alternatives, and we can calculate only the relative probability of ending up in any one of them. Moreover, unlike CM where the trajectory from initial to final configuration is well defined, in QM we cannot exclude the possibility of the SUO taking any of the countless trajectories to go from initial to final states. The Feynman path integral gives us the rules for taking *all* of their contributions into account [Feynman & Hibbs, 1965]. This is the final nail in the coffin for classical determinism.

Teleology

Teleology is the principle that *final causes* exist in nature, that is, that there is something or someone that has established a purpose or goal to which a system such as the universe develops in time. Teleology makes an implicit appeal to some absolute primary observer, which is why it finds favour with religious people.

Teleology is regarded in science as an unscientific principle because either it is stated in a contextually incomplete form designed to lead to an acceptance that there is an ultimate primary observer (the designer of the universe), or else it is stated with explicit reference to a metaphysical primary observer such as the God concept.

Chaos

As science and mathematics develop, there arises, surprisingly frequently, popular interest in various new or rediscovered ideas. One of these is chaos theory. It is of interest to non-scientists because their lives can be affected by chance events, the consequences of even the smallest change can be magnified over time to have devastating unforeseen consequences. Chaos theory is a mathematical discussion of such phenomena.

Dynamical systems exhibiting chaotic behaviour occur in both classical and quantum theory. It is classical chaos that is generally referred to in popular debate, because that is where it seems most surprising to the non-mathematician. After all, quantum mechanics is well known for its random outcomes, whilst standard classical mechanics was, until relatively recently, always thought of as fully deterministic. In fact, chaos and strict determinism are not inconsistent. Far from it: chaotic phenomena are most spectacular when discussed within a strictly deterministic context.

We said above that Newtonian mechanics is based on a process time perspective: given an initial state Σ_0 of an SUO, the equations of motion can be used to determine Σ_t, the final state at a later time $t > 0$. That is a strictly deterministic picture.

Now suppose we alter the initial conditions, from Σ_0 to say Σ_0^*, reminiscent of Reichenbach's mark method discussed above. Then the equations of motion applied to Σ_0^* will evolve it to a unique final state Σ_t^*. The question is, what is the relationship between the ordered pairs (Σ_0, Σ_t) and (Σ_0^*, Σ_t^*)?

To quantify this, we need some way of describing differences between states. A typical method would be to use the properties of real numbers as follows. Given the two initial states Σ_0 and Σ_0^*, construct in some agreed way a positive number $d(\Sigma_0, \Sigma_0^*)$, with the property that $d(\Sigma_0, \Sigma_0) = 0$ and $d(\Sigma_0, \Sigma_0^*) > 0$ if $\Sigma_0^* \neq \Sigma_0$. This will serve as a definition of 'distance' between the states.

The question then is, what is the relationship between $d(\Sigma_0, \Sigma_0^*)$ and $d(\Sigma_t, \Sigma^*(t))$ for any t greater than zero? This question can be discussed in several ways. We give now a heuristic (hand waving) account of one, involving the so-called *Lyapunov exponent*.

A Lyapunov exponent λ is a specific measure of sensitivity to initial conditions that gives a measure of the timescale over which chaos sets in. It can be estimated for various realistic SUOs. For such systems, there is an approximate relationship of the form

$$d(\Sigma_t, \Sigma_t^*) \approx e^{\lambda t} d(\Sigma_0, \Sigma_0^*), \tag{9.1}$$

for large enough time t and small enough $d(\Sigma_0, \Sigma_0^*)$. The exponent λ may itself depend on the initial states. If the maximal Lyapunov exponent is positive, then the system is taken to display chaos: given $d(\Sigma_0, \Sigma_0^*) > 0$, then $d(\Sigma_t, \Sigma_t^*)$ will eventually exceed a given bound. For example, if we want to ensure $d(\Sigma_t, \Sigma_t^*) < \delta$, where δ is positive, then if (9.1) is taken to be exact, then by a time $T = \lambda^{-1} \ln \left\{ \delta / d(\Sigma_0, \Sigma_0^*) \right\}$ we have reached the limit.

The point here is that if an SUO demonstrates chaos, then there will be a characteristic timescale beyond which predictions become meaningless. An essential feature here is that chaos cannot be attributed to an SUO alone: a deterministic SUO evolves without any concern for predictability. Chaos is a discussion of a relationship between an SUO, time, and the observer.

10

Physics and time

Physics and persistence

Humans are conditioned by experience to think that the world around them is stable and subject only to gradual change. In fact that is a dangerous illusion based on *persistence*, the apparent endurance in geometrical shape of vast patterns of atoms. This is due to the relatively enormous differences in various scales of time: we do not live long enough to see mountains rise or church window glass slowly creep downwards under gravity. Our natural perceptions such as vision are based on physical processes that take tiny fractions of a second to register in our brains, so that processes that take days or longer can appear to be static. The point was well put by the physicist Chew:

> There is, nevertheless, a possibly unavoidable reason for a difference in status between electromagnetic and strong interactions. Electromagnetism provides the tools that make feasible the measurements on which physics is based. For example, the existence of solid (or pseudosolid) matter, which can maintain a shape in which different components are distinguishable, is essential to the concept of measurement. It is difficult to see how solids could exist without two key features of electromagnetism: (1) the zero (or very small) photon mass, which gives long range forces; and (2) the small fine-structure constant, which allows the size of the atom to be much larger than that of the nucleus, while adding very little mass.[20] Purely nuclear matter, even if unrestricted by electromagnetism, seems unlikely to develop the characteristics of a solid.[21] If such a view is correct, the photon mass and the fine-structure constant are closely interlocked with the problem of measurement, perhaps even with the meaning of macroscopic space-time . . .   [Chew, 1966]

The relative persistence of patterns in our environment not only conditions us to think of those environments as relatively stable but makes the concept of a system under observation (SUO) a practical and economical way of dealing

---

[20] Assuming that the small mass of the electron is associated with the smallness of the fine-structure constant.

[21] Pulsars, or rotating stars consisting of tightly compressed neutrons with a mass density comparable to purely nuclear matter, were first observed in 1967 [Hewish *et al.*, 1968].

Images of Time. First Edition. George Jaroszkiewicz.
© George Jaroszkiewicz 2016. Published in 2016 by Oxford University Press.

with the vicissitudes of life. The illusions of permanence that we create in our minds guide us so well that we easily forget how fragile those illusions are, until an earthquake reminds that nothing endures forever.

These illusions have had an enormous influence on the development of science: Aristotelians believed that the Earth was fixed and that everything else moved relative to it. It took two thousand years to break this conditioning, and the effects still linger. We like to reminisce about our old school or our old home town, and imagine that we might return one day to relive past glories. It is sobering to realize that none of that things 'exist' exactly as they were. They will have changed, just as we have: we are not quite the same people we were then. The Earth will have rotated thousands of times on its axis since then, moved around the Sun many times since then, and travelled with it on its huge orbit around the centre of the galaxy many millions of miles since the old days.[22] Since then, the old school may have been torn down and replaced by a supermarket, or as in the author's case, entire streets razed and communities relocated. When our past is no longer there, when only our memories remain, there comes the most frightening thought of all: when we too have passed and all our civilization with it, there might be left no trace in the universe that we ever existed. Not even persistence is permanent. There is only one fundamental law of time: *everything changes, nothing endures forever.*

Scales of time

To put our remarks on persistence into perspective, let us now review the temporal scales associated with various physical processes.

The Planck time

As physicists probed elementary particles with greater and greater energies, they realized that as energies increased, the timescales for the processes that they were investigating decreased.

There are several ways of understanding this reciprocity. For instance, de Broglie's relation $E = h\nu$ shows this explicitly. Here E is the energy of a particle, h is Planck's constant, and ν is the frequency (a reciprocal time) of the associated de Broglie wave. We shall explore this reciprocity in more detail later on in this book. Suffice it for now to say that most physicists would agree that there is a limit to shortness of duration, to how quickly we can perform any action. Below a certain scale, there seems to be no physical or theoretical meaning to intervals of time. This limit is generally called the Planck time, denoted T_P. It is a heuristic concept used extensively to motivate research in speculative areas such as string theory and quantum gravity.

---

[22] The Sun moves around the galactic centre at a speed over half a million miles per hour, taking 240 million years to complete one orbit as defined by line of sight with the distant stars.

The Planck time is assigned a value from dimensional analysis via the following heuristic argument. Modern physics is based on several fundamental constants, three of which are c, the speed of light, $\hbar$, Planck's (reduced) constant, and G, the Newtonian constant of gravitation. From dimensional analysis, it is easy to work out that $\hbar G/c^5$ has the physical dimensions of a time squared. Dimensional analysis does not give us any numerical factors, so taking the overall numerical factor to be unity, we define the Planck time as

$$T_P \equiv \sqrt{\frac{\hbar G}{c^5}} \approx 5.4 \times 10^{-44} \text{sec} \tag{10.1}$$

Such a timescale is far, far smaller than any timescale encountered in the laboratory (these are discussed next). It is commonly believed that the Big Bang, the origin of the universe, took place over such a timescale.

The concept of Planck time and the related Planck scales of length and mass have been criticized by Meschini [Meschini, 2006].

The natural chronon

There are a number of scenarios where a timescale of about 10^{-24} sec. appears naturally in physics. It is convenient to use this as a base unit, called here the *natural chronon*, T_N, a yardstick to compare similar timescales. We caution against thinking of such a timescale as the most fundamental unit of time. In particular, we should not think of a chronon as a 'quantized particle' of time, unless we have a specific, contextually complete model to explain what we mean by this. Not every quantity that comes in integral units is derived from a quantum theory, examples being days, pennies, students in a class, and galaxies in the universe.

Our first example of a natural chronon comes from hadronic physics, the study of processes involving protons, neutrons, and mesons. In 1956, there came the first direct evidence that the proton is not a fundamental pointlike object: it has an effective electric charge radius of about 0.8768 fermi, that is, just under 10^{-15} metres [McAllister & Hofstadter, 1956]. The speed of light is almost 3×10^8 metres per second, and so light takes about six natural chronons to cross the diameter of a proton.

Another example comes from high-energy particle collisions, where particle resonances have been observed with inferred lifetimes of the order of a natural chronon: this is about a hundred million million million times longer than the Planck time.

Notable theoretical chronons are Caldirola's chronon and Finkelstein's chronon. The former comes from Caldirola's theory of discrete time [Caldirola, 1978], which features a timescale $T_C \equiv e^2/(6\pi\varepsilon_0 m_e c^3) \approx 6T_N$, where e is the electronic charge, ε_0 is the permittivity of free space, m_e is the mass of the electron, and c is the speed of light. In a series of papers labelled *space-time code*, Finkelstein gave

arguments for the empirical significance of a similar timescale in high-energy hadronic collisions [Finkelstein, 1968]. The current shortest hadronic collision time,
associated with the Large Hadron Collider, is of the order one tenth of a natural
chronon [Antchev, 2012].

Time for light to cross an atom

Bohr's model of the atom [Bohr, 1913] pictures each hydrogen atom as a central
point-like particle (the proton) orbited by a much lighter particle (the electron). The diameter d of the electron's orbit in the model is approximately one
Ångstrom, or one tenth of a thousand-millionth of a metre. The relevant timescale here is the time T_H for light to cross the electron's orbit, given by $T_H = d/c$,
where c is the speed of light. Light travels at a speed approximately three hundred
million metres per second and so T_H is approximately one third of an *attosecond*.
An attosecond is one billionth of a billionth of a second, that is, 10^{-18} seconds,
approximately the time for light to cross three hydrogen atoms.

Hydrogen is the lightest element, with one electron per atom. Atoms of heavier
elements have more electrons and correspondingly more protons in their nuclear
centres. The electric forces that bind the electrons to the protons are proportional to the number of protons, and so get larger as we go to heavier elements.
It might be expected therefore that electrons in heavier elements would be forced
into tighter orbits than that of hydrogen. Whilst this is true of the next element in
the periodic table, helium, which has a diameter about half that of hydrogen, there
are notable factors conspiring to undermine the simplicity of the Bohr model.
First, electrons obey the Pauli exclusion principle, a veto on particles such as electrons sharing the same quantum numbers.[23] Then there is electron shielding: the
other electrons in a multi-electron atom repel any given electron and cancel out
to some extent the attractive force of the protons in the nucleus. Finally, atoms
cannot be accurately modelled by classical mechanics but by quantum mechanics,
which raises numerous technical issues. The net effect is that atomic diameters
range from about one Ångstrom for hydrogen to about six Ångstrom for caesium,
which is not the heaviest naturally occurring element. The attosecond turns out
to set a suitable temporal scale for many atomic processes, leading to a branch of
physics known as *attophysics*, based on pulses of light or electrons of the duration
of an attosecond.

Period of an electron in a Bohr atom

In Bohr's 1913 theory of the atom [Bohr, 1913], an electron in the lowest circular
orbit has a period of approximately 1.5×10^{-16} seconds.

---

[23] The analogue of classical properties.

Fastest electronics

The fastest electronics available today have a time resolution of about 10^{-11} seconds.

Period of insect wing

Insect flight is based on muscle power and its efficiency depends primarily on the species of insect. The characteristic period of oscillation of the wings ranges from a few times per second for some butterflies to over a thousand times a second for some mosquitoes. The fastest flying insects generally have the longest and thinnest wings.

In insects with *asynchronous flight muscles*, a single nerve impulse can cause a muscle fibre to contract multiple times, allowing the frequency of wing beats to exceed the rate at which the nervous system can send impulses. Such insects can hover, fly backward, and outperform insects with *synchronous flight muscles*, which beat once per nerve impulse. The period of asynchronous wing beats can be faster than 10^{-3} seconds, compared with the slower range of 2×10^{-1} to 5×10^{-2} seconds for synchronous wing beats.

Period of wing of humming bird

Hummingbird wings beat at a typical rate of 80 times per second, but can range from as low as 12 times per second up to 200 times per second. Hummingbird heart beats have been recorded as high as 1000 times per second.

Average human heartbeat

The average period of the human heartbeat is about 0.86 seconds ['t Hooft & Vandoren, 2014].

The second

Defined conventionally as 1/60th of a minute, which is 1/60th of an hour, which is 1/24th of an Earth day, the second is too important a scientific unit to be so defined, given the variations in the period of rotation of our planet. By international treaty, the unit of time known as the *second* is defined as exactly 9,192,631,770 cycles of the electromagnetic radiation emitted when an electron makes a transition between the two hyperfine levels of the ground state of the caesium 133 atom [International des Poids et Mesures, 2006].

The hundred metres

The current world record [2015] for the shortest time for a human to run 100 metres is 9.58 seconds held by Usain Bolt. In comparison, the fastest land mammal, the cheetah, would take about 3 seconds. The fastest bird in powered flight, the spine-tailed swift, would fly that distance in about 2 seconds whilst the fastest fish, the sail fish, would take 3 seconds in water.

The Minute Waltz

Chopin's masterpiece for the piano, written in 1847, takes between 1.5 to 2.5 minutes to play depending on the pianist.

The period of rotation of the Earth

The *true siderial day* is the period of rotation of the Earth relative to the distant (fixed) stars and is 86164.09053 seconds.

The lunar month

Although the Moon appears relatively stable in its orbit about the Earth, its motion is complex and there are several different definitions of lunar months. The *siderial month* is approximately 27.32166 days, the time for it to return to a fixed point relative to the distant stars. The *tropical lunar month* is the time for the Moon to return to the same point on the ecliptic (the imaginary point on the celestial equator intersected by the plane of the Earth's orbit, discussed below). Because of the precession of the equinoxes, the tropical lunar month is approximately 27.32158 days, slightly shorter than the siderial lunar month.

The year

Whilst Newtonian orbit theory predicts that the Earth's orbit is a perfect ellipse with a definite period known as a *year*, a full description of the motion of the Earth around the Sun is complicated by many factors. There are, consequently, several different definitions of an Earth 'year'. The *sidereal year* is the time required for the Earth to complete an orbit of the Sun relative to the distant stars and is 365 days, 6 hours, 9 minutes, and 9.5 seconds. This would be the length of the year if there were no complicating factors.

The *ecliptic plane*, the plane of the Earth's orbit, is inclined by just over 23 degrees relative to the *equatorial plane*, the plane perpendicular to the axis of the Earth's daily rotation. Over a year, observers on the Earth see the Sun moving slowly relative to the distant stars on a path known as the *ecliptic*. This path intersects the equatorial plane at two points known as the *equinoxes*. One of the equinoxes takes place approximately on 20 March (the vernal equinox) at a point on the ecliptic known as the First point of Aries. The other equinox is on 22 September (the autumnal equinox).

A significant complicating factor is that the Earth is not a perfect sphere and on that account there is a 'wobble' or precession of its axis of rotation. One of the triumphs of Newtonian mechanics is that it can account for such wobbles in the motion of a rotating mass. It was noticed in Antiquity that the equinoxes move gradually eastwards, a phenomenon known as the *precession of the equinoxes*. Every year, the First Point of Aries crosses the equatorial plane just over 20 minutes earlier than the previous year. The astronomer Hipparchus estimated that the period of the Earth's 'wobble' was about 36 thousand years. Modern measurements give a figure of 25,722 years. A *tropical* or *solar* year is the time for the Sun to return to the same position, relative to the cycle of seasons, as seen from Earth. The *mean tropical year* on 1 January 2000 was about 20 minutes and 25 seconds shorter than the sidereal year.

The galactic year

Our galaxy is a vast rotating system of stars and gas. This includes our star, the Sun. Like all stars, the Sun is orbiting the galactic centre with a period estimated between 220 and 250 million years. This period is convenient in a number of respects and serves as the definition of a galactic 'year', denoted GY in the following.

According to the most recent estimates, the universe (and hence time) started just over 60 GY ago. It is commonly believed that following a period of rapid expansion known as inflation, our galaxy formed just over 50 GY. Our star, the Sun, is a relatively latecomer in galactic chronology, appearing just over 18 GY ago. Therefore, the Sun has orbited the galactic centre close to 20 times in its lifetime so far.

The decay lifetime of the proton

The proton is one of the fundamental particles that define our observable world. Protons are, as far as is known, completely stable. However, various proposed 'Grand Unified Theories' (GUTs) predict that protons can decay into other particles. The essential fact that permits us to discuss such a concept is that protons are no longer believed to be 'elementary', but bound states of more elementary objects called quarks and gluons. GUTs describe the interactions of those more elementary objects and not protons *per se*.

Motivated by GUT predictions, physicists have attempted to observe proton decay. What helps is Avogadro's number, about 6×10^{23}, which is relatively colossal and a measure of how many atoms there are in a handful of matter. The sheer scale of Avogadro's number, when factored in with the predicted enormously long lifetime of the proton, means that vast numbers of protons can be monitored over accessible timescales such as years. If just one of these protons does something unusual, such as decay during the course of being monitored, then hopefully that could be observed and allow meaningful limits to be placed on that lifetime. No such events have been observed, and so currently, the best estimate is that the proton half-life cannot be less than about 10^{34} years. This is about 10^{26}GY or about 10^{24} times the current estimated age of the universe.

11

Biological time

Introduction

The discussion of the scales of time in Chapter 10 is important if we want to understand what time means. The relative persistence of structures in the universe simultaneously defines time and is defined by time. Without persistence in one form or another, the concept of system under observation (SUO) makes no sense. But change is the counterbalance to persistence: we would not have evolved into observers without changes in the past.

The time concept as we discuss it in this book, therefore, is predicated on change in two fundamental respects. We will call them *relatively external* (RE) changes and *relatively internal* (RI) changes respectively. RE changes are those changes in the universe that were involved in the evolution of living organisms, some of which became primary observers such as physicists and astrophysicists. RI changes are changes in SUOs that those primary observers observe going on.

The expansion of the universe is an RE change because, whilst astronomers 'observe it', they have no control over it, no more than they can alter the fossil record in geological strata. They are simply finding out about their own place in the universe. RE changes provide evidence that real observers are *endophysical*, that is, inside the universe. A theory of *observers*, therefore, will deal with RE changes, whilst RI changes will allow those observers to pretend FAPP (for all practical purposes) that they are external to SUOs and can observe their properties. RE changes affect relative external context (discussed in Chapter 2) whilst relative internal context is needed for the interpretation of RI changes.

Traditionally, physics has focused on RI changes: experiments are not done *on* observers but *by* observers. But as science progressed and widened its scope, it started to look beyond RI changes and explored the meaning of RE change, hitherto the domain of metaphysics and religious-based mythology. This advance of science was particularly evident in the nineteenth century, when the sciences of geology and biology started to explore our planet in depth. It became clear not only that our planet had changed, but so had we: humans had evolved over many millions of years.

Images of Time. First Edition. George Jaroszkiewicz.
© George Jaroszkiewicz 2016. Published in 2016 by Oxford University Press.

Evolution, which played a crucial role in the transition from potential SUO into observer, requires a breakdown of persistence in a particular way. In this chapter we look at the processes of life that underpin this transition.

The solar model

As astronomers and astrophysicists discover more and more exoplanets,[24] it becomes more and more likely that the *solar model* of life what we see in our own solar system will be found throughout the universe. We do not know if it is the only model on which life can be based. Given the discovery of organisms that can exist in the most extreme environments, such as hydrothermal vents on our planet, and the indications that there is an internal ocean of water on Europa, one of the moons of Jupiter, we should not be surprised to find life in many diverse corners of the universe and in many forms.

The solar model consists of a star with a relatively stable configuration of planets in orbit, ranging from small Mercury-like objects to giant failed star objects such as Jupiter. Within that range of planets, there will be one or more planets capable of supporting life of some kind.

There is no merit in speculating on different stellar models: we have evidence currently only for what happened on our planet. Given the timescales of evolution that have been observed on Earth, the central star in a solar model system should be relatively stable over timescales measured in several billions of Earth years.

The Sun today is roughly halfway through the most stable part of its life. It has not changed dramatically for 4 billion years and will remain fairly stable for 4 billion more. However, even before hydrogen fusion in its core has stopped, the Sun will undergo severe changes, both internally and externally that will make like on Earth impossible in about a billion years' time.

Assuming that the physical constants c (the speed of light), G (the Newtonian constant of gravitation), and $\hbar$ (Planck's reduced constant) have remained constant as the universe expanded, scientist have estimated that the Sun formed about 4.57 thousand million years ago, or about 20 galactic years, from the collapse of a gas cloud. The trigger for this collapse is believed to have been shock waves from nearby supernova explosions, enormous outpourings of energy created by the collapse of stars when their nuclear furnaces ran out of fuel.

The earliest life on Earth

A recent study has concluded that there is evidence for life on Earth as far back as 3.7 thousand million years ago [Ohtomo *et al.*, 2013]. Therefore, life started less than a billion years after the formation of the planet.

---

[24] Close to 2,000 exoplanets have been reported to date [early 2015]. A recent analysis [Petigura *et al.*, 2013] has led to speculation that there may be several *billion* habitable Earth-size planets in our galaxy.

Assuming such a relatively brief timescale for life to have developed on Earth-type planets, there has been opportunity for it to have occurred several times over since the Big Bang. It should be kept in mind that the idea that life was brought to the Earth from other planets may be a valid statement, but by itself does not explain how *that* life on those other planets started.[25]

Lifetimes of organisms

A definition of life is problematic. As with time, everyone thinks they know what it means, but an exact definition is impossible. According to some perspectives, life is a process that involves organized units called cells that can absorb energy and nutrients from their environment and reproduce. Another perspective is that *life is any relatively persistent pattern of particles that has a finite existence, with the ability to create replicas of itself before that pattern disappears.*[26] But where then is the dividing line between life and inanimate life? Could we say that stars were alive? They are complex patterns of atoms that, like our Sun, were created when other stars died in supernova explosions. Some scientists have considered viruses to be life in that they reproduce. But others disagree, noting that viruses need host organisms in order to do that. Viruses therefore may be classified as a form of endophysical life. Other forms of life do not have this limitation and can be classified as exophysical life: they can reproduce autonomously. Of course, resources are required in all cases, so this classification may be unhelpful.

An interesting question is: how long do lifeforms persist in time? Is there any limit? If an organism never died, the proverbial immortal being, then there would be no need for reproduction. Time itself would have little meaning to such a creature. There are various science fiction stories touching on this theme, such as the *Star Trek* Original Series 'Requiem for Methuselah'. In that episode, Flint, the immortal man, laments the price paid for his immortality: '*I have married a hundred times, Captain. Selected, loved, cherished. Caressed a smoothness, inhaled a brief fragrance. Then age, death, the taste of dust.*'

The great variation in species lifespan has preoccupied many scientists for decades. For instance, a mouse has a maximum lifespan of about four years whilst some species of whale may exceed 200 years. There are also some animals that appear to be immortal, such as flatworms and lobsters. Plants too show great variation in lifespan: *annuals* live one year whilst some perennials such as brittlecone pine may live as much as 5,000 years.

For over a century, a popular belief amongst bioscientists has been that there is statistical correlation between animal lifespan, metabolic rate, and mass. In 1908, Max Rubner formulated his *rate of living theory*, which posited that slow metabolism was correlated with longer lifespan and with larger mass. Later, work

---

[25] This is absolutely not an argument for a Creator, but a warning against complacency.
[26] As defined by Dr David Sinclair, private communication.

by Kleiber on possible correlation of surface area to metabolic rate led him to formulate *Kleiber's law*. This asserts that for many animal species, their individual metabolic rate, μ, scales as some power of the animal's mass m: specifically, Kleiber found a relationship $\mu \approx m^{3/4}$ [Kleiber, 1932]. However, recent studies have shown a more complex picture: '*We confirm the idea that age at maturity is typically proportional to adult life span, and show that mammals that live longer for their body size, such as bats and primates, also tend to have a longer developmental time for their body size*' [De Magalhães et al., 2007].

Chronobiology

The study of time in living organisms is known as *chronobiology*. It focuses on the numerous temporal cycles that all organisms employ in their individual and species survival strategies.

These cycles are necessary because of some basic facts of life. First, all organisms are liable to die: even those species that seem immortal, such as lobsters, can have accidents. Therefore, there has to be a reproductive cycle in order to ensure species survival. Second, all known species live on our planet, which rotates once a day and orbits the Sun once a year, processes that impinge directly on the fortunes of all organisms. There are therefore cycles that are conditioned on a daily basis, known as *circadian rhythms*, and cycles that are conditioned by the seasons.

Another cycle predicated on the dynamics of our planet are lunar cycles. We note that '*Lunar cycles had, and continue to have, an influence upon human culture, though despite a persistent belief that our mental health and other behaviours are modulated by the phase of the moon, there is no solid evidence that human biology is in any way regulated by the lunar cycle*' [Foster & Roenneberg, 2008]. On the other hand, there is evidence of lunar cycles in fish, birds, and mice [Zimecki, 2006].

Biological time travel

In a sense, all biological organisms travel in time, that is, the forwards direction. But some species of plant seem to do it rather spectacularly. Plant species such as foxtail (*Setaria*) produce seeds that do not automatically germinate the next year but remain dormant until conditions are conducive for survival. In other words, '... *there is an active selection of the final destination of this time travel, as opposed to the passive travel through space, exhibited by tumbleweeds, sycamore seeds, or plant seeds ingested by animals*' [Smith, 2002].

12

The dimensions of time

Introduction

In this chapter we discuss the question: *why is time modelled by a single real parameter rather than two or more?*

As with all things to do with time, the answer is contextual: it depends on our image of time, observers and their protocols of observation, and the systems under observation (SUOs). In Absolute Time architecture for example, we normally use a single real-time parameter to describe a chorus of observers collecting information from an SUO, regardless of how many particles there are in that SUO. Although each member of the chorus carries their own clock, these clocks are standardized and synchronized, so that in effect any one of them represents any and all of the others. That is what is meant by 'laboratory time'.

Even with Absolute Time architecture, however, there are situations where we might have to consider two or more separate time parameters. This happens in the case of a superobserver, discussed in Chapter 6. Consider a superobserver S receiving, at their own clock time t_E^*, information about an event E that had happened at laboratory (chorus) time t_E as measured by the clock of that chorus member O_E who happened to be coincident with that event. Suppose this information is carried from O_E to S at the fastest possible speed, the speed of light c. If d is the Euclidean distance between O_E and S (assumed fixed normally) then t_E^* is necessarily always *later* than t_E because of the time delay in getting the information from E to S, the specific relation being $t_E^* = t_E + d/c$.

Because of retardation, superobservers have to do a lot of unravelling of the signals they receive if they want to theoretically reconstruct what actually happened throughout the laboratory. Otherwise, a distorted image will be obtained. An interesting example of general unawareness of the problems faced by superobservers in special relativity (SR) is the so-called Terrel–Penrose effect, originally discussed by Lampa [1924]: because of retardation, a cube moving at relativistic speed has the appearance of a normal cube but spatially rotated, rather than a Fitzgerald contracted unrotated cube. Fitzgerald length contraction is discussed in Chapter 16.

Images of Time. First Edition. George Jaroszkiewicz.
© George Jaroszkiewicz 2016. Published in 2016 by Oxford University Press.

A related problem arises in observational cosmology involving the concept of *lookback time*: images of galaxies observed today show what those galaxies were like many millions of years ago [Hobson *et al.*, 2006].

With the advent of SR and general relativity (GR), the situation regarding the number of time parameters changed dramatically on a number of fronts. First, relatively moving inertial frames in SR do not use Absolute Time (we refer the reader at this point to Chapter 18, where we discuss Generalized Transformations and Tangherlini's approach to relativity). Second, superobservers move along timelike worldlines in spacetime. The time recorded by the clock moving with each observer, known as the *proper time*, depends crucially on the worldline. Different superobservers will usually register different intervals of proper time even if they start together at one event, move apart, and finally meet up at a later event in spacetime. This is the origin of the so-called Twin Paradox, discussed in Chapter 17.

The issue of multiple times becomes critical when quantum mechanics (QM) and relativity are used together. For instance, when theorists attempted to discuss bound states in relativistic quantum field theory (RQFT), they naturally encountered a form of multi-dimensional time, a 'multi-fingered' time, because there is a separate time coordinate associated with each one of the particle-field operators involved in a bound-state calculation. Unravelling the multi-time aspects of relativistic bound states has been a severe technical problem for decades. A fundamental complicating factor is that the bosonic quantum fields (the photons) that bind the fermions (the electrons) in quantum electrodynamics propagate with retardation effects, so that what may appear to be a two-body problem is in fact an intractable field-theoretic nightmare. The Bethe–Salpeter equation is widely used in this context [Salpeter & Bethe, 1951].

What really helps the interpretation here is that because of the assumed homogeneity of special relativistic spacetime, centre of mass coordinates can be used to define four spacetime coordinates for the centre of mass, the other spacetime coordinates, including the additional times, being hidden in some way as internal to the bound state. Then the centre of mass time coordinate can serve as an effective overall single time parameter for the bound state. Multiple times in RQFT are discussed further in Chapter 22.

The theoretical consequences of having two or more fundamental temporal dimensions, as opposed to two or more arising in bound-state problems, were analysed by Tegmark [Tegmark, 1997]. He concluded that there are inconsistencies in the flow of information over spacetime when fundamental time has two or more dimensions. This is discussed in more detail towards the end of this chapter. His analysis is based on standard models of classical and quantum fields evolving over spacetime manifolds with various initial and boundary conditions.

Important implicit assumptions in our discussion involve the following choices: *process time* versus *manifold time*, *endophysics* versus *exophysics*, and *active transformations* versus *passive transformations*. We discuss these elsewhere in this book. Assuming that we are taking a standard exophysical perspective, we turn now to the role of partial differential equations (PDEs) in the study of time.

Partial differential equations and the flow of information

A fundamental but generally unstated principle in physics is that all things happen locally, action at a distance is nonsense, and there is no such thing as magic. What all this means is that whenever some event A is deemed to have caused event B, there has to be some identifiable agent or agency that transmitted the signal, force, interaction, or influence from A to B in a well-defined process with a finite speed, relative to any frame of observers.

To illustrate the point, consider Isaac Newton, who stated his law of universal gravitation in *The Principia* [Newton, 1687]. This law asserts that the gravitational force on one mass due to another mass depends on the instantaneous distance between them. Therefore it seems at first sight that Newton believed that gravitational forces act instantaneously over any distance. To most physicists, this is anathema. In fact, Newton also rejected action-at-a-distance, writing in a letter to Bentley that [Newton, 2006]

> Tis unconceivable that inanimate brute matter should (without the mediation of something else which is not material) operate upon & affect other matter without mutual contact; as it must if gravitation in the sense of Epicurus be essential & inherent in it. And this is one reason why I desired you would not ascribe {innate} gravity to me. That gravity should be innate inherent & {essential} to matter so that one body may act upon another at a distance through a vacuum without the mediation of any thing else by & through which their action or force {may} be conveyed from one to another is to me so great an absurdity that I believe no man who has in philosophical matters any competent faculty of thinking can ever fall into it. Gravity must be caused by an agent {acting} consta{ntl}y according to certain laws, but whether this agent be material or immaterial is a question I have left to the consideration of my readers.

When he formulated the law of gravitation, Newton had no way of measuring the speed of propagation of the forces of gravity, which to all intents and purposes appears to be infinite. It is currently commonly believed that gravitational disturbances are metrical (distance) perturbations in spacetime that propagate at the speed of light.

The way that physicists encode this locality principle into the laws of physics is by formulating them in differential equations. Differential equations represent the principle of locality in its purest form. Most of the great equations of physics are partial differential equations, such as Newton's laws of motion, Maxwell's equations of electromagnetism, Einstein's field equation in GR, Schrödinger's wave equation, Dirac's electron equation, and many more.

Differential equations come in many varieties: ordinary, partial, homogeneous, inhomogeneous, linear, nonlinear, coupled, real, complex, and so on. Physicists are fortunate that the structure of spacetime seems to restrict the class of differential equations that they need to solve. Almost all the fundamental equations of physics are second-order or first-order in time PDEs. Some are linear, others are nonlinear: some are classical, others need quantum mechanical interpretation. We discuss now a class of PDE fundamental to all branches of science.

Classification of second-order PDEs in two independent variables

We can learn a great deal about the structure of time and causality by studying second-order PDEs in a two-dimensional space. Consider the general second-order PDE for one field φ in two independent variables x, y of the form

$$A\frac{\partial^2}{\partial x^2}\varphi + 2B\frac{\partial^2}{\partial x\partial y}\varphi + C\frac{\partial^2}{\partial y^2}\varphi + \dots \text{ (lower order terms)} = 0, \qquad (12.1)$$

where A, B, and C are given functions of x and y, and φ is the x, y dependent field that has to be solved for. If $A^2 + B^2 + C^2 > 0$ over a region $\mathcal{R}$ of the $x-y$ plane[27] then the PDE is *second-order* in that region, so we shall assume that condition holds in the following.

A critical quantity at each point in $\mathcal{R}$ is the *discriminant* Δ, defined by $\Delta \equiv B^2 - AC$. This will vary over $\mathcal{R}$ as A, B, and C vary and dictates the structure of causality and the flow of information throughout $\mathcal{R}$. Now in physics, spacetime coordinates are real, particle masses are real, and the constants of physics are real. Therefore, we may take A, B, and C to be real functions over $\mathcal{R}$. The same restriction does *not* apply to the field φ: fields in physics can be real, complex, vector valued, and so on. They could even be quantum field operators.

Given this, the discriminant Δ is a real-valued function over $\mathcal{R}$. Therefore, at any given point, Δ can be either positive, zero, or negative. These three cases give a fundamental classification scheme to such PDEs: if $\Delta < 0$ over $\mathcal{R}$ the PDE is said to be *elliptic*, if $\Delta = 0$ over $\mathcal{R}$ the PDE is *parabolic*, whilst if $\Delta > 0$ the PDE is *hyperbolic*. We discuss now each case separately for the restricted but nevertheless important class of PDEs for which the coefficients A, B, and C are constant over $\mathcal{R}$.

The elliptic regime

Suppose we are given that over the region $\mathcal{R}$, a field $\psi(u, v)$ satisfies the PDE

$$\left(\frac{\partial^2}{\partial u^2} + \frac{\partial^2}{\partial v^2}\right)\psi(u, v) = 0, \qquad (12.2)$$

where u, v are real coordinates. Define new coordinates x, y by

$$x = \alpha u + \beta v, \quad y = \gamma u + \delta v, \qquad (12.3)$$

---

[27] This condition means that at least one of the functions A, B, and C is non-zero at every point in the region.

where α, β, γ, and δ are real constants. Then the partial differential operators transform according to the chain rule and are given by

$$\frac{\partial}{\partial u} = \alpha \frac{\partial}{\partial x} + \gamma \frac{\partial}{\partial y}, \quad \frac{\partial}{\partial v} = \beta \frac{\partial}{\partial x} + \delta \frac{\partial}{\partial y}. \tag{12.4}$$

If now we define $\varphi(x, y) \equiv \psi(u, v)$ we readily convert equation (12.2) into the more general form

$$\underbrace{(\alpha^2 + \beta^2)}_{A} \frac{\partial^2}{\partial x^2} \varphi + \underbrace{2(\alpha\gamma + \beta\delta)}_{B} \frac{\partial^2}{\partial x \partial y} \varphi + \underbrace{(\gamma^2 + \delta^2)}_{C} \frac{\partial^2}{\partial y^2} \varphi = 0. \tag{12.5}$$

From this we read off $A = \alpha^2 + \beta^2$, $B = \alpha\gamma + \beta\delta$, $C = \gamma^2 + \delta^2$, from which we find $\Delta = -(\alpha\delta - \beta\gamma)^2 < 0$, confirming that the PDE is elliptic. Of course, we could have found this directly from the original equation (12.2). The point is, could we have started off with (12.5) and transformed it into (12.2), which does not have a term containing mixed partial derivatives and so is easier to solve? The answer comes from the transformation (12.3): the transformation is invertible if and only if $\alpha\delta - \beta\gamma \neq 0$.

Note that there is no change in sign between the two second partial derivatives in (12.2). Mathematicians assign to this a classification called *signature*, discussed in the Appendix. In this case, the signature is $(1, 1)$.[28] Sylvester's law of inertia (a theorem in linear algebra, not physics) tells us that a coordinate change such as (12.3) does not change signature.

Euclidean signature plays a crucial role in speculations about the origin of the universe and time. Following earlier work by Hartle and Hawking [Hartle & Hawking, 1983], it was posited by Hawking and Turok in their 'pea' model of the very early universe that there was no singularity at the Big Bang [Hawking & Turok, 1998]. Instead, the signature of the metric of spacetime (impacting on the signature of PDEs) was Euclidean, that is, did not have a sign change associated with a time coordinate, right where a sign change was supposed to be. Another way of seeing what is going on in the Hawking–Turok model is to think of the real-time parameter t relevant to the later universe being the analytic continuation from *imaginary time* τ, such that $\tau \rightarrow it$, at which point the signature changed from Euclidean to hyperbolic. Imaginary time is discussed in Chapter 21.

The parabolic regime

Consider the PDE

$$\frac{\partial^2}{\partial u^2} \psi(u, v) = 0. \tag{12.6}$$

---

[28] We could multiply all terms in the equation by -1 and lose no content, so signature could just as well be given in the form $(-1, -1)$. What is important is the relative number of positive and negative terms.

With the same coordinate transformation (12.3), equation (12.6) becomes

$$\alpha^2 \frac{\partial^2}{\partial x^2}\varphi + 2\alpha\gamma\frac{\partial^2}{\partial x\partial y}\varphi + \gamma^2\frac{\partial^2}{\partial y^2}\varphi = 0, \tag{12.7}$$

so we read off $A = \alpha^2$, $B = \alpha\gamma$, $C = \gamma^2$, which gives $\Delta = 0$.

The general solution to (12.6) is $\psi(u, v) = \psi_0(v) + \psi_1(v)u$, where ψ_0 and ψ_1 are arbitrary functions of v. This shows that the PDE (12.6) is in fact equivalent to an ordinary differential equation.

The hyperbolic regime

Given the PDE

$$\left(\frac{\partial^2}{\partial u^2} - \frac{\partial^2}{\partial v^2}\right)\psi(u, v) = 0, \tag{12.8}$$

then the transformation (12.3) gives

$$(\alpha^2 - \beta^2)\frac{\partial^2}{\partial x^2}\varphi + 2(\alpha\gamma - \beta\delta)\frac{\partial^2}{\partial x\partial y}\varphi + (\gamma^2 - \delta^2)\frac{\partial^2}{\partial y^2}\varphi = 0 \tag{12.9}$$

so $A = \alpha^2 - \beta^2$, $B = \alpha\gamma - \beta\delta$, $C = \gamma^2 - \delta^2$ giving $\Delta = (\alpha\delta - \beta\gamma)^2 > 0$. In this case the signature is $(1,-1)$: the sign change is fundamental to time and causality, as we shall see below.

Classification of second-order PDEs in n independent variables

The above discussion is readily extended to more than two dimensions or independent variables. Given coordinates $\mathbf{x} \equiv (x_1, x_2, \ldots, x_n)$ consider the second-order PDE

$$\sum_{i=1}^{n}\sum_{j=1}^{n} A_{ij}\frac{\partial^2}{\partial x_i\partial x_j}\varphi(\mathbf{x}) = 0, \tag{12.10}$$

where the coefficient matrix $[A_{ij}]$ is constant. Then we have the following classification determined by the set of eigenvalues of this matrix:

1. If the eigenvalues are all positive or all negative, then the PDE is *elliptic*.
2. If the eigenvalues are all positive or all negative, except for one that is zero, the PDE is *parabolic*.
3. If one eigenvalue is positive and the rest negative, or one negative and the rest positive, then the PDE is *hyperbolic*.
4. If more than one eigenvalue is positive and more than one eigenvalue is negative, the PDE is *ultrahyperbolic*.

It turns out that the fundamental PDEs used in physics to encode the properties of time as we know it, such as causality, are all hyperbolic.

Boundary conditions

It is not sufficient to have equations of motion that can propagate information throughout spacetime: we need *boundary conditions*, that is, the information that is propagated from one region of spacetime to another. There are three important types of boundary condition that lead to different effects, depending on the kind of PDE involved.

Cauchy boundary conditions are such that the value of the field φ and its normal derivative are specified on part of the boundary of a region. In spacetime physics, such a boundary will usually be a so-called *spacelike hypersurface*. Cauchy boundary conditions are the field theoretic analogues of initial position and velocity that are required to solve the second-order equations of motion in Newtonian mechanics.

Dirichlet boundary conditions appear to be more limited, being a specification of the value of the field over a complete boundary of a region. These conditions are the analogue of the specification of initial and final positions in the Calculus of Variations.

Neumann boundary conditions are a variant of the Dirichlet conditions: the normal derivative of the field is specified over the boundary of the region concerned.

The following table shows how the various boundary conditions match up with the various classes of PDE [Arfken, 1985].

Table 12.1 *Boundary conditions versus PDE classification.*

Boundary condition	Elliptic	Parabolic	Hyperbolic
Cauchy open surface	unstable	too restrictive	unique, stable solution
Cauchy closed surface	too restrictive	too restrictive	too restrictive
Dirichlet open surface	insufficient	unique, stable in one direction	insufficient
Dirichlet closed surface	unique, stable solution	too restrictive	solution not unique
Neumann open surface	insufficient	unique, stable in one solution	insufficient
Neumann closed surface	unique, stable solution	too restrictive	solution not unique

The signature of spacetime

We shall see in Chapter 17 that Minkowski spacetime, the space and time of special relativity, has a distance structure or metric with a *Lorentzian* signature, meaning that the distance s between two events with relative Cartesian coordinates $x \equiv (ct, x_1, x_2, x_3)$ is given by $s^2 = c^2t^2 - x_1^2 - x_2^2 - x_3^2$. Here t is the time as measured by observers at rest in the inertial frame concerned and is known as a *timelike* coordinate. The other coordinates x_1, x_2, and x_3 correspond to the standard (x, y, z) spatial Cartesian coordinates and are accordingly known as *spacelike* coordinates.

The term *signature* is appropriate because PDEs based on this spacetime generally have the same signature. For example, the Klein–Gordon equation for a free spinless particle of mass m in SR is given by

$$\hbar^2 \left\{ \frac{\partial^2}{c^2\partial t^2} - \frac{\partial^2}{\partial x_1^2} - \frac{\partial^2}{\partial x_2^2} - \frac{\partial^2}{\partial x_3^2} \right\} \varphi(x) + m^2 c^2 \varphi(x) = 0, \tag{12.11}$$

where $\hbar$ is Planck's reduced constant and c is the speed of light.

The theorist Tegmark gave an analysis of the properties of a spacetime with signature $p - q$, that is, one where we could find a coordinate patch where there were p timelike coordinates and q spacelike coordinates [Tegmark, 1997]. Then the equivalent of the Klein–Gordon equation for this spacetime would be

$$\hbar^2 \left\{ \underbrace{\frac{\partial^2}{c^2\partial t_1^2} + \ldots + \frac{\partial^2}{c^2\partial t_p^2}}_{p \text{ terms}} \underbrace{\frac{\partial^2}{\partial x_1^2} - \ldots - \frac{\partial^2}{\partial x_q^2}}_{q \text{ terms}} \right\} \varphi(x) + m^2 c^2 \varphi(x) = 0. \tag{12.12}$$

In Figure 12.1 we show the conclusions of Tegmark's analysis, with the number p of timelike dimensions and the number q of spacelike dimensions parametrizing the rows and columns respectively. This table shows the unique property of Minkowski spacetime, the spacetime we live in: it has one time dimension and three spatial dimensions. All other signatures have something unphysical about them.

Empirical studies

There have been some empirical studies in various fields where the observers attempted to fit the data to multi-dimensional time models. Two of these are mentioned briefly here now.

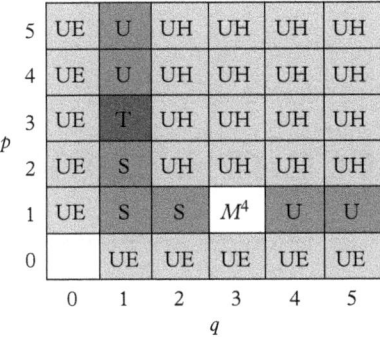

Fig. 12.1 *Conclusions from Tegmark [1997]: UE = unstable elliptic, U = unstable, T = tachyonic, S = too simple, UH = unstable ultrahyperbolic, M^4 = Minkowski spacetime (our local spacetime).*

Tifft's analysis of red shifts

The astrophysicist W. G. Tifft measured the red shift of distant galaxies and believed that there were some unusual patterns in the data. He attempted to fit this data with a model of three-dimensional time [Tifft, 1996]. Subsequent independent observation did not verify Tifft's data so his model is no longer viable.

The two-dimensional arrow of biological time

Recent studies by Bailly and collaborators on multiple biological rhythms such as heart beat, circadian rhythms, and so on, came to the conclusion that biological time should not be invariably modelled by a linear concept of time but could be better understood in terms of a two-dimensional time 'parameter'. One of these dimensions would represent the ordinary 'flow' of linear time whilst the other would be cyclic, 'rotating' around the other time [Bailly *et al.*, 2011].

13

The architecture of time

What is temporal architecture?

The subject of time is vast and complex, generally benefiting from explicit commentary on several key aspects. Some of these tend to be ignored or neglected by many authors on the subject, an oversight that we attribute to the implicit assumption that 'time' has a well-defined structure accepted by one and all.

We disagree. The indications are that time means different things in different contexts: each context has its particular aspects, including the implicit architecture that we focus on in this chapter. Our strategy in this book is to cover as many bases within our means, giving as many insights, opinion, and perspectives on the subject as we can. The title of this book is not *Image of Time*.

When the diversity of images of time is examined, one thing becomes apparent: there are almost as many conceptual models of time as there are authors. We should not claim any model is inherently better than any other unless we have empirical evidence to that effect. What we can do is to point out the mathematical assumptions inherent to each model. These assumptions define the *temporal architecture* of a model, a conceptual structure that we can usually describe in words, much as we could describe the very different architecture of buildings such as Stonehenge in Britain, the Pyramids in Egypt, the Parthenon in Athens, the Pantheon in Rome, the Great Wall of China, or the Taj Mahal in India.

Before we discuss any temporal architecture, we should ask three questions: *What is time?*, *What is the structure of time?*, and *How are observers related to time?* These questions are explained as follows.

Process or manifold?

In McTaggart's influential analysis of time [McTaggart, 1908], *A-Series* propositions are tensed, meaning they have a relative temporal label, such as *yesterday*, whilst *B-Series* propositions are untensed, meaning they have an absolute temporal label, such as a date.

Presentism is the view that only the moment of the *now* exists, time is not an object but a process, and only A-Series propositions are fundamental. This is the process time perspective.

Images of Time. First Edition. George Jaroszkiewicz.

On the other hand, if we believe in the manifold time perspective, we are prepared to believe in an observer-independent objective time structure, possibly intermingled with an objective physical space in the form of spacetime. This was Minkowski's view of relativistic spacetime [Petkov, 2012]. In the manifold time perspective, sometimes referred to as the *Block Universe* [Price, 1997], A-Series propositions are secondary and illusionary: only B-Series propositions are meaningful.

Points or intervals?

Does time consist of *points* or does time come in *intervals* of duration? This question was debated in Antiquity. One important difference between these images of time is that points of time cannot be added together to give a resulting point, but intervals can be added together to give longer intervals. In Chapter 23, Discrete Time, we shall discuss a model of Maxwell's electromagnetic theory where the magnetic field is defined on temporal *nodes*, or end points of intervals, whilst the electric field is defined on temporal *links*, that is, on the successive intervals themselves. Therefore, that version of discrete time mechanics requires time to be associated with intervals, rather than just points.

Endo or exo?

Recognizing that the role and status of observers is crucial to the debate about time, we may decide that the concept of observers standing outside spacetime and looking in makes sense. Then we will be adopting the exophysical perspective. On the other hand, recognizing that we are part of the universe and therefore always inside it, we will be adopting the endophysical perspective. The two perspectives have very different architectures.

Which of these we adopt will profoundly affect the logic of how we discuss time: for instance, an endophysical perspective cannot be meaningfully applied to time before the universe was created, not unless we are extremely careful in our interpretation of data. By this we mean the following. Suppose we accept that we are endophysical. Then we can only comment on what we have observed directly, in the form of information collected by astronomers at this time. Does this mean we cannot comment on what the past was like?

The answer is that we can, but we should really know what we are doing. For instance, some theorists have discussed pre-Big Bang physics. Is this meaningful? To help us answer this question, it may help to discuss an analogy with *beachcombing*, or the leisurely art of looking for interesting washed up objects on beaches.

The analogy goes like this. Suppose we find ourselves walking alone along a deserted beach after the overnight tide has gone out. As we walk on the still-wet sand we suddenly encounter another person's footprints impressed on the surface. Clearly, they are not our footprints. Can we deduce something of what might have happened before we got there?

With the knowledge available to us, we can assume with reason that another person had been there not long before us, but certainly *after* the last tide had gone out. We deduce that, because in our experience, footprints in sand would not survive any tide: tides have the power to wash over all traces of footprints, leaving really smooth sand. Therefore, we reason that the footprints were created by another human who had been there that morning before us. We cannot believe that any animal or natural process could create footprints like the ones we can see.[29]

But suppose now we went further along the beach and saw some strange depressions and mounds, laid out in a recognizable square pattern. We recognize this as the tenuous remains of the very large sandcastle that we ourselves had built on that beach yesterday, *before* the overnight tide had come in. A single tide will not always remove traces of very large sandcastles. So we deduce that we are in fact seeing traces of a genuine artefact from the previous day, a surviving memory of reality before last night's tide.

The analogy here is with the Big Bang, the conjectured origin of the universe and the start of time. No one knows whether it was a one-off event or whether it is the latest in a series of expansions and contractions. It is good science, not metaphysics, to look at the possibilities of residual patterns, faint memories from potential previous cycles of the universe, that we could detect today. That justifies pre-Big Bang theorists in their endeavours. But we would have to make sure that we were not being led by wishful thinking, as was the case with the 'Face on Mars' photographs.

Examples of temporal architectures

Temporal architectures can be very different to each other: it is sufficient to compare linear time with cyclic time to see this. In the following subsections we elaborate on specific aspects of temporal architecture that seem most important to us.

Aristotelian space-time architecture

In ancient times, humans had limited experience of travel, and no real understanding of the planet on which they lived. It is not surprising that the relative stability of our planet created the impression that the Earth defines an absolute frame of reference in which objects are located and move about in. This generates an architecture that we can call Aristotelian space-time, in which points of three-dimensional space have an identity throughout time. In Aristotle's view of mechanics, the natural state of an object is at rest relative to this space. In this temporal architecture, when we revisit old places such as our former home or school, we imagine we are returning to them.

---

[29] Strange things can happen naturally. In 1976 the Viking 1 spacecraft orbiting Mars photographed a geological feature that looked like a human face in some photographs, causing much wild speculation.

Galilean–Newtonian space-time

In this architecture, there is an acknowledgement that there is no absolute state of rest: everything is moving from some or other perspective. The natural state of a free object now is uniform motion relative to the distant stars.

Minkowski spacetime

In his radical vision of space and time, Minkowski thought of them as part of a single, absolute, four-dimensional continuum that we refer to as spacetime (no hyphen) [Petkov, 2012]. Different inertial frames of reference cut across this structure in different 'directions', each frame giving its own space and time coordinates to an event, or point in spacetime.

Multi-fingered time

Different observers in Minkowski's spacetime carry their own clocks: these tick away according to their individual 'proper' times, rates determined by the specific worldlines or tracks that each clock traces out in spacetime. There are as many separate, individual times as there are clocks, giving a truly 'multi-fingered', endophysical architecture of time. This is the origin of the Twin Paradox, discussed in Chapter 17. Inertial frames of reference are idealizations composed of a 'chorus', a continuum of observers moving in a coordinated way with their clocks synchronized and their relative positions in 'space'[30] fixed.

Continuous or discrete time

The choice of mathematical model of the time parameter plays a central role in all theories. If we think of time as continuous, then we can formulate concepts of velocity and acceleration, fundamental to most modern theories of the universe. On the other hand, if time is discontinuous or discrete, we will have different mathematical structures, such as difference equations rather than differential equations.

Particle decays

Many so-called 'elementary' particles are unstable, decaying into two or more other particles. Usually there is a characteristic timescale for each type of decay. The measurement of these decay lifetimes is a fundamental weapon in the validation of modern particle theories, as these usually give predictions for such

---

[30] 'Space' being the usual three-dimensional position space that each chorus imagines it is sitting at rest in.

lifetimes. Normally, particle decay lifetimes are modelled in terms of time dependent decaying exponentials of the form $e^{-\Gamma t}$, where Γ is the theoretical lifetime being investigated. Although this seems simple and straightforward, the architecture associated with particle decay experiments is actually complicated and very different to the simplistic Block Universe architecture usually assumed to faithfully model the universe. Such exponentials represent data acquired from many repetitions of a decay, with measurements taken at the end of each run, each of which has its own stopping time. It was pointed out by Misra and Sudarshen that decay experiments involved several different architectures that depend on the type of questions asked by the observers of the decay processes being investigated [Misra & Sudarshan, 1977]. A spectacular demonstration of this is in the class of experiments demonstrating the so-called 'Quantum Zeno' effect [Itano *et al.*, 1990], where the observational protocol (specific techniques of measurement) imposes an architecture in which an otherwise unstable SUO appears to be stable. Such experiments can be discussed using discrete time rather than continuous time [Jaroszkiewicz, 2008b].

Multiple universes

There is no empirical evidence for multiple universes, but some paradigms suggest otherwise. A logical question to ask in such cases is: *how is time correlated between universes, if they can interact in some way?* There is an implicit exophysical architecture that is necessary in those conjectured models to coordinate each universe with the others, so as to account for quantum interference for instance: such a process is asserted to occur in the Multiverse paradigm [Deutsch, 1997]. Such models are invariably contextually incomplete in their architecture and are therefore metaphysical.

Time travel

Time travel stories invariably require some form of postulated architecture that departs from the linear time Block Universe model. As with the Multiverse paradigm, lack of detail prevents counterarguments to be presented. Novel forms of temporal architectures such as wormholes, ansibles,[31] and hyperspace are posited with no details. A century earlier, such stories would have been classified as fairy stories.

There is one reputable scenario where the architecture of time travel can be discussed in a meaningful way, and that is in general relativity. There we find spacetimes that allow for closed timelike curves (CTCs), or loops in spacetime that an endophysical observer could follow. An important paper on this topic was given by the logician Gödel in 1949: we discuss that paper in Chapter 20.

---

[31] Devices that permit instantaneous communication across vast distances.

Architectural levels of observation

A commonly ignored yet fundamental aspect of time is what we call the *architecture level of observation,* viz., a non-mathematical description of the principles and processes underpinning any experiment. It is essential to state which architectural level is being used in any discussion involving time. Failure to do so can contribute greatly to confusion and cross-purpose communication, especially in the interpretation of quantum mechanics. We identify at least four distinct observational architectural levels, listed here in increasing order of sophistication.

Level 1: classical observation

In this the most basic level of observation, detailed modelling of observers and their apparatus is ignored, with a godlike, exophysical perspective being assumed for observers. Different observers are differentiated by their individual rest frames, which are identified with laboratories and apparatus. In this level of architecture there is a belief, sometimes stated as the principle of general covariance, that physical processes are independent of observers and their frames of reference. Therefore, states of SUOs are assumed to evolve independently of any processes of observation: however, how each observer describes these states is frame dependent. Observation is cost free at this level, so observers can in principle know everything possible about the current state of an SUO without affecting it in any way. Newtonian classical mechanics operates at this level of architecture. It was a ubiquitous paradigm of observation until advanced technology caught it out and a more sophisticated level of observation, quantum mechanics, was found necessary.

Newtonian mechanics, special relativity (SR), and general relativity (GR) employ this lowest level of architecture, differing only in how different observers' frames of reference are related. Although SR and GR are based on more sophisticated models of spacetime and its geometry than the Newtonian paradigm, they operate at the same level as far as the processes of observation are concerned.

By definition, metaphysics is conducted below this level of architecture. Some sophisticated philosophical discussions do recognize the importance of observers, but do not go beyond discussion. A characteristic of level 1 statements is that they are contextually incomplete, but not necessarily metaphysical. Our classification scheme discussed at the end of Chapter 2 may help identify this level.

Level 2: standard quantum mechanical observation

Schrödinger's wave mechanics and Heisenberg's matrix mechanics operate at the next level of observation, referred to as standard quantum mechanics (SQM). SQM has commonly been misinterpreted as a statement of what the properties of SUOs *are,* but our view is that it is a theory of how SUOs can be *perceived* by observers. SQM without observers is physically meaningless, a severe criticism of quantum cosmology [Fink & Leschke, 2000].

In SQM, primary observers are exophysical and their apparatus is not modelled directly *per se* but indirectly via the concept of an *observable*. SQM architecture is based on the pattern of state preparation followed by outcome detection, the possible outcomes being related to a chosen observable. Each observable represents some experimental context that *could* be implemented, such as a measurement of position or momentum. The choice of observable is made by the exophysical observer and is not contextual, in that there is nothing in SQM that dictates this choice.

The choices made by the observer in this architecture are not trivial in several respects:

1. A choice of one observable can exclude a simultaneous choice of another, if these are incompatible observables. For instance, a position measurement is incompatible with a momentum measurement. This does not happen in level 1 architecture, where it is assumed that everything can be known contemporaneously about states of SUOs.

2. Processes of observation can affect prepared states of SUOs. There remains a longstanding debate about what happens during the process of observation. In a common scenario, observed states 'collapse' or reduce to eigenstates of the observables involved. Lüders gave a mathematical rule for such a situation [Lüders, 1951]. Our view is that state reduction is a manifestation of change in an observer's information about an SUO. Whether anything physical is done to an SUO depends on the form of detection employed. Recent developments in quantum measurement theory support the notion that weak and strong forms of measurement can give different information about states of SUOs [Svensson, 2013].

3. Heisenberg's uncertainty principle [Heisenberg, 1927] can be interpreted as the statement that *the temporal order in which two incompatible observations are made is physically significant*. There is no such principle in level 1 architecture.

Living as they do in the real world, theorists are naturally conditioned to think in terms of level 1 architecture, which is also known as a *realist* perspective. Without some effort to overcome this conditioning, this can lead to attempts to interpret level 2 quantum architectural scenarios as level 1 classical scenarios. Three notable examples are hidden variables theory (Bohmian mechanics), the Multiverse paradigm [Deutsch, 1997], and early forms of decoherence theory that posit a universal wavefunction. Each of these approaches leaves unanswered various crucial questions, such as where the observers are. A common feature is a lack of contextuality. Indeed, such approaches purport to give an absolutist view of reality, so it is never made clear exactly for whom any observational information is significant, that is, there is no concept of a primary observer in any of these approaches.

It is noteworthy that the great pioneers of QM such as Heisenberg, Bohr, Born, and Dirac understood the dangers of level 1 conditioning, whilst equally great theorists such as Einstein and Schrödinger did not.

Level 3: quantum detection

In this level of quantum architecture, an explicit attempt is made to factor the processes of detection into quantum mechanical calculations. This level is much more ubiquitous than might be imagined, for a very good reason: quantum mechanics is really a theory of *observation*, not about the strange particle-wave properties of SUOs. We list here a number of scenarios that we classify as being based on this architecture.

Planck's quantization of energy: When he introduced quanta in 1900, Planck did *not* think of this as the quantization of the black-body radiation field energy itself, that is, he did not think in terms of photons [Planck, 1900, 1901]. Planck quantized the energy levels of the atomic oscillators in the container walls, and these behave as quantized sources and detectors of energy.

Mott's cloud chamber analysis: In 1929, Mott gave a prescient account of how quantum theory could account for linear alpha particle tracks in a cloud chamber even when the source wavefunction has spherical symmetry. With hindsight, we may readily interpret Mott's analysis as a quantum register description of multi-detector apparatus [Mott, 1929].

Feynman's attempt to detectorize physics: According to Brown, Feynman attempted early in his career to discuss quantum electrodynamics solely in terms of signals in apparatus [Brown, 2005].

Schwinger's source theory: Schwinger devised a remarkable method of interfacing an observer with quantum state evolution: he introduced a classical source term into the quantum action and showed how physical information could be extracted by suitable external manipulation of the source by the observer [Schwinger, 1969].

The LSZ formalism: At first sight, the calculation of scattering cross-sections in relativistic quantum field theory (RQFT) appears to have level 2 architecture. However, the LSZ formalism posits a separate Hilbert space for the 'in' states of the SUO, another one for the 'out' states, and yet another one for state evolution in between [Lehmann *et al.*, 1955]. An explicit acknowledgement that the physics of state preparation and detection is different to evolution in vacuo is marked by the use of possibly divergent renormalization constants bridging the worlds of preparation, vacuum, and detectors.

The Ludwig–Kraus POVM approach to observation: Although the POVM approach of Ludwig and Kraus [Ludwig, 1983a,b; Kraus, 1983] is centred on the Hilbert space of a prepared state, the number of detector outcomes can exceed the dimensionality of that Hilbert space, something that does not occur with level 2 architecture.

Glauber's approach to quantum optics: Glauber used quantum field theoretic n-point functions to discuss correlations in quantum optics experiments. He emphasized the irreversibility of observation by focusing on matrix elements of $\mathbf{E}^{(+)}(\mathbf{x}, t)$, the positive frequency part of the electric field operator, rather than the complete field operator $\mathbf{E}(\mathbf{x}, t)$ itself [Glauber, 1963a,b].

Detector pointer bases: In various models of quantum observation, an attempt is made to describe SUOs and detectors within the same framework by tensoring SUO states with detector pointer basis states [Peres, 2000a,b]. This is a halfway house between a level 2 description and a level 3 description, in that whilst a single quantum state representing the combined SUO–detector system is discussed, the individual SUO and detector components retain their identities, with the former being traced out in the calculation of detector outcome probabilities.

The Unruh effect: In this scenario, a non-inertial observer detects particles in what would be empty space when viewed from the perspective of an inertial observer. An explicit attempt is made to model the accelerating detector in a non-local way that does not cover the whole of spacetime, resulting in the observation of what looks like a thermal bath of particles [Unruh, 1976].

The Quantum Zeno effect: The processes of observation play a central role in the Quantum Zeno effect [Misra & Sudarshan, 1977; Jaroszkiewicz, 2008b]. The effect arises when there is a specific relationship between two timescales: the intrinsic decay lifetime of some unstable state of an SUO and the timescale associated with the process of observation. The effect cannot be discussed without the presence of the observer.

Quantum teleportation: In this form of experiment, a quantum state is first prepared at some initial time t_0. The state is looked at partially at a later time t_1 and the results of that observation are transmitted classically to a final stage of the experiment at an even later time.

There are several other scenarios in quantum theory that we do not have scope to list here. It is clear that this level of architecture is the focus of great attention currently, particularly in the field known as relativistic quantum information (RQI). Whilst RQI does not currently address the effect of observation on the background spacetimes discussed, it does represent the recognition that the detailed processes of detection cannot be ignored if the long-sought-for unification of relativity and quantum mechanics is desired.

Level 4: theories of observation

This architectural level of observation should give an account of the origin of observers and their apparatus. It is virtually unexplored at this time. Even describing it risks the accusation from practitioners of level 2 quantum mechanics that it is metaphysics. There is some validity in that comment at this time, but the situation may change with more attention being given to it.

In a pioneering account of quantum computation, Feynman implicitly acknowledged that there was something missing in our current formulation of QM, a general lack of any understanding of this level, when he wrote

> I mentioned something about the possibility of time – of things being affected not just by the past, but also by the future, and therefore that our probabilities are in some sense 'illusory'. We only have the information from the past, and we try to predict the next step, but in reality it depends upon the near future which we can't get at, or something like that. A very interesting question is the origin of probabilities in quantum mechanics. Another way of putting things is this: we have an illusion that we can do any experiment that we want. We all, however, come from the same universe, have evolved with it, and don't really have any 'real' freedom. For we obey certain laws and have come from a certain past. Is it somehow that we are correlated to the experiments that we do, so that apparent probabilities don't look like they ought to look if you assume that they are all random.   [Feynman, 1982]

Any attempt to address these issues would required us to move to a far deeper level of mathematical modelling of physical systems than we have access to currently. The problem is not just having to deal with vastly increased complexity, generated by vastly many more degrees of freedom than our computers can handle. Complexity itself is not a well-defined concept and we simply do not have a proper understanding of the differences between the sort of complexity induced by vast numbers of elementary systems acting together (mechanical complexity if you will) and the sort of complexity that biological organisms demonstrate (such as human thought) [Clayton, 2013].

As with most great problems, the best strategy seems to be to move progressively, small step by small step, modularizing the discussion and tackling each module separately. Perhaps we will never fully understand why we do certain experiments and not others, as Feynman would ask, but it may be possible to go some way towards that goal. For instance, it is possible to model the construction and destruction of apparatus in an otherwise empty laboratory, once we have decided that we want to perform a specific experiment [Jaroszkiewicz, 2010]. This takes us from a level 3 architecture, where apparatus is given, to level 4, where the construction of apparatus would be part of the natural evolution of the laboratory in which the observer is based.

At this fourth level of architecture, it would be explicitly acknowledged that experiments are contextual on the apparatus so constructed, and that there is a physical reality that is not being described but which is necessary for the existence of the observer. We speculate that a level 4 description should involve the tensor product of *three* Hilbert spaces: one space describing quantum states of the SUO, another space describing quantized signal (i.e. pointer) states of the apparatus, and a third one describing the states of existence of the apparatus itself relative to the primary observer's laboratory. A proper dynamical theory of observation that aimed to include time-dependent apparatus could not be constructed with just the first two Hilbert spaces.

In such a three-component framework, temporal evolution would have a reversible component describing the unitary evolution of SUO states in the *information void*[32] and two irreversible components. The first irreversible component would be generated by the processes of detection in apparatus, along the lines discussed by Mott [Mott, 1929] and Glauber [Glauber, 1963a,b], whilst the second irreversible component would be generated by the observers themselves as they created, operated, and decommissioned their apparatus [Jaroszkiewicz, 2010]. In this scenario, reversibility and irreversibility would be contextually defined relative to the memories of the primary observers concerned. Even if such a three-component framework proved satisfactory, we would still be a very long way from a comprehensive theory of observation and reality along the lines implied by Feynman's remarks quoted above [Feynman, 1982].

In the three-component scenario envisaged by us, the unitary evolution part would be generated via conventional quantum evolution operators, constructed from empirically validated Hamiltonians. In order to make predictions in terms of signals in apparatus, the SUO state component would be traced out, leaving behind a time-dependent quantum register network or 'Heisenberg net' of detector states, connected by transition amplitudes that result from the aforementioned tracing out. Essentially, evolution in the information void generates the signal state amplitudes that the observer is interested in. This means that there is an aspect of physical reality, the existence of which cannot be detected directly but inferred from the behaviour of our detecting apparatus. A good analogy is with the discovery of dark matter: it cannot be seen directly by optical means but its presence is inferred by the behaviour of objects that can be seen optically.

The Heisenberg net approach to observation has a number of advantages, such as allowing for multiple, time-dependent detectors [Jaroszkiewicz, 2008a].

---

[32] A spacetime region where no detectors exist to influence dynamical evolution, often referred to as the vacuum in relativistic quantum field theory.

14

Absolute time

Introduction

In this chapter we look at *Absolute Time,* the model of time used in classical mechanics (CM) until the advent of special relativity (SR) in 1905. Absolute Time was described by Newton in his book *Philosophiae Naturalis Principia Mathematica,* otherwise known as *The Principia.* He wrote:

> Absolute, true and mathematical time, of itself, and from its own nature, flows equably without relation to anything external, and by another name is called duration: relative, apparent, and common time, is sensible and external (whether accurate or unequable) measure of duration by the means of motion, which is commonly used instead of true time; such as an hour, a day, a month, a year.   [Newton, 1687]

Newton's statement is contextually incomplete. In common with most people at that time, Newton believed in God: it is reasonable to interpret his statement as how he imagined time ran relative to that hypothesized absolute primary observer.

In *The Principia,* Newton aimed to provide a usable, applicable system of mechanics. He must have understood the unsatisfactory metaphysical nature of his statement on Absolute Time, because he supplemented his statement with reference to *relative,* or *apparent,* or *common,* time, the time used by observers. He gave an empirical method of measuring this relative time in terms of motion.

Absolute Time is implicitly assumed to be an ordered set of abstract elements called *absolute times.* Ordered sets are defined in Chapter 7. This means that there is a relationship denoted $\leqslant$ between any two absolute times α, β, such that one of two conditions holds: either $\alpha \leqslant \beta$ or else $\beta \leqslant \alpha$. If $\alpha \leqslant \beta$ and $\alpha \neq \beta$ then we write $\alpha < \beta$ and say that α is *earlier than* β (or equivalently that β is *later than* α).

Images of Time. First Edition. George Jaroszkiewicz.
© George Jaroszkiewicz 2016. Published in 2016 by Oxford University Press.

An important additional property is that, given three absolute times α, β, and γ such that $\alpha \leqslant \beta$ and $\beta \leqslant \gamma$, then $\alpha \leqslant \gamma$. We shall call this *temporal transitivity*.

Clocks

In order to make use of Absolute Time we need to translate it into mathematical terms. A near universal assumption is that Absolute Time is a one-dimensional manifold (q.v. Appendix) $\mathcal{A}$. In a given laboratory and motivated by some physical mechanism or process, a *clock function* is a special mapping, a bijection T from a given open subset $\mathcal{T}$ of $\mathcal{A}$ into some open subset $T(\mathcal{T})$ of the real numbers $\mathbb{R}$. Given any absolute time α in $\mathcal{T}$ then its clock value $T(\alpha)$ is a real number called the *time of α relative to the clock function T*. The significance here is that observers, that is, humans, cannot deal directly with the metaphysical, abstract elements of $\mathcal{A}$ but only with the clock times, the real numbers associated with those elements.

Clock functions satisfy certain conditions:

1. Bijections are single valued functions, so an absolute time always has a unique clock value relative to a given clock function.

2. Bijections are one-to-one invertible mappings, so any two different absolute times do not have the same clock time relative to a given clock.

3. Clock functions respect the ordering property of Absolute Time: the conventional rule is that if absolute time α is earlier than absolute time β then $T(\alpha) < T(\beta)$. It is always possible, however, to choose the reverse convention, that is, to have that $T(\beta) < T(\alpha)$, but there is no particular advantage in this and it will be generally avoided.

4. Clock functions respect *temporal transitivity*: if $\alpha \leqslant \beta \leqslant \gamma$ then relative to a given clock, we have $T(\alpha) \leqslant T(\beta) \leqslant T(\gamma)$.

The reparametrization of time

Clocks parametrize the temporal evolution of the primary observer, but they are not unique. Suppose we have two clocks T and T' each mapping a given open subset $\mathcal{T}$ of Absolute Time into $\mathbb{R}$. If α is some absolute time in $\mathcal{T}$ then $T(\alpha)$ and $T'(\alpha)$ are each well-defined real numbers, but not necessarily equal to each other. We may use the properties of clocks to define a *reparametrization* of clock values as follows.

First, choose a T clock value t_α in $T(\mathcal{T})$. By the injective properties of clocks, we can identify a unique absolute time α in $\mathcal{T}$ such that $T(\alpha) = t_\alpha$. Then we may formally write $\alpha = T^{-1}(t_\alpha)$, that is, we can assume that there is an inverse function mapping us from $T(\mathcal{T})$ back into $\mathcal{T}$. We can now use clock T' to calculate its clock value $t'_\alpha \equiv T'(\alpha)$. This defines a bijection $f : T(\mathcal{T}) \to T'(\mathcal{T})$ from $T(\mathcal{T})$ to $T'(\mathcal{T})$ which we write in the form

$$t \underset{f}{\to} t'(t) \equiv T'(T^{-1}(t)). \tag{14.1}$$

Such a bijection is called a *reparametrization of time*, or *temporal reparametrization*.

Three additional conditions are normally imposed on temporal reparametrizations:

1. Each clock has the same temporal direction, that is, if $t_\alpha \leqslant t_\beta$ then $t'_\alpha \leqslant t'_\beta$.

2. Absolute Time is generally assumed to be a linear continuum, discussed in Chapter 7, so the function $t'(t)$ is assumed to be a continuous function of t.

3. It is assumed that $t'(t)$ is a differentiable function of t satisfying the rule

$$\frac{dt'}{dt} > 0. \tag{14.2}$$

The point about Newton's comments on Absolute and Relative Time is that the former is a metaphysical concept whilst the latter is relevant to the real world of experimentation. This begs the question: which of our physical clocks are reliable indicators of Absolute Time?

At the time of Newton, it was generally believed that the motions of the planets were regular enough to provide good clock standards. However, it was observed subsequently that all astronomical bodies have variable motions, including the Earth: we cannot rely on the daily rotation of our planet to be constant. For example, earthquakes can alter the mass distribution of our planet and hence change its rotation rate. This comes about because of the conservation of total angular momentum (the product of the Earth's moment of inertia about a given axis and its angular velocity about that axis). Redistribution of mass due to an earthquake changes the moment of inertia and consequently the angular velocity has to change in order to conserve total angular momentum. The NASA scientist Richard Gross has calculated that recent changes to the length of the day caused by known earthquakes are of the order of millionths of a second per day [Gross & Chao, 2006].

The Earth's rotation sets the standard of time known as *Universal Time*, monitored by the International Earth Rotation and Reference Systems Service (IERS). Complete daily rotations can be defined using distant celestial objects such as stars as a baseline. Given the observed irregularities in Universal Time caused by numerous factors such as earthquakes, the creation of reservoirs (these raise masses of water), and gravitational interaction with the Moon, astronomical movement is no longer used to provide the best scientific clock standards. Instead, the regularity of atomic processes is used [International des Poids et Mesures, 2006]. For example, the second is now defined as the duration of 9,192,631,770 oscillations of the light emitted when a caesium 133 atom makes a transition between two specified energy configurations, rather than a fraction 1/86,500 of a mean solar day.

Absolute space

In *The Principia*, Newton also commented upon Absolute Space, the physical space in which we move about:

> Absolute space, in its own nature, without relation to anything external, remains always similar and immovable. Relative space is some movable dimension or measure of the absolute spaces . . .
> [Newton, 1687]

It was appreciated in Antiquity that Absolute Space has what appears to be intrinsic properties: it has three dimensions, it appears unbounded, it seems directly linked to the surface of the Earth, and there is a distance between any two points in it. Because it has three dimensions, we shall denote the abstract 'thing' described by Newton by $\mathcal{S}^3$. In the same way as we represent Absolute Time $\mathcal{A}$ by the real numbers $\mathbb{R}$, we represent Absolute Space by three-dimensional Euclidean space, $\mathbb{E}^3$: this is the set of points in the Cartesian product $\mathbb{R}^3 \equiv \mathbb{R} \times \mathbb{R} \times \mathbb{R}$ plus the Euclidean distance rule between any two points in $\mathbb{R}^3$.

The Euclidean distance rule states that given two points A, B in $\mathcal{S}^3$, then the distance $d(A, B)$ is given by the rule

$$d(A, B) = \sqrt{(x_A - x_B)^2 + (y_A - y_B)^2 + (z_A - z_B)^2},$$

where (x_A, y_A, z_A) are the coordinates of A in $\mathbb{R}^3$ and similarly for B. Such a rule has all the properties of a metric (q.v. Appendix).

Aristotelian space-time versus Galilean–Newtonian space-time

Long before Galileo and Newton, great minds like Aristotle looked at the universe and created mental models of what they observed. In Antiquity, the relatively low level of technology severely constrained observations and, in consequence, the mental models invoked turned out to have limited value and potential. In particular, observers were constrained to move over the surface of what appeared to be a small part of a relatively flat horizontal surface called the Earth with very limited movement possible in the vertical direction. This surface appeared immensely stable, apart from isolated moments when earthquakes occurred, whilst up in the distant sky, the Sun, Moon, and small points of light moved around in relatively regular patterns. All of this conditioned the ancients to suppose that the surface that they lived on, the Earth, was a fundamental or absolute frame of reference, and that changes in events could be parametrized by an absolute time parameter.

The Ancient Greeks were not all Flat Earthers: they had long speculated that the Earth was round. Around 240 BCE, Eratosthenes gave an estimate for the Earth's circumference that was out by about 5–15 per cent. Nevertheless, the view that became standard by the time of Galileo was that the Earth was held fixed relative to some absolute frame.

We can represent this architecture of space-time mathematically by *Aristotelian space-time* $\mathbb{A}^4$, the Cartesian product $\mathbb{E}^3 \times \mathbb{R}$ of two sets: three-dimensional Euclidean space $\mathbb{E}^3$ and one-dimensional Absolute Time $\mathbb{R}$. Setting up a Cartesian coordinate frame in $\mathbb{E}^3$ then allows us to assign four coordinates (x_P, y_P, z_P, t_P) to any event P in Aristotelian space-time. In this paradigm, points of space have a real existence, are labelled by coordinates, and their properties (such as being occupied by matter) change with Absolute Time: Absolute Space is the amber of the Universe in which the insects (matter) are embedded.

Not long after Newton formulated the laws of his mechanics in the Principia, it was pointed out by thinkers such as George Berkeley [1685–1753] that absolute motion could not be detected. Berkeley wrote

> The laws of motion . . . hold without bringing absolute motion into account. As is plain from this that since according to the principles of those who introduce absolute motion we cannot know by any indication whether the whole frame of things is at rest, or moved uniformly in a direction, clearly we cannot know the absolute motion of any body.   [Berkeley, 1721]

This was a reminder to followers of Newton of Galileo's relativity principle stated in 1632 [Galileo, 1632]: *the laws of motion are the same in all inertial frames*.

The following scenario motivates this principle. You are an important passenger on a sailing ship bound for the New World. You go on board late at night, are given a fine cabin with a bed, and you fall asleep. You wake up long after dawn, convinced that the ship is still in port. You open the windows and find that you, the bed, and everything else on the ship are moving serenely downriver towards the sea: in fact, you discover later that the ship had been moving since sunrise, the crew having been ordered not to make any noise that would disturb your sleep.

The impossibility of detecting absolute motion is responsible for hindering the development of mechanics and hence the study of time: for thousands of years humans were unaware that they were on a planet rotating them at a speed of about 1,000 miles an hour (relative to line of sight of the stars) about the centre of the planet, carrying them in orbit around the Sun at a speed of about 67,000 miles an hour, and following that star around its orbit about the centre of our galaxy at a speed of about half a million miles per hour. We are also on a collision course with the Andromeda Galaxy in about 4 billion years' time, with a relative speed of about a quarter of a million miles per hour.

Galileo's relativity principle creates difficulties in the mathematical modelling of space-time. The traditional physics way is to invoke the intuitive concept of *inertial frame*. Inertial frames are idealized structures that can be approximated by sets of synchronized observers coasting along together in space-time, each set or chorus being reasonably extended and capable of acting as a laboratory in which observations of events in space-time can be made and recorded. The chorus concept is discussed in Chapter 6. There is a theoretical infinity, a continuum, of inertial frames, each with a chorus and, in principle, some superobserver associated with them that acts as a primary observer, collecting after-event data from the members of the chorus and using it to build a conceptual account of what had gone on in their frame.

If that seems bizarre, you have not seen the worst. It is assumed without further ado that each chorus can coast through any of the others without any effects whatsoever. This is a very subtle and dangerous assumption that comes about because of the particular conditions prevailing in our corner of the universe: during the day, SUOs are illuminated by sunlight and give off vast amounts of electromagnetic radiation. Different observers see different parts of that radiation and generally agree on what they think they see. Each observer frame will believe that it is seeing the same image of a given SUO. This is the source of two erroneous beliefs about 'reality': first that SUOs have properties that can be detected by all observers; second, that those properties 'exist' even in the absence of those observers. Hence we can dispense with the observers.

This illusion was shattered by quantum mechanics, as we shall discuss in later chapters, but it has entered the conditioned thought patterns of humans so effectively that it seems impossible to eradicate it. It is responsible for the famous debate known as EPR,[33] in which the contextually incomplete notion of 'element of reality' played a central role.

To the mathematician, a sound description of the inertial frame concept requires more sophisticated technology than mere Cartesian products. $\mathcal{N}^4$, Galilean–Newtonian space-time, is described mathematically as a *fibre bundle* [Schutz, 1980]. The time aspect of $\mathcal{N}^4$, Newton's Absolute Time, is modelled by $\mathbb{R}$ just as with Aristotelian space-time $\mathcal{A}^4$. Now, however, this $\mathbb{R}$ is interpreted as the *base space* of the fibre bundle, at each point of which is attached a *fibre*, a separate copy of Euclidean space $\mathbb{E}^3$.

The point here is that separate fibres, that is, copies of Euclidean space at different times, are in principle uncorrelated: they are separate spaces. What is missing is a mechanism to link structural information across all the different copies of space, such as the direction of coordinates axes in each copy, where the origin of coordinates in each actually is the system of units to be used, and so on.

The problem can be better understood by the following analogy. Consider yourself going into a completely darkened room, such as your bedroom, when all the lights in the house have gone off and you have no torch. You open the door by touch and go in. You have a memory of where objects such as your bed are, but you cannot see anything. Slowly you make your way in the dark, hoping to reach the window to open the curtains, so that the light from outside might at least provide some illumination. You step gingerly forwards towards where you think the window is. Suddenly, you bump into a wall. You have gone in the wrong direction.

---

[33] After Einstein, Podolsky and Rosen, authors of a paper that triggered a famous debate with Bohr [Einstein *et al.*, 1935; Bohr, 1935].

The point is, without any information provided by light or touch to tell you where you are, you have no intrinsic way of determining your position and direction of motion in your bedroom. To navigate successfully, you need information in order to correlate your movements at different times.

Another analogy is with a person stepping out of a shower and drying their hair with a rough towel. When it is dry but before being combed, the hair is a mess, with all the individual fibres (hairs) randomly oriented. After combing, the hairs are aligned in some well-defined pattern.

The necessary information that correlates different copies of $\mathbb{E}^3$ in Galilean space-time is given by Newton's first law of motion, otherwise known as Galileo's Principle of Inertia. This states that a body moving on a level surface will continue in the same direction at constant speed unless disturbed.[34] This correlation supposes that a 'free' particle can be empirically identified and monitored by observers in some Galilean frame of reference. This gives the chorus a working rule on how to align coordinates in separate copies of $\mathbb{E}^3$: they have to be coordinated so that a 'free' particle is described as moving uniformly along a straight line. In this respect, Eddington's parody of the first law of motion was entirely unjustified: the first law is needed to tell us what is meant by an inertial frame.

Of course, the entire procedure rests on us finding a 'free' particle in the first place, and that is the rub. Two practical matters need to be sorted out. First, can we be sure that all external forces that might be acting on a particle have been eliminated; and second, what if the frame is rotating?

The first question is answered by understanding the laws of physics: by identifying all the known forces on a particle we can try to eliminate them in practical terms. As for the problem of rotation, Newton gave an answer: look at the distant stars. If the spatial axes of your frame are rotating with respect to the line of sight with those stars, then you are not sitting in an inertial frame of reference.

The problem of correlating axes in successive copies of Euclidean space is not particular to Galilean space-time. It occurs in GR, where it is known as the *parallel transport* problem. It is solved in GR by the so-called *metric connection*, discussed in Chapter 19.

Particle worldlines

In Newtonian mechanics, it is imagined that particles are idealized points, occupying position in physical space modelled by $\mathbb{E}^3$ and moving around it as Absolute Time advances. The set of all positions that a particle occupies throughout its history or lifetime is called its *worldline*. Worldlines may be thought of as spaghetti-like lines threading their way through a four-dimensional box called *space-time*. Three of these dimensions are those of physical space and the fourth

---

[34] Parodied by Eddington as '*Every body continues in its state of rest or uniform motion in a straight line, except insofar as it doesn't.*' [Eddington, 1929]

is time itself. This view is the one described by H. G. Wells in his book *The Time Machine* [Wells, 1895].

In Newtonian mechanics, the assumption is made that space and time are continuous and that particle worldlines have certain definite properties. Without these properties, Newtonian mechanics would not take the form that we are used to.

First, it is assumed that if a particle starts at some position in physical space $\mathbf{r}_i$ at initial absolute time t_i, and is subsequently found at final position $\mathbf{r}_f$ at final time $t_f > t_i$, then for any time t between t_i and t_f, the particle certainly has to be *somewhere* in physical space. It cannot disappear or not exist anywhere in physical space at any intermediate time. This is a fundamental CM assumption directly rejected by the Copenhagen interpretation of QM, discussed later, particularly in Chapter 25. It is not that QM says that a particle 'does not exist' if it is not being observed: rather it declines to comment, for the good reason that to say something exists or does not exist requires evidence to support either assertion. Lack of observation for a proposition implies inability to validate that proposition one way or the other: it does not mean evidence against that proposition.

Other implicit assumptions made in Newtonian mechanics are that no particle can exist in two or more in places in physical space at the same time, that particles have mass, and that particle coordinates are normally differentiable functions of time. We sum up these assumptions by the statement: *particle worldlines are single-valued, differentiable functions of time, except at isolated points, where impulses (discontinuities in velocity) may occur.*

The Newtonian mechanical paradigm

The Newtonian paradigm of mechanics corresponds closely to the way humans experience time: we have a single consciousness operating in the present, the past is gone and beyond our control, whilst the future can be influenced by what we do in the present.

A game of football provides a useful illustration of this paradigm. If their team is one-nil down, as a striker approaches the opposition's goal mouth with the ball, nothing can alter the fact that the opposition have already scored a goal. The past is fixed as far as the players, the referee, and the spectators are concerned. The striker hopes to equalize before the end of the match. In order for that to happen, the striker has to do two things more or less simultaneously: (i) make sure the ball is in a suitable position just outside the opposition's goal mouth and (ii) make sure the ball has been given the right initial velocity for it to go into the back of the net subsequently. Newtonian mechanics does the rest, because once the ball has been kicked, its trajectory is completely outside the control of the striker.[35]

---

[35] Apart from any 'Hand of God'.

The Euler–Lagrange mechanical paradigm

There is another way of discussing mechanics, which we can illustrate by the game of football discussed above. As the striker approaches the opposition's goal mouth, their action in kicking the ball into the goal is not done meaninglessly. The striker has a great deal of visual information about the dispositions of all the other players in the match and how they are running. The striker can imagine where those players might be if the ball came their way, so the kick has to be such that the trajectory of the ball is clear to the back of the net. In some currently not understood way, the striker will analyse, in their brain and before the kick, potential future trajectories of the ball. Then they will decide almost spontaneously which potential trajectory is best or optimal in some sense, and then arrange to kick the ball so as to make it follow that trajectory. A good player will often succeed. Pelé usually succeeded.

According to this way of looking at it, football could be regarded as an exercise in *teleology*: the performance of certain actions so as to ensure a desired final result.

In the Euler–Lagrange approach to CM, an SUO is analysed non-locally in time via the Calculus of Variations. At any time between initial and final times, the instantaneous state of a classical SUO is fully specified by the current values of a number n of configuration degrees of freedom $\mathbf{q} \equiv (q_1, q_2, \ldots, q_n)$ and their time derivatives (velocities) $\dot{\mathbf{q}} \equiv (\dot{q}_1, \dot{q}_2, \ldots, \dot{q}_n)$. These are fed into the Lagrangian $L \equiv L(\mathbf{q}, \dot{\mathbf{q}}, t)$, a chosen function of the coordinates, their velocities, and time, specified over some trajectory Γ running from initial configuration at initial time t_i to final configuration at final time t_f. The *action integral* $A[\Gamma]$ is then the integral of the Lagrangian over the time interval of interest, that is,

$$A[\Gamma] \equiv \int_{t_i}^{t_f} L(\mathbf{q}, \dot{\mathbf{q}}, t)\,dt. \tag{14.3}$$

The Calculus of Variations explores variants of the trajectory Γ and selects an optimum one, according to some principle [Goldstein *et al.*, 2002]. For fixed end points, the chosen principle is Hamilton's principle, whilst for variable end points we use the Weiss Action Principle [Weiss, 1936; Sudarshan & Mukunda, 1983]. The resulting equations of motion are known as the Euler–Lagrange equations of motion:

$$\frac{d}{dt}\left(\frac{\partial L}{\partial \dot{q}_i}\right) = \frac{\partial L}{\partial q_i}, \quad i = 1, 2, \ldots, n. \tag{14.4}$$

One of the great mysteries of science crops up here. No one knows why all useful Lagrangians involve only the coordinates q^i and their velocities $\dot{q}^i$ and no higher time derivatives.[36] The fact is, canonical Lagrangians, those of the form

---

[36] It is possible to use Lagrangians containing higher time derivatives of the coordinates [Tapia, 1988], but quantization becomes problematical.

$L \equiv L(\mathbf{q}, \dot{\mathbf{q}}, t)$, are used in all known fundamental theories, including those of elementary particle physics and general relativity. This has the following consequence: the Euler–Lagrange equations (14.4) give second-order in time differential equations. This impacts directly on all theories of time, because the only viable way we have of investigating time is through observation, and all observations are based on the laws of physics consistent with canonical Lagrangians.

Paul Dirac [1902–84] found third-order in time differential equations of motion for extended classical electric charges [Dirac, 1933]. He took into account the possibility of the dissipation of electromagnetic field energy whenever an electric charge accelerates. In consequence, his equations of motion have a built-in temporal asymmetry, in that they are not time-reversal invariant. Time reversal is discussed in Chapter 26.

Phase space

The Euler–Lagrangian paradigm remains a powerful approach to CM. However, a crucial development occurred in 1834, when Hamilton published an alternative approach, one that had immense influence on the further development of CM, the discovery of QM, and our understanding of time.

In Hamilton's approach, the $2n$ variables $(\mathbf{q}, \dot{\mathbf{q}})$ used to construct the Lagrangian are first replaced by an equivalent set of variables $(\mathbf{q}, \mathbf{p})$, where the p_i are known as *conjugate momenta*. These are obtained from the Lagrangian by the rule

$$p_i \equiv \frac{\partial L}{\partial \dot{q}_i}, \quad i = 1, 2, \ldots, n. \tag{14.5}$$

Whereas the original variables $(\mathbf{q}, \dot{\mathbf{q}})$ can be thought of as points in *configuration-velocity space*, or some variant concept such as fibre bundles [Schutz, 1980; Abraham & Marsden, 2008], $(\mathbf{q}, \mathbf{p})$ is now regarded as the coordinates of a point in *phase space*, a $2n$-dimensional space.

Assuming that the relationships (14.5) can be inverted or reorganized so that the velocities can be found as explicit functions of the phase-space coordinates, the next step is to define the Hamiltonian $H(\mathbf{p}, \mathbf{q}, t)$, a time-dependent function over phase space, by the rule

$$H(\mathbf{p}, \mathbf{q}, t) \equiv \mathbf{p} \cdot \dot{\mathbf{q}} - L(\mathbf{q}, \dot{\mathbf{q}}, t). \tag{14.6}$$

At this point, all notion of trajectory has disappeared. It reappears with Hamilton's equations of motion, which are given by

$$\frac{dq_i}{dt} \overset{c}{=} \frac{\partial H}{\partial p_i}, \quad \frac{dp_i}{dt} \overset{c}{=} -\frac{\partial H}{\partial q_i}, \quad i - 1, 2, \ldots, n, \tag{14.7}$$

where the 'dynamical equality' $\overset{c}{=}$ denotes an equality holding over a true (i.e., classical) phase-space trajectory.

Several comments are relevant here.

1. The inversion of the relations (14.5) is easy for most applications in Newtonian mechanics. However, that is not the case in relativistic particle
   mechanics and, indeed, for all modern theories of space, time, and matter
   including GR. This fact is behind the so-called 'problem of time' in quantum cosmology, which has led some theorists such as Barbour to assert that
   'time does not exist' [Barbour, 1999]. Dirac developed a formalism called
   constraint mechanics to deal with this inversion issue [Dirac, 1964] because
   he wanted to quantize GR. We shall have more to say on this topic in the
   next chapter.

2. The Hamiltonian is related to the concept of energy, but is more subtle.
   Consider an SUO consisting of a bead on a smooth wire that is forced to
   rotate at constant angular speed in a plane about a fixed point. The Hamiltonian for this SUO is constant over any dynamical trajectory[37] but the
   'energy' is not. Such examples suggest that energy can involve the interface
   between endophysics and exophysics. In the case of the rotating bead, energy from outside the SUO is absorbed by the bead as it is pushed further
   and further out by inertial forces.

3. Given a phase space and a Hamiltonian, we can evaluate Hamilton's equations of motion at any point in that phase space. These are first-order in
   time differential equations. Therefore, given a starting point in that phase
   space, these equations can transport the state of the SUO unambiguously
   and without further information from that point. It is almost as if the Hamiltonian gives us a set of fingerprints or grooves in phase space, the paths
   that would be followed by particles subject to the equations of motion.
   Moreover, different paths would never cross.

Poisson brackets

In Chapter 3 we mentioned Hamilton and his view that time could be discussed
in terms of algebra. Hamilton's equations allow us to do just that and vindicate his
assertion. Suppose $f(\mathbf{p}, \mathbf{q}, t)$ is some time-dependent function over phase space.
Then as we move along a dynamical trajectory, we find

$$\frac{df}{dt}\bigg|_c = \{f, H\}_{PB} + \frac{\partial f}{\partial t}, \tag{14.8}$$

where $\{f, g\}_{PB}$ is the *Poisson bracket* of any two functions f, g over phase space,
define by

$$\{f, g\}_{PB} \equiv \sum_{i=1}^{n} \left\{ \frac{\partial f}{\partial q_i} \frac{\partial g}{\partial p_i} - \frac{\partial f}{\partial p_i} \frac{\partial g}{\partial q_i} \right\}. \tag{14.9}$$

---

[37] A dynamical trajectory is one that satisfies Hamilton's equations of motion.

An immediate conclusion is that if we choose f to be the Hamiltonian itself, then we deduce $\dfrac{dH}{dt} \underset{c}{=} \dfrac{\partial H}{\partial t}$: if a Hamiltonian is explicitly independent of time, that is, $\dfrac{\partial H}{\partial t} = 0$, then the Hamiltonian is constant over any dynamical trajectory.

The use of Poisson brackets to discuss evolution in time has an enormous advantage. Just after QM was discovered at the end of the first quarter of the twentieth century, Dirac showed that canonical quantization, the process of finding quantum mechanical version of a classical mechanical theory, could be understood as a replacement of Poisson brackets in the classical theory by commutators of operators in the quantum theory [Dirac, 1925], that is, $\hat{p}_i\hat{q}_j - \hat{q}_j\hat{p}_i = i\hbar\left\{p_i, q_j\right\}_{PB}$, and so on, where $\hat{p}_i$ is the quantum operator corresponding to the classical variable p_i, and so forth.

The algebraic approach using Poisson brackets was also the technology that Dirac used in his development of constraint mechanics [Dirac, 1964]: this has been applied to all fundamental theories in particle physics, general relativity, and quantum cosmology.

An important refinement of these ideas was the formulation of the so-called *equation of small disturbances* [DeWitt, 1965] and unequal-time Poisson brackets [Peierls, 1952]. This set the scene for unequal-time commutation relations in relativistic quantum field theory, currently the most successful technology available to study the physics of time and space.

Canonical transformation theory

The Hamiltonian phase space approach to mechanics has many fundamental advantages over Lagrangians in addition to providing a path to canonical quantization. One of these is the insight it gives into coordinate transformations in phase space.

In many situation, the initial phase-space coordinates $(\mathbf{p}, \mathbf{q})$, referred to as the *old coordinates,* are not the best coordinates to use. Often the observer will consider transforming to *new phase-space coordinates* $(\mathbf{P}, \mathbf{Q})$ related in some specific way to the old coordinates by a rule of the form

$$P_i = P_i(\mathbf{q}, \mathbf{p}, t), \quad Q_i = Q_i(\mathbf{q}, \mathbf{p}, t), \quad i = 1, 2, \ldots, n, \qquad (14.10)$$

where we include the possibility of time dependence in the coordinate transformation.

Not all such transformations are useful. *Canonical transformations* are those for which Hamilton's equations of motion (14.7) are satisfied in the new coordinates as well as in the old, and these will be of primary interest to us for the following reasons:

1. Restricting transformations to be canonical ensures that we have a unified approach to mechanics.

2. Quantum mechanics is often obtained from a classical mechanical description based on Hamilton's equations of motion.

Restricted canonical transformations are such that there is no explicit time dependence in the transformation, that is, the phase-space coordinate transformations takes the form

$$P_i = P_i(\mathbf{q}, \mathbf{p}), \quad Q_i = Q_i(\mathbf{q}, \mathbf{p}), \quad i = 1, 2, \ldots, n. \tag{14.11}$$

Point transformations are even more restrictive, being transformations of the form

$$P_i = P_i(\mathbf{q}, \mathbf{p}), \quad Q_i = Q_i(\mathbf{q}), \quad i = 1, 2, \ldots, n. \tag{14.12}$$

Although Lagrangians and Hamiltonians are related, they are fundamentally different objects. In special relativistic mechanics, for example, a point particle Lagrangian is usually a Lorentz scalar,[38] whereas the corresponding Hamiltonian is the zeroth (or time) component of the energy–momentum four-vector. With this in mind, we shall assume that Lagrangians are invariant to canonical transformations. If L' is the Lagrangian at a point on a trajectory as seen in the new coordinates, we assume

$$L' = L'\left(\mathbf{Q}, \frac{d\mathbf{Q}}{dt}, t\right) = L\left(\mathbf{q}, \frac{d\mathbf{q}}{dt}, t\right). \tag{14.13}$$

We note that in this discussion, time is being treated as absolute, that is, it is the same parameter in old and new coordinate descriptions.

Again, taking a lead from special relativity, we cannot insist that the Hamiltonian $H' \equiv H'(\mathbf{P}, \mathbf{Q}, t)$ in the new phase-space coordinates is the same as the Hamiltonian in the old coordinates. However, since we are using a canonical transformation, we can say that

$$H' \equiv H'(\mathbf{P}, \mathbf{Q}, t) \equiv \mathbf{P} \cdot \frac{d\mathbf{Q}}{dt} - L'\left(\mathbf{Q}, \frac{d}{dt}\mathbf{Q}, t\right), \tag{14.14}$$

where

$$\mathbf{P} \equiv \frac{\partial L'}{\partial \frac{d\mathbf{Q}}{dt}}. \tag{14.15}$$

Imposing the condition that the transformation is canonical means that

$$\frac{d\mathbf{Q}}{dt} = \frac{\partial H'}{\partial \mathbf{P}}, \quad \frac{d\mathbf{P}}{dt} = -\frac{\partial H'}{\partial \mathbf{Q}}. \tag{14.16}$$

---

[38] A Lorentz scalar is a quantity in SR, the value of which is independent of inertial frame.

Standard analysis [Goldstein *et al.*, 2002; Leech, 1965] then leads to the relation

$$\mathbf{p} \cdot \frac{d\mathbf{q}}{dt} - \mathbf{P} \cdot \frac{d\mathbf{Q}}{dt} - H(\mathbf{p}, \mathbf{q}, t) + H'(\mathbf{P}, \mathbf{Q}, t) = \frac{dF}{dt}, \tag{14.17}$$

for some function F of the variables $\mathbf{q}, \mathbf{Q}, \mathbf{p}, \mathbf{P}$, and t. Because of the transformation equations (14.10), there are only $2n + 1$ variables. Standard transformation theory classifies four main transformation function types. We shall be interested in the type known as $F_2 \equiv F_2(\mathbf{q}, \mathbf{P}, t)$, for which the rules are

$$\mathbf{p} = \frac{\partial F_2}{\partial \mathbf{q}}, \quad \mathbf{Q} = \frac{\partial F_2}{\partial \mathbf{P}}. \tag{14.18}$$

Infinitesimal transformations

We can use the above transformation theory to develop an approach to symmetries that sheds light on the significance of time evolution in classical mechanics. Consider an infinitesimal transformation function of type F_2 of the form

$$F_2(\mathbf{q}, \mathbf{P}, t) \equiv \mathbf{q} \cdot \mathbf{P} + \varepsilon G(\mathbf{q}, \mathbf{P}, t), \tag{14.19}$$

where ε is an infinitesimal parameter and G is some chosen function of $\mathbf{q}, \mathbf{P}$, and t. Then we find from (14.18) that

$$\mathbf{Q} = \frac{\partial F_2}{\partial \mathbf{P}} = \mathbf{q} + \varepsilon \frac{\partial G}{\partial \mathbf{P}}, \quad \mathbf{p} = \frac{\partial F_2}{\partial \mathbf{q}} = \mathbf{P} + \varepsilon \frac{\partial G}{\partial \mathbf{q}}. \tag{14.20}$$

Given that ε is infinitesimal, we deduce that, to lowest order in ε,

$$\delta q^a \equiv Q^a - q^a = \varepsilon \frac{\partial G(\mathbf{q}, \mathbf{p}, t)}{\partial p_a} + O(\varepsilon^2)$$

$$\delta p_a \equiv P_a - p_a = -\varepsilon \frac{\partial G(\mathbf{q}, \mathbf{p}, t)}{\partial q^a} + O(\varepsilon^2), \tag{14.21}$$

which can we rewritten in terms of the Poisson brackets as follows:

$$\delta \mathbf{q} = \varepsilon \{\mathbf{q}, G\}_{PB} + O(\varepsilon^2), \quad \delta \mathbf{p} = \varepsilon \{\mathbf{p}, G\}_{PB} + O(\varepsilon^2). \tag{14.22}$$

When a transformation is expressed in this way, we say that the function εG is a *generator* of the infinitesimal canonical transformation (14.22).

Two important results follow from this analysis.

1. Conserved quantities

Suppose G has no explicit time dependence, that is, $G = G(\mathbf{q}, \mathbf{p})$. Further, suppose that G is a symmetry of the Hamiltonian, that is,

$$\delta H = \varepsilon \{H, G\}_{PB} = 0. \tag{14.23}$$

Now the Hamilton–Poisson equation of motion for G is

$$\frac{d}{dt}G = \{G, H\}_{PB} + \frac{\partial}{\partial t}G. \tag{14.24}$$

But both terms on the right-hand side of (14.24) are zero. We then deduce that if G is explicitly independent of time and generates a symmetry of the Hamiltonian, then G is conserved, that is, is constant along true trajectories in phase space.

2. Translation in time

Suppose we choose the infinitesimal generator εG to be εH, where H is the Hamiltonian itself. Then we have

$$\delta\mathbf{q} = \varepsilon \{\mathbf{q}, H\}_{PB} + O(\varepsilon^2), \quad \delta\mathbf{p} = \varepsilon \{\mathbf{p}, H\}_{PB} + O(\varepsilon^2). \tag{14.25}$$

If now we take $\varepsilon \to 0$, we deduce

$$\lim_{\varepsilon \to 0} \frac{\delta\mathbf{q}}{\varepsilon} = \{\mathbf{q}, H\}_{PB}, \quad \lim_{\varepsilon \to 0} \frac{\delta\mathbf{p}}{\varepsilon} = \{\mathbf{p}, H\}_{PB}, \tag{14.26}$$

but we also know that the Hamilton–Poisson equations of motion are

$$\frac{d\mathbf{q}}{dt} = \{\mathbf{q}, H\}_{PB}, \quad \frac{d\mathbf{p}}{dt} = \{\mathbf{p}, H\}_{PB}. \tag{14.27}$$

This leads us to interpret the Hamiltonian as *the generator of translations in time*, a fundamental insight into the relationship between time and energy.

15

The reparametrization of time

Introduction

We saw in the previous chapter that Newton's Absolute Time can be used to parametrize successive classical mechanical (CM) states of systems under observation (SUOs). We also commented on the fact that Newton was aware that Absolute Time is an idealization, so he defined measured time as '*relative, apparent, and common time*', which is a '. . . *sensible and external (whether accurate or unequable) measure of duration*' [Newton, 1687]. In this chapter we look more carefully at this issue. Specifically, following on from our discussion in Chapter 12 on the single dimensionality of the time parameter, we discuss in this chapter how variable this parameter can be in mechanics, that is, investigate how precisely we need to pin time down.

This discussion is of importance to us because it leads to perhaps the deepest question of them all: what is the physical status of time? Is time real or not?

Surprisingly, an important perspective on this question is to be found in the branch of mechanics known as *constraint mechanics*. Constraints arise naturally in CM whenever there is a mathematical redundancy in the description of some physical system. This happens surprisingly frequently. Suppose a theorist knows that a given SUO has a certain number n of physically observable degrees of freedom, but for some reason that theorist finds it necessary to employ a greater number m of mathematical degrees of freedom (coordinates) to model that system. This happens in all modern field theories. In such a case, constraint mechanics generates just the right number $m - n$ of constraints that effectively eliminate the redundant, unphysical degrees of freedom, resulting in a theory that has the right number n of physical degrees of freedom.

An important example of such a situation occurs in the special relativistic (SR) formulation of the classical point particle. The logic goes as follows. If we started discussing a point particle in non-relativistic (Newtonian) CM, we would use *three* spatial degrees of freedom, such as the Cartesian position coordinates x, y, and z in an inertial frame $\mathcal{F}$. We would use laboratory time t to label the particle's instantaneous position in space, relative to our frame of reference $\mathcal{F}$. This is a space-time perspective (note the hyphen), where time t is a *parameter*.

Images of Time. First Edition. George Jaroszkiewicz.
© George Jaroszkiewicz 2016. Published in 2016 by Oxford University Press.

We shall see in later chapters that Minkowski's approach to SR [Minkowski, 1908] led to the idea that space and time should be merged into a single entity called *Minkowski spacetime* (no hyphen), a four-dimensional continuum with a pseudo-Euclidean distance structure or metric. According to Minkowski's approach, the particle discussed above should be described by a worldline in this four-dimensional continuum. A worldline is a set of events or points in spacetime with *four* spacetime coordinates $\{t, x, y, z\}$, parametrized by some new parameter τ that plays the role of an internal time. From this perspective, the SUO appears now to have *four* degrees of freedom, not three, because t is now regarded not as a parameter but as a dynamical degree of freedom.

This raises the obvious question: how can a Newtonian space-time description that involves *three* mechanical degrees of freedom x, y, z describe the same physics as an SR description that involves *four* mechanical degrees of freedom t, x, y, z?

The answer is provided by constraint mechanics: it shows how to accommodate both the space-time perspective and the spacetime perspective. Constraint mechanics is relevant to us in this chapter because we shall need it when we attempt to reparametrize time. We do not need to be discussing SR: the method works equally well for purely Newtonian, non-relativistic systems.

We will follow the analysis given in an influential book on constraint mechanics by Dirac [Dirac, 1964]. It was Dirac's intention in that book to lay down an approach to quantum gravity that would permit the so-called canonical quantization of Einstein's GR. Although Dirac did not succeed in giving a consistent quantization scheme for GR, his approach to constraints has been applied successfully to many other situations, including the one we are concerned with now.

Temporal parametrization

Before we discuss the reparametrization of time, we should review briefly what *temporal parametrization* means.

A parameter is generally an independent real variable t defined over some interval $[a, b]$ which labels distinct elements of some set, referred to as the *dependent* variables. The dependent variables could be position coordinates, quantum state vectors, or whatever is considered to be changing with respect to the parameter.

To illustrate what we mean, consider a real-valued function f of a parameter t defined over the interval $[a, b]$. Then for any value t of the parameter in the interval $[a, b]$, the value of the function is denoted by $f(t)$.

We note two important points about this example that apply to more general situations:

1. For each value t of the independent parameter in the interval $[a, b]$, there is always one and only one function value $f(t)$.
2. There is nothing in the definition which rules out the possibility $f(t_1) = f(t_2)$ for $t_1 \neq t_2$. In such a case, the different values of the independent parameter serve to distinguish the two function values, even though numerically they are equal.

The second point above reflects a fundamental feature of the physics of time: if a localized SUO takes on mathematically identical states at different times, we should not say that those states represent the same physical reality. The time parameter is really a measure of the physical evolution of the observer, their apparatus, and the SUO in the physical universe. Therefore, because any discussion of 'physical reality' should include observers as well as the SUOS, mathematically identical states of SUOs at different times are in fact components of physically different states of the total system, which includes the observers, their apparatus, and the SUOs. Essentially, a holistic view of reality has to be taken. We recall here Ray Cumming's dictum 'Time . . . is what keeps everything from happening at once' [Cummings, 1922]. The Leibniz principle of the identity of indiscernibles says much the same thing differently: '*objects that have absolutely every property in common are the same object*'.

Temporal reparametrization

The reparametrization of time is a passive transformation, corresponding to a coordinate change involving the real line $\mathbb{R}$. Given an initial temporal coordinate $a \leqslant t \leqslant b$ over the interval $I \equiv [a, b]$, and a final temporal coordinate $\tilde{a} \leqslant \tilde{t} \leqslant \tilde{b}$ over interval $\tilde{I} \equiv [\tilde{a}, \tilde{b}]$, we will assume that $\tilde{t}$ is an invertible function of t, mapping points in I to points in $\tilde{I}$, viz. $t \rightarrow \tilde{t}(t)$, such that $\tilde{a} = \tilde{t}(a), \tilde{b} = \tilde{t}(b)$. If we wanted to, we could choose to make the new time parameter $\tilde{t}$ run in the opposite direction, that is, $\tilde{a} = \tilde{t}(b), \tilde{b} = \tilde{t}(a)$, but there is no advantage in this so we shall not make this choice.

If we believe in the continuity of time, then we require $\tilde{t}$ to be a continuous function of t. Additionally, since we want to deal with differentiable functions, we will require $\tilde{t}$ to be a differentiable function of t over the open interval (a, b). All of these conditions means that we must have $\dfrac{d\tilde{t}}{dt} > 0$ over the interval (a, b). If this condition failed to hold anywhere over that interval then that would mean that the new time parameter $\tilde{t}$ appeared to stop for some value of the old time parameter t. Whilst we are prepared to accept that our new time parameter $\tilde{t}$ could vary non-uniformly with the old (Absolute) time parameter t in the way Newton commented on above, we should not be prepared to use a clock that appears to stop at some point.

In the following analysis, we shall use the inverse transformation, writing $t = t(\tilde{t})$ and taking $\dfrac{dt}{d\tilde{t}} > 0$.

Action integrals

The reparametrization of time is best studied from the action integral perspective. Recall that in the Euler–Lagrange formulation of CM, an SUO is described classically by a finite number n of real degrees of freedom $\mathbf{q} \equiv \{q_1, q_2, \ldots, q_n\}$:

these are configuration space coordinates over some coordinate patch chosen by the observer. We shall start off assuming in the first instance that the observer is describing successive configurations of the system using their laboratory time, denoted by t.

At any value t of laboratory time, the observer represents the instantaneous spatial configuration of the SUO by $\mathbf{q}(t) \equiv \{q_1(t), q_2(t), \ldots, q_n(t)\}$, a single, well-defined point in the configuration space manifold Q. We shall call it the *system point*. As the time parameter goes from initial time t_i to final time t_f, the system point traces out a path in configuration space denoted by Γ. The objective in the Calculus of Variations approach to mechanics is to find an equation for Γ_c, the true or classical path taken by the system point in reality.

In Lagrange's approach to mechanics, the coordinates q_i could represent standard Cartesian coordinates, angles, or even ratios, so the discussion here is quite general. The main implicit assumption is that the coordinates $q^i(t)$ are suitably differentiable functions of time t. This means that we can calculate the instantaneous velocities

$$\mathbf{v} \equiv \frac{d\mathbf{q}}{dt} = \left\{ \frac{dq_1}{dt}, \frac{dq_2}{dt}, \ldots, \frac{dq_n}{dt} \right\} \tag{15.1}$$

at every point along the trajectory Γ.

Given a differentiable trajectory Γ, the next step is to calculate the value of the Lagrangian $L = L(\mathbf{q}, \mathbf{v}, t)$ at every point along it. After that, we construct the action integral $A_{if}[\Gamma]$, defined by

$$A_{if}[\Gamma] = \int_{t_i}^{t_f} L(\mathbf{q}, \mathbf{v}, t)\, dt, \quad t_f > t_i. \tag{15.2}$$

The Calculus of Variations and Hamilton's principle applied to this action integral give the standard Euler–Lagrange equations of motion that are taken to hold over Γ_c for each coordinate, that is,

$$\frac{d}{dt}\left(\frac{\partial L}{\partial v_i} \right) \underset{c}{=} \frac{\partial L}{\partial q_i}, \quad i = 1, 2, \ldots, n, \quad t_i < t < t_f, \tag{15.3}$$

where the symbol $\underset{c}{=}$ denotes an equality holding only over the true or classical trajectory Γ_c.

For convenience we shall rewrite (15.3) in the form

$$\frac{d\mathbf{s}}{dt} \underset{c}{=} \frac{\partial L}{\partial \mathbf{q}}, \tag{15.4}$$

where the *conjugate variables* $\mathbf{s}$ are defined by

$$\mathbf{s}(\mathbf{q}, \mathbf{v}, t) \equiv \frac{\partial L}{\partial \mathbf{v}}, \quad \text{that is,} \quad s_i \equiv \frac{\partial L}{\partial \left(\dfrac{dq_i}{dt} \right)}, \quad i = 1, 2, \ldots, n. \tag{15.5}$$

At this stage we are working entirely with Lagrangians and velocity-configuration space. The s_i are not to be regarded as phase-space 'canonical' momenta. This only happens once we perform a Legendre transformation taking us from Lagrangian mechanics to Hamiltonian mechanics. This is an important point, as the *primary constraints* that arise in Dirac's constraint mechanics live in phase space. The analogues of primary constraints are what we call *primary identities*, and these live in velocity-configuration space.

Temporal reparametrization in detail

We are now in a position to consider the reparametrization of time. In the action integral (15.2) consider replacing the variable of integration t by some alternative real parameter λ such that

$$t = t(\lambda): \quad t_i = t(0), \quad t_f = t(1). \tag{15.6}$$

For convenience and without any loss of content, the new temporal evolution parameter λ is chosen to run from $\lambda = 0$ to $\lambda = 1$.

Following our earlier discussion on reparametrization, we will assume that

$$\dot{t} \equiv \frac{dt}{d\lambda} > 0 \tag{15.7}$$

for $0 < \lambda < 1$, where we reserve the Newtonian 'dot' or *fluxion* notation for derivatives with respect to λ.

Turning to the coordinates $q_i(t)$ along a given trajectory Γ, we shall use the notation

$$\mathbf{Q}(\lambda) \equiv \mathbf{q}(t(\lambda)). \tag{15.8}$$

Then the original velocities $\mathbf{v}$ are given by

$$\mathbf{v} \equiv \dot{\mathbf{Q}}/\dot{t}, \quad 0 < \lambda < 1. \tag{15.9}$$

Hence the action integral (15.2) now takes the form

$$A_{if}[\Gamma] = \int_0^1 \tilde{L}(\mathbf{Q}, t, \dot{\mathbf{Q}}, \dot{t})\, d\lambda, \tag{15.10}$$

where $\tilde{L}(\mathbf{Q}, t, \dot{\mathbf{Q}}, \dot{t}) \equiv \dot{t} L(\mathbf{Q}, \dot{\mathbf{Q}}/\dot{t}, t)$.

An important point to note in the context of this discussion is that whilst the original Lagrangian L could have an *explicit* dependence on the original (Absolute) time parameter t, the reparametrized Lagrangian $\tilde{L}$ has only an *implicit* dependence on its time parameter λ. This has everything to do with the so-called 'timelessness' encountered in quantum cosmology. This timelessness is no more than an artefact of the formalism but has been given, in our opinion, undue physical significance [Barbour, 1999]. A fundamental but generally unstated precept that we advise is that *mathematics can describe physics, but is not physics*.

To illustrate the above, let us apply it to the Newtonian point particle. Given the standard Lagrangian

$$L(\mathbf{q}, \mathbf{v}, t) = \frac{1}{2} m \mathbf{v}^2 - V(\mathbf{q}, t) \tag{15.11}$$

for a particle of mass m in a time dependent force potential V, we have

$$\tilde{L}(\mathbf{Q}, t, \dot{\mathbf{Q}}, \dot{t}) \equiv \frac{m \dot{\mathbf{Q}}^2}{2 \dot{t}} - \dot{t} V(\mathbf{Q}, t), \quad \dot{t} > 0. \tag{15.12}$$

Reparametrization form invariance

The original action integral (15.2) and the reparametrized version (15.10) look manifestly different. However, the reparametrized version has an extraordinary property: it is *form invariant* to further reparametrizations.

A function F of n independent variables $\mathbf{q} \equiv (q_1, q_2, \ldots, q_n)$ is *form invariant* to the transformation $q_i \to q_i'(\mathbf{q})$, $i = 1, 2, \ldots, n$, if $F(\mathbf{q}') = F(\mathbf{q})$.

Form invariance has been invoked as a fundamental principle in the development of relativistic equations. The basic idea is that all 'reasonable' frames of reference are 'as good as each other' in the description of physics, and therefore the laws of physics should take the same functional form in each of those frames.

We should take the following points into account when applying this idea:

1. What constitutes a 'reasonable' frame of reference is contextual. For example, if we wish to ignore gravitation and/or curvature of spacetime, then we generally restrict our attention to transformations between inertial frames of reference. This is the SR scenario. If on the other hand we want to discuss the GR scenario involving curvature, we have to use generalized coordinates with suitable conditions on transformations between frames.

2. Too much faith in the principle of general covariance might be misplaced. For example, the idea that all inertial frames of reference are equivalent in SR is at odds with the existence of the symmetry frame of the cosmic background radiation field at any given point. By this we mean that we can use the cosmic background radiation field to determine a preferred class of frames at any point in the universe. Since this is believed to be a product of the evolution of the universe, it is hard to accept the argument that such a frame is 'accidental', or not really special.

3. We should be wary of any seductive arguments based on symmetry and/or intuitive logic. We have referred in previous chapters to the suggestion that the Ancient Greeks did not develop physics because they considered experiments, which are artificial, to be the wrong way to discover intrinsic truth. A more recent example of misplaced confidence in symmetry occurred in the 1950s, when experiments proved that the universe does not respect left–right symmetry, the so-called overthrow of parity conservation experiment [Lee & Yang, 1956; Wu *et al.*, 1957].

4. Form invariance as a principle of physics is based on a classical notion that SUOs have intrinsic properties independent of observers: on that basis, descriptions of their properties should be independent of any particular observers. However, there is a theorem in quantum mechanics due to Kochen and Specker [Kochen & Specker, 1967] that states that quantum states of SUOs cannot 'have' classical properties in the manner of classical states in CM. Observers *must* be taken into account. Therefore, it seems to us that form invariance is at best a property that we should build into *classical* descriptions of SUOs.

5. There is no proof that the laws of physics have to be 'beautiful', a common theme of many mathematically inspired but as yet unverified approaches to physics, such as string theory. We should be prepared to find that the laws of physics are cumbersome and ugly, *if* that is what experiment eventually supports. In cosmology, for example, the *cosmological principle* states that there is no special place or direction in space, over large enough distance scales. Observations of galaxies at the furthest limit of observation may yet reveal this to be false. Form invariance to arbitrary coordinate transformations, otherwise known in GR as *general covariance*, is at best a guide for the development of potential theory.

6. It is not the case that general covariance implies GR. SR can be rewritten in a generally covariant form but that theory does not discuss gravitation. True gravity comes about from spacetime curvature, which cannot be induced by mere coordinate transformations.

7. Time is a phenomenon that cannot be separated from the observers of SUOs. It seems to us that too much adherence to the classical principle of general covariance, which does not take into account specific observers, might well be at the root of the failure to 'quantize' GR. We take the view that quantum mechanics is *not* a theory about bizarre particle/wave-like properties of SUOS, but a theory of observation. Form invariance is an emphasis on the properties of SUOs. Therefore, theorists who tend to focus too readily and too much on the construction of a generally covariant formulation of quantum gravity at the expense of looking at the observational physics of gravity stand of good chance of getting nowhere, as the history of the last seven decades has shown. We shall return to this point in later chapters.

To demonstrate the form invariance of (15.10) we consider a second reparametrization, from λ to λ'. We take $\lambda \equiv \lambda(\lambda')$ to be a differentiable function of λ' such that $d\lambda/d\lambda' > 0$ and define

$$Q'(\lambda') \equiv Q(\lambda(\lambda')), \quad t'(\lambda') \equiv t(\lambda(\lambda')), \tag{15.13}$$

with λ' running from 0 to 1. Then we find

$$A_{if}[\Gamma] = \int_0^1 \tilde{L}(Q, t, \dot{Q}, \dot{t})\,d\lambda = \int_0^1 \tilde{L}(Q', t', \dot{Q}', \dot{t}')\,d\lambda', \tag{15.14}$$

which demonstrates explicitly the form invariance of the action integral under this particular class of reparametrizations of time. Note that on the right-hand side of (15.14), $\dot{t} \equiv dt/d\lambda$ whilst $\dot{t}' \equiv dt'/d\lambda'$, and so on.

The extended equations of motion

In the reparametrized action integral (15.10), the dependent variables appear to have increased by one and now include t, giving us an extended set of dependent variables, (Q, t). It turns out that we can treat all of these variables as if they were dynamical variables. The reason is that arbitrary variations of reparametrization function $t(\lambda)$ can be interpreted as temporal reparametrizations, and these do not change the value of the action integral (by form invariance under temporal reparametrization).

In the following, we shall revert to the notation (q, t) in place of (Q, t), it being understood that all the variables are functions of the temporal path parameter λ.

In the Calculus of Variations, variations of the original coordinates q are essentially *active* transformations of the path, that is, Hamilton's principle involves a comparison of different ways of going from A to B. Turning to our extended coordinates, consider a variation of path consisting of an active variation of the original coordinates q and a passive temporal reparametrization in the action integral (15.10), given by

$$t \to t' \equiv t + \delta t, \quad q \to q' \equiv q + \delta q \tag{15.15}$$

for fixed λ. A careful application of the Calculus of Variations to first-order in the variations gives the $n + 1$ equations of motion

$$\frac{ds}{dt} \underset{c}{=} \frac{\partial L}{\partial q}, \quad \frac{dE}{dt} \underset{c}{=} -\frac{\partial L}{\partial t}, \tag{15.16}$$

where $s \equiv \partial L/\partial v$ as before and $E(q, v, t) \equiv s \cdot v - L$ is interpreted as the canonical energy. In the phase-space reformulation of these equations, we make the transition $s \to p$, and then the canonical energy becomes identified with the instantaneous value of the original Hamiltonian $H(\mathbf{p}, \mathbf{q}, t)$ at the position of the system point on the trajectory. The first set of equations in (15.16) are the usual Euler–Lagrange equations of motion, whilst the second equation involving the energy expresses the standard result that energy is conserved if the Lagrangian does not depend explicitly on time.

Transformation to phase space

At this point we need to briefly review the Hamiltonian formulation of mechanics, because the above reparametrization introduces a constraint.

The canonical Lagrange formulation of CM is based on velocity-configuration space, that is, the tangent bundle TQ, where Q is configuration space. The action principle applied to canonical Lagrangians then leads to the Euler–Lagrange equations of motion, which are second-order in time differential equations for the true trajectory Γ_c from initial to final time.

On the other hand, the Hamiltonian formulation is based on the cotangent bundle T^*Q, which is referred to in this context as *phase space*. If Q has dimension n then the n second-order equations of motion in the Lagrangian formulation are replaced in the Hamiltonian formulation by $2n$ first-order in time differential equations. Although Feynman showed how to use Lagrangians in his path integral approach to quantization [Feynman & Hibbs, 1965], quantization has generally been attempted via phase space. Therefore, we need to discuss how to go from TQ to T^*Q.

The tradition method is to set up a coordinate patch $\mathbf{q} \equiv (q_1, q_2, \ldots, q_n)$ covering a region of interest in Q, construct a canonical Lagrangian $L(\mathbf{q}, \mathbf{v}, t)$, and then map into the momentum coordinates $\mathbf{p} \equiv (p_1, p_2, \ldots, p_n)$ via the rule

$$p_i \equiv \frac{\partial L}{\partial v_i}, \quad i = 1, 2, \ldots, n. \tag{15.17}$$

At this point we may hit a fundamental problem: it may be impossible to invert the map from the momentum coordinates back into velocity space. This occurs whenever we encounter a *primary constraint* in the language of Dirac [Dirac, 1964]. A primary constraint is a relation of the form $\Phi(\mathbf{q}, \mathbf{p}) = 0$ whenever the momenta are given by the canonical map (15.17).

The interpretation of primary constraints is that they indicate that the $2n$ dimensional phase space T^*Q that we are using is too large to describe the true physics of the SUO that the chosen Lagrangian is intended to model. A primary constraint effectively reduces the dimensionality of phase space.

We shall not go over the phase-space analysis of primary constraints, as this is quite involved and is discussed in detail by Dirac [Dirac, 1964]. Instead, we shall investigate what is happening from the position–velocity space perspective.

The key to our analysis is the Lagrangian. Recall that this is chosen by the observer to reflect the physics believed to apply to the SUO. Consider a canonical

Lagrangian $L(\mathbf{q}, \mathbf{v}, t)$ that happens to have the property that there exist one or more *primary identities*. A primary identity is a relation involving the configuration coordinates $\mathbf{q}$, the conjugate variables $\mathbf{s} \equiv \partial L / \partial \mathbf{v}$, and possibly the time t, of the form

$$\Phi(\mathbf{q}, \mathbf{s}, t) = 0, \tag{15.18}$$

for *all* possible values of the $\mathbf{q}$, the $\mathbf{s}$, and all time. In other words, a primary identity is a true identity.

To illustrate, consider an SUO described by two configuration degrees of freedom q_1, q_2 and Lagrangian

$$L(q_1, q_2, \dot{q}_1, \dot{q}_2) = \frac{1}{2}(\dot{q}_1 - \dot{q}_2)^2. \tag{15.19}$$

Then

$$s_1 \equiv \frac{\partial L}{\partial \dot{q}_1} = \dot{q}_1 - \dot{q}_2, \quad s_2 \equiv \frac{\partial L}{\partial \dot{q}_2} = \dot{q}_2 - \dot{q}_1. \tag{15.20}$$

By inspection, we see there is a primary identity given by $\Phi \equiv s_1 + s_2 = 0$.

We investigate the conditions for primary identities to arise as follows. Given an identity of the form (15.18), we note it holds for all values of configuration coordinates and conjugate variables. Consider an infinitesimal change in these. Then we may write

$$d\Phi = \sum_{i=1}^{n} \left\{ \frac{\partial \Phi}{\partial q_i} dq_i + \frac{\partial \Phi}{\partial s_i} ds_i \right\} = 0. \tag{15.21}$$

But the conjugate variables $\mathbf{s}$ are themselves functions of $\mathbf{q}$ and $\mathbf{v}$, so we may write

$$ds_i = \sum_{j=1}^{n} \left\{ \frac{\partial^2 L}{\partial v_j \partial v_i} dv_j + \frac{\partial^2 L}{\partial v_i \partial q_j} dq_j \right\}. \tag{15.22}$$

Since the variations in $\mathbf{q}$ and $\mathbf{v}$ can be taken to be independent, we deduce

$$\frac{\partial \Phi}{\partial q_i} + \sum_{j=1}^{n} \frac{\partial^2 L}{\partial q_i \partial v_j} \frac{\partial \Phi}{\partial s_j} = 0, \tag{15.23}$$

$$\sum_{j=1}^{n} \frac{\partial^2 L}{\partial v_i \partial v_j} \frac{\partial \Phi}{\partial s_j} = 0. \tag{15.24}$$

Now from (15.24), if the matrix $\left[\dfrac{\partial^2 L}{\partial v_i \partial v_j} \right]$ is invertible we deduce $\partial \Phi / \partial \mathbf{s} = 0$, and then (15.23) gives $\partial \Phi / \partial \mathbf{q} = 0$.

We conclude that there are $n - \text{rank} \left(\dfrac{\partial^2 L}{\partial v_i \partial v_j} \right)$ primary identities.

Reparametrized primary identity

We are now in position to understand the phenomenon of timelessness in quantum gravity. Consider the temporally reparametrized Lagrangian

$$\tilde{L} \equiv \dot{q}_0 L\left(\mathbf{q}, \frac{\dot{\mathbf{q}}}{\dot{q}_0}, q_0\right), \tag{15.25}$$

where for convenience we define $q_0 \equiv t$ and work with the extended coordinates $q_A \equiv (q_0, \mathbf{q})$, where capital Latin subscripts run from 0 to n.

It is straightforward to show that the matrix $\left[\dfrac{\partial^2 \tilde{L}}{\partial \dot{q}_A \partial \dot{q}_B}\right]$ is singular, proving that a temporarily reparametrized canonical Lagrangian, of whatever kind, has at least one primary identity. We can find this directly as follows. Using the reparametrized Lagrangian $\tilde{L}$, we calculate the conjugate variables $\tilde{s}_A \equiv \partial \tilde{L}/\partial \dot{q}_A$. We find

$$\tilde{s}_0 \equiv \frac{\partial \tilde{L}}{\partial \dot{q}^0} = L - \mathbf{s} \cdot \mathbf{v}, \tag{15.26}$$

$$\tilde{\mathbf{s}} \equiv \frac{\partial \tilde{L}}{\partial \dot{\mathbf{q}}} = \frac{\partial L}{\partial \mathbf{v}} \equiv \mathbf{s}. \tag{15.27}$$

Remarkably, equation (15.27) shows that the original conjugate momenta $\mathbf{p}$ are invariant to temporal reparametrization. This is consistent with the fact that temporal reparametrization is a passive process that affects no physics. Equation (15.26) is a primary identity in extended state space. The interpretation is that the momentum conjugate to time in extended phase space is minus the canonical Hamiltonian $H(\mathbf{q}, \mathbf{p}, t)$.

The end point of our analysis is to calculate the extended Hamiltonian $\tilde{H}$, defined by the Legendre transformation

$$\tilde{H}(\tilde{q}_A, \tilde{p}_B) \equiv \sum_{A=0}^{n} \tilde{s}_A \dot{q}_A - \tilde{L}. \tag{15.28}$$

Using (15.26) we immediately find that $\tilde{H} \underset{c}{=} 0$, which is the primary constraint we are looking for.

This is the source of 'timelessness' in quantum cosmology. According to the transformation theory [Goldstein *et al.*, 2002] discussed in the previous chapter, a Hamiltonian is a generator of translations in time. Since the above classical extended Hamiltonian $\tilde{H}$ vanishes on the classical trajectory, this suggests that its quantized operator version vanishes: therefore there should be no meaning to translation in the corresponding temporal parameter, so *time appears to have no dynamical content*.

Actually, the situation is more subtle than that, since primary constraints vanish only on the so-called 'surface of constraints' [Dirac, 1964]. Quantum mechanically, it means that the quantized extended Hamiltonian is not a generator of unitary translations in time but an operator that restricts physical states to those that it annihilates, that is, equivalent to the Wheeler–DeWitt equation (8.2).

Our final thoughts in this chapter are the following:

1. 'Timelessness' relates to reparametrized temporal parameters and not to the original time coordinate t. The universe still evolves as that parameter changes.

2. Dirac's constraint theory tells us that the existence of primary constraints/identities indicates a redundancy in a mathematical description of an SUO. This redundancy is eliminated by choosing one representative state out of each set of physically equivalent (i.e., mostly redundant) states, and that choice no longer has any redundancy. Such a choice is referred to as a choice of *gauge*, for historical reasons. In our case, once we make a choice of temporal gauge, then we lose temporal reparametrization invariance and timelessness goes away. But then the theory no longer looks as elegant as it did before the choice of gauge. This is consistent with the fact that attempts to include observers directly into the formalism of any modern theory generally makes those theories much less elegant and invariably much more complicated.

3. The Wheeler–DeWitt equation has inspired work in quantum cosmology but has itself been the subject of recent criticism [Unger & Smolin, 2015] that we agree with. The Wheeler–DeWitt equation

$$\hat{\tilde{H}}|\Psi\rangle = 0 \tag{15.29}$$

for the so-called 'wavefunction for the universe' is not a valid quantum equation for the simple reason it is contextually incomplete, and therefore any claim that it represents physics is unjustified.

16

Origins of relativity

Inertial frames

In this chapter we discuss the origins of special relativity (SR). Although SR is usually associated with the name of Einstein and his seminal paper in the year 1905 [Einstein, 1905b], a number of theorists and experimentalists had worked out many of the mathematical details of SR by that date. The limits of Newtonian classical mechanics (NCM) had been detected empirically well before 1900, so by 1905 many physicists were ready to accept a rapid paradigm shift away from the Newtonian space-time of NCM towards a radically new view of time and space.

Before SR, physicists believed that the laws of physics should look the same in every inertial frame, that is, the laws of physics should be invariant to standard Galilean transformations.

Such transformations between inertial frames can be encapsulated by the simplified transformation

$$t'_P = t_P, \quad x'_P = x_P + vt_P, \quad y'_P = y_P, \quad z'_P = z_P, \tag{16.1}$$

which incorporates Absolute Time in an obvious way. Here the subscript P refers to any event P in Newtonian space-time. The property of such a transformation is that whilst velocities are not invariant to Galilean transformations, accelerations are invariant, and that means that Newton's laws of motion are form-invariant to Galilean transformations, provided mass is an invariant. If a particle has instantaneous position $\mathbf{x}(t) \equiv (x(t), y(t), z(t))$ in one inertial frame and in another inertial frame it has instantaneous position $\mathbf{x}'(t') \equiv (x'(t'), y'(t'), z'(t'))$ then under the Galilean transformation (16.1) we find

$$m\frac{d^2\mathbf{x}}{dt^2} = m\frac{d^2\mathbf{x}'}{dt'^2} = \mathbf{f}, \tag{16.2}$$

where $\mathbf{f}$ is the instantaneous force on the particle.

Images of Time. First Edition. George Jaroszkiewicz.
© George Jaroszkiewicz 2016. Published in 2016 by Oxford University Press.

The speed of light and Galilean transformations

What motivated the move from NCM to SR was a conflict of principles involving Maxwell's equations of electromagnetism and Galilean relativity. Consider the following three statements:

1. The laws of NCM are form-invariant to Galilean transformations, that is, they take the same form in every standard inertial frame.

2. Maxwell's equations have been empirically validated, supporting the view that light is an electromagnetic wave phenomenon propagating with a speed $c \approx 3 \times 10^{10}$ metres per second.

3. Velocity is not invariant to Galilean transformations, that is, the velocity of a moving object depends on the frame in which that velocity is being measured.

Until about the end of the nineteenth century, most physicists had no trouble accepting (1): all NCM calculations gave excellent agreement with observations except for a number of technical issues. These included:

1. **The electrodynamic stability of atoms**: Maxwell's equations predict that classical electric charges will radiate electromagnetic energy into the surrounding space whenever they undergo acceleration. This is the principle behind radio broadcasting and reception. If atoms are bound states of electrically charged particles, then those particles would be accelerating and hence dissipating energy into their environments. NCM therefore predicts that atoms should collapse relatively quickly (small fractions of a second), contrary to the observed stability of atoms.

2. **The ultraviolet catastrophe**: All dynamical quantities, such as linear momentum and energy of particles, are continuous in NCM, meaning they could in principle take on any permitted value. Based on this assertion, Rayleigh [Rayleigh, 1900] and Jeans [Jeans, 1905] discussed collections of classical charged harmonic oscillators in the walls of a container in thermal equilibrium with an electromagnetic radiation field inside that container. Rayleigh and Jeans predicted that, at a given temperature T, these oscillators would be exchanging energy with the radiation field by amounts proportional to T and to the square of the frequency of the field electromagnetic vibrations. Experiments showed exactly the opposite, that there was a sharp fall-off in the energy density of the cavity radiation field at high frequencies.

3. **The orbit of the planet Mercury**: Newtonian mechanics and Newton's law of universal gravitation could account for all the deviations in the orbit of Mercury around the Sun, away from a closed elliptic orbit, except for a tiny but measurable discrepancy between theory and observation. The calculations took into account the perturbations of Mercury's orbit by other planets, but this residual discrepancy could not be explained by Newtonian gravitation.

These issues were eventually addressed by quantum mechanics in the case of (1) and (2) and by Einstein's theory of general relativity in the case of (3). Otherwise, close to the end of the nineteenth century, most physicists would have believed that their subject was in an excellent state of health.

However, there was an obvious problem with the speed of light and the *principle of Galilean relativity*. Stated in 1632, this principle asserts that the laws of motion are the same in all inertial frames [Galileo, 1632]. Newton's laws of motion, enunciated in *The Principia* [Newton, 1687] obey Galileo's principle. By the time Maxwell had formulated his equations of electrodynamics, c. 1862, most physicists would have expected Galileo's principle to hold for them as well.

The problem was this. Maxwell's equations led to the prediction that light propagates through space with a finite speed, the predicted value of which was in excellent agreement with experiment. But speed is frame dependent, according to the Galilean transformation rules. If the speed of light is c in one inertial frame, then according to Galilean transformations of the form (16.1), the speed of light in a relatively moving inertial frame will be different.

These facts could not be reconciled: Maxwell's equations were inconsistent with Galileo's principle. For several decades, theorists and experimentalists tried unsuccessfully to understand the relationship between Maxwell's equations and the Galilean transformations.

A widespread belief at that time was that, in much the same way that sound is a vibration of a physical medium, the air, electromagnetic waves had to be vibrations in an actual physical medium called the *Aether*. No one knew what sort of medium it could be. Rather like the dark matter that most physicists currently believe exists but cannot be seen, the Aether was assumed to exist but to be undetectable, apart from transmitting electromagnetic vibrations through space. Unlike sound waves, which are caused by vibrations in air pressure along the direction of a sound wave, Maxwell's theory predicted that electromagnetic waves vibrate in directions transverse to the direction of the wave.

Whatever the mechanics, it was believe that the Aether was a real substance, possibly unlike ordinary matter, that defined a unique inertial frame, its own rest frame, $\mathcal{F}_{\text{Aether}}$.[39] Relative to this frame, light would move at the speed $c = 1/\sqrt{\varepsilon_0 \mu_0}$ as predicted by Maxwell's theory, where ε_0 and μ_0 are fundamental, measurable constants. Relative to any inertial frame of reference that was moving relative to $\mathcal{F}_{\text{Aether}}$, it was expected that light would be observed to move at a different speed fully in accordance with the rules of Galilean transformations.

The Michelson–Morley experiment

Motivated by the idea that the Aether existed, physicists undertook experiments to detect its influence on the speed of light. One of the most famous experiments

---

[39] Unique up to choice of origin of coordinates and rotations of spatial axes: this is regarded as an inessential complication.

in physics is the Michelson–Morley experiment, first conducted in 1881 and repeated subsequently many times [Michelson & Morley, 1887]. In that experiment, conducted in a laboratory on Earth, light from a source was split by a beam-splitter into two transverse directions then reflected back onto a detector. Depending on the time spent in transit, there would be a pattern of interference fringes observable in the detector: electromagnetic waves arriving in phase at the same point would add constructively whilst those arriving out out phase would add destructively.

In addition, if the light travelling in one direction had a different speed to the other, then when the two returning components were superposed, there would be additional interference effects observable. It was calculated that if the Aether were stationary relative to the Sun, then the Earth's motion during the course of a year would produce an observable effect: an easily measurable interference fringe shift of 1/25 of a fringe was predicted.

Michelson and Morley's equipment was good enough to detect effects of the order 1/100 of a fringe. Their results did not show any expected effects: their experimental errors were consistent with a fringe shift of zero. There appeared to be no effect of motion relative to the Aether

Subsequently various other experiments were done but generally the results were the same. No Aether could be detected. This is the situation to date.

FitzGerald length contraction

There were various attempts to explain Michelson and Morley's failure to detect the Aether. One of these was that perhaps the Aether was at rest locally, that is, at rest relative to the Earth. Most of these ideas were found to be inadequate for one reason or another.

One of the more interesting ideas was due to George Francis FitzGerald [1851–1901]. In 1889 he wrote a short letter, given below, to a science journal suggesting that perhaps the Michelson–Morley experiment had failed to detect any effects because objects moving with speed v relative to the Aether rest frame shrank in length, relative to those at rest, by a factor which depended on v^2/c^2 and *in the direction of motion*. FitzGerald suggested that such an effect could be just enough to compensate for the predicted effects of a fringe shift in the Michelson-Morley experiment, resulting in a net effect of zero. He wrote:

The Ether and the Earth's Atmosphere

Letters to the Editor, *Science*, 1889, **13**, page 390:
I HAVE read with much interest Messrs. Michelson and Morley's wonderfully delicate experiment attempting to decide the important question as to how far the ether is carried along by the earth. Their result seems opposed to other experiments showing that the ether in the air can be carried along only to an inappreciable extent.

I would suggest that almost the only hypothesis that can reconcile this opposition is that the length of material bodies changes, according as they are moving through the ether or across it, by an amount depending on the square of the ratio of their velocity to that of light. We know that electric forces are affected by the motion of the electrified bodies relative to the ether, and it seems a not improbable supposition that the molecular forces are affected by the motion, and that the size of a body alters consequently. It would be very important if secular experiments on electrical attractions between permanently electrified bodies, such as in a very delicate quadrant electrometer, were instituted in some of the equatorial parts of the earth to observe whether there is any diurnal and annual variation of attraction, — diurnal due to the rotation of the earth being added and subtracted from its orbital velocity; and annual similarly for its orbital velocity and the motion of the solar system.   [FitzGerald, 1889]

Working with Maxwell's equations, the physicist Hendrik Lorentz [1853–1928] managed to account for such a shrinking effect by 1891. By 1899 he had derived a revised version of the Galilean transformation and on that account these new transformations are called *Lorentz transformations*. In fact, Joseph Larmor [1857–1942] had derived these transformations and predicted relativistic time dilation by 1897 [Larmor, 1897].

Relativistic length contraction is frequently referred to as *FitzGerald length contraction* and the SR transformation between inertial frames is referred to as a *Lorentz transformation*.

Derivation of FitzGerald length contraction

We give now a derivation of FitzGerald's length contraction hypothesis based on an idealized experiment. The same approach predicts *relativistic time dilation*, or the temporal slowing down of clocks moving relative to the Aether. This is a phenomenon that FitzGerald does not mention in his letter of 1889. This may be either because he did not reach that conclusion or perhaps because he realized that time dilation would be something too difficult for Newtonian theorists to take seriously.

In order to explain the null result of the Michelson–Morley experiment, Fitz-Gerald proposed that an object moving relative to the Aether rest frame suffers a length contraction along the direction of motion only: transverse to the direction of motion, he assumed that there was no change in physical lengths. This transversality assumption is a reasonable one and it works, but in principle should be examined empirically.

To derive FitzGerald's result, consider the Michelson–Morley experiment from the perspective of the Aether rest frame $\mathcal{F}_{\text{Aether}}$, in which light moves with speed c in all directions (Figure 16.1)

The diagram shows the spatial Cartesian coordinates x, y in the Aether rest frame. These are the spatial coordinates used by an observer chorus at

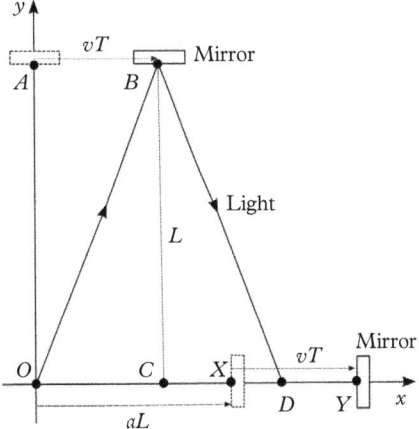

Fig. 16.1 *FitzGerald contraction factor apparatus as seen in the Aether frame.*

rest in $\mathcal{F}_{\text{Aether}}$. Each member of the chorus carries a standard clock and these have all been previously synchronized to Absolute Time.

The experiment involves a rigid apparatus, moving with constant speed v along the positive x direction and consisting of two perpendicular rods of length L, as measured by the chorus in the rest frame $\mathcal{F}_{\text{Apparatus}}$ of the apparatus. One rod is aligned along the x-direction whilst the other is aligned along the y-direction. At the end of each rod is a mirror. At Absolute Time $t = 0$, a sharp pulse of light from the source O is sent towards each mirror, travels with speed c according to the Aether chorus, and is reflected at events B and Y. The reflected pulses intersect at D, the event where the source has finally moved to.

Coordinates in the Aether frame $\mathcal{F}_{\text{Aether}}$ of an event P are given by $P \rightleftharpoons (t_P, x_P, y_P, z_P)$ and are as follows for the various events shown in the diagram:

O: origin of coordinates $O \rightleftharpoons (0, 0, 0, 0)$

A: y-axis mirror at Absolute Time zero $A \rightleftharpoons (0, 0, L, 0)$

B: y-axis mirror when light from origin reaches it $B \rightleftharpoons (T, vT, L, 0)$

C: source has moved here by Absolute Time T $C \rightleftharpoons (T, vT, 0, 0)$

D: source has moved here by Absolute Time $2T$ $D \rightleftharpoons (2T, 2vT, 0)$

X: the x-axis mirror at Absolute Time zero $X \rightleftharpoons (0, \alpha L, 0, 0)$
   (where α is the factor by which the x-axis rod has shrunk, as measured in the Aether frame, according to FitzGerald's hypothesis)

Y: the x-axis mirror at the Absolute Time $\overline{T}$ when the light reaches it:
$$Y \rightleftharpoons (\overline{T}, \alpha L + v\overline{T}, 0, 0)$$

In such discussions, *empirical protocol* is a critical issue, that is, the specific details about how things are measured is significant. Given two events P, Q connected by light, and measuring all times and distances in the same inertial frame $\mathcal{F}$, we use the rule $c = d_{\mathcal{F}}(P, Q)/t_{\mathcal{F}}(P, Q)$, where c is the speed of light, $d_{\mathcal{F}}(P, Q)$ is the spatial interval, and $t_{\mathcal{F}}(P, Q)$ is the temporal interval, between the two events, as measured in the given frame $\mathcal{F}$.

This experiment does not make sense if the apparatus is moving faster than the speed of light c according to frame $\mathcal{F}_{\mathrm{Aether}}$, because then a light signal from event O would never catch up with the mirror at event Y. Therefore, we may safely assume $|v| < c$.

Now consider the following speed of light relations:

1. Light travelling from O to the transverse mirror, hitting it at B, according to frame $\mathcal{F}_{\mathrm{Aether}}$ travels at speed c, hence

$$c = \frac{d_{\mathrm{Aether}}(O, B)}{t_{\mathrm{Aether}}(O, B)} = \frac{\sqrt{v^2 T^2 + L^2}}{T}, \tag{16.3}$$

from which we find

$$T = \frac{L}{\sqrt{c^2 - v^2}}. \tag{16.4}$$

We see here confirmation of the requirement that $|v| < c$, for otherwise T would be imaginary, which is unphysical and hence rejected.

2. Light travelling from O to the longitudinal mirror, hitting it at Y, according to frame $\mathcal{F}_{\mathrm{Aether}}$ also travels at speed c, hence

$$c = \frac{d_{\mathrm{Aether}}(O, Y)}{t_{\mathrm{Aether}}(O, Y)} = \frac{\alpha L + v\overline{T}}{\overline{T}}, \tag{16.5}$$

which gives

$$\overline{T} = \frac{\alpha L}{(c - v)}. \tag{16.6}$$

We rule out $|v| > c$ in this last equation because otherwise we would have $\overline{T} < 0$, which would violate the causality condition that the light signal *from* O reaches the mirror at Y *after* it is sent.

3. Light bouncing from the longitudinal mirror at Y and returning to the source, now moved to D, according to frame $\mathcal{F}_{\mathrm{Aether}}$ gives

$$c = \frac{d_{\mathrm{Aether}}(Y, D)}{T(Y, D)} = \frac{\alpha L + v\overline{T} - 2vT}{2T - \overline{T}}, \tag{16.7}$$

which gives

$$2T - \bar{T} = \frac{\alpha L}{(c + v)}. \tag{16.8}$$

Using (16.4) and (16.6) in (16.8) gives

$$2\frac{L}{\sqrt{c^2 - v^2}} - \frac{\alpha L}{c - v} = \frac{\alpha L}{c + v}, \tag{16.9}$$

which leads to $\alpha = \sqrt{1 - \dfrac{v^2}{c^2}}$, the required contraction factor.

Although FitzGerald did not give this contraction factor explicitly in his letter, it seems reasonable to believe that he did just such a calculation: he was certainly capable of it.

There are several remarks worth making here.

1. All the above measurements are made in the Aether frame, $\mathcal{F}_{Aether}$, with the only reference made to the apparatus frame $\mathcal{F}_{Apparatus}$ being the assumption that each rod had length L in that frame, *as far as the observers moving with that apparatus are concerned*.

2. The Michelson–Morley experiment deals with *closed* light paths, so only the average round-trip speed is being discussed. The Michelson–Morley experiment does *not* show, therefore, that the speed of light is the same in each direction, whereas the above calculation explicitly assumes this. This fact was pointed out by Tangherlini, Chang, Rembieliński, and others in various attempts to find alternative relativistic transformation rules. These Generalized Transformations are discussed in Chapter 18.

3. The same thought experiment leads to the conclusion that the clocks carried by the apparatus must go slow relative to the Absolute Time rate of the clocks in the Aether frame $\mathcal{F}_{Aether}$. We show this in the next section. FitzGerald does not mention this in his letter.

FitzGerald time dilation

On the assumption that observers in the rest frame $\mathcal{F}_{Apparatus}$ of the apparatus could not detect any change in the speed of light, it is easy to show that clocks at rest in that frame ($\mathcal{F}_{Apparatus}$) will appear to run slow relative to Absolute Time, when said clocks are monitored by observers at rest in the Aether rest frame $\mathcal{F}_{Aether}$.

With reference to the same experiment discussed above, consider light travelling from O to B from the point of view of both frames. Frame $\mathcal{F}_{Aether}$ registers a time interval

$$t_{Aether}(B, O) = T - 0 = \frac{L}{\sqrt{c^2 - v^2}} \tag{16.10}$$

at B. Now the apparatus rest frame $\mathcal{F}_{\text{Apparatus}}$ must have its clocks working in such a way that events O and A are synchronized to read zero. According to frame $\mathcal{F}_{\text{Apparatus}}$, the light has moved over a length L, and so its clock at B must read a time $t_{\text{Apparatus}}(B) = \dfrac{L}{c}$. Hence

$$t_{\text{Apparatus}}(B, O) = \frac{L}{c} - 0 = \frac{L}{c}. \tag{16.11}$$

Hence we immediately deduce

$$t_{\text{Apparatus}}(B, O) = \sqrt{1 - \frac{v^2}{c^2}} \, t_{\text{Aether}}(B, O), \tag{16.12}$$

which is the famous relativistic time dilation effect.

The interpretation of this effect requires a more detailed analysis of the relationship between inertial frames in relative motion.

Lorentz transformations

In the years before Einstein's 1905 paper, Lorentz attempted to resolve the problem by showing that Newtonian physics would indeed lead to length contractions as proposed by FitzGerald. By looking carefully at Maxwell's equations, he came up with the so-called Lorentz transformations, which are the correct transformation rules between two standard inertial frames, assuming their coordinate axes are aligned and the origins of spatial and temporal coordinates coincide. Given two such frames, $\mathcal{F}$ and $\mathcal{F}'$, the Lorentz transformation from $\mathcal{F}$ to $\mathcal{F}'$ is

$$t' = \gamma(v)\left(t - \frac{vx}{c^2}\right), \quad x' = \gamma(v)(x - vt), \quad y' = y, \quad z' = z, \tag{16.13}$$

where $\gamma(v)$ is the *Lorentz factor* $\gamma(v) = 1/\sqrt{1 - v^2/c^2}$. We shall discuss this transformation further in the next chapter. Points to note are:

1. Transformations in physics come in two varieties: active and passive. Mathematically, the above Lorentz transformation appears to be a mere relabelling of events in space-time, so it can be considered to be a passive transformation. On the other hand, FitzGerald, Larmor, and Lorentz thought that length contraction was a real phenomenon, so from their perspective, the transformation is very much an active one. In other words, they believed objects moving relative to the Aether actually shrink. However, from the perspective given by Einstein, discussed in the next chapter, objects do not shrink: it is the protocols of observation in different inertial frames that gives that impression. According to Einstein, observers in

every inertial frame say that objects moving relative to them have shrunk: Einstein's SR is symmetrical in that respect.

2. The significance of the parameter v is seen as follows. Consider the origin of spatial coordinates $(x', y', z')' = (0, 0, 0)'$ in $\mathcal{F}'$. This corresponds to

$$\gamma(v)(x - vt) = 0, \quad y = 0, \quad z = 0, \tag{16.14}$$

or $\mathbf{x} = \mathbf{v}t$, the equation of the worldline of an object moving uniformly with respect to $\mathcal{F}$ with velocity $\mathbf{v} = (v, 0, 0)$.

3. The form of the Lorentz factor tells us that we shall run into mathematical problems if the magnitude of the relative velocity between inertial frames is equal to c or greater. Then the Lorentz factor either diverges or is imaginary. Both situations are regarded as unphysical. This leads to the conclusion in SR that nothing can travel faster than the speed of light.

4. We may rewrite the Lorentz transformation in matrix terms, viz.,

$$\begin{bmatrix} ct' \\ x' \\ y' \\ z' \end{bmatrix} = \begin{bmatrix} \gamma(v) & -\gamma(v)\beta & 0 & 0 \\ -\gamma(v)\beta & \gamma(v) & 0 & 0 \\ 0 & 0 & 1 & 0 \\ 0 & 0 & 0 & 1 \end{bmatrix} \begin{bmatrix} ct \\ x \\ y \\ z \end{bmatrix}, \tag{16.15}$$

where $\beta \equiv v/c$.

5. The inverse transformation, from $\mathcal{F}'$ to $\mathcal{F}$ exists provided the determinant of the above 4×4 matrix is non-zero. A calculation gives this determinant to be unity, so the inverse transformation always exists. It is easy to obtain the inverse Lorentz transformation. We just recall that if frame $\mathcal{F}$ sees frame $\mathcal{F}'$ moving with velocity $\mathbf{v}$, then frame $\mathcal{F}'$ will see $\mathcal{F}$ moving with velocity $-\mathbf{v}$.

From this we deduce that the inverse transformation is

$$\begin{bmatrix} ct \\ x \\ y \\ z \end{bmatrix} = \begin{bmatrix} \gamma(v) & \gamma(v)\beta & 0 & 0 \\ \gamma(v)\beta & \gamma(v) & 0 & 0 \\ 0 & 0 & 1 & 0 \\ 0 & 0 & 0 & 1 \end{bmatrix} \begin{bmatrix} ct' \\ x' \\ y' \\ z' \end{bmatrix}, \tag{16.16}$$

since $\gamma(-v) = \gamma(v)$.

6. If the frames $\mathcal{F}$ and $\mathcal{F}'$ are not in standard configuration (spatial axes aligned), then it is a straightforward matter to introduce purely spatial rotations which take this into account. Likewise, we can shift the origin of coordinates in space and time suitably.

7. Lorentz transformations involve c in various places. In ordinary terms, that is, in Standard International (SI) units, c is an enormous number relative to unity. This gives us a way to understand the relationship between Galilean

and Lorentz transformations. First, expand the Lorentz factor in powers of β:

$$\gamma(v) = 1 + \frac{1}{2}\beta^2 + O(\beta^4).\tag{16.17}$$

Then the original Lorentz transformation (16.13) looks like

$$t' = t + O(\beta), \quad x' = x - vt + O(\beta^2), \quad y' = y, \quad z' = z,\tag{16.18}$$

where $O(\beta)$ means 'terms of the order β, β^2, etc.'. For ordinary speeds, β is entirely negligible and so we recover the standard Galilean transformation (16.1) in the limit $\beta \to 0$. Note that this limit could be reached in two ways: either $v \to 0$ or $c \to \infty$.

17

Special relativity

Lorentz transformations

In special relativity (SR), each standard inertial frame[40] $\mathcal{F}$ is associated with a set $\{x^0, x^1, x^2, x^3\}$ of four coordinates that covers the whole of Minkowski spacetime, M^4. Here $x^0 \equiv ct$, where t is the laboratory time of clocks at rest in the frame and x^1, x^2, x^3 are identified with standard spatial Cartesian coordinates commonly denoted by x, y, and z.[41] SR deals with the transformation rules between different standard inertial frames. Note that all references to observations by $\mathcal{F}$ really mean the data collected by the observer chorus at rest in $\mathcal{F}$, and so on.

There are two kinds of SR transformation. The more general are the *Poincaré transformations*. These take the form

$$x'^{\mu} = \Lambda^{\mu}{}_{\nu} x^{\nu} + a^{\mu}, \quad \mu = 0, 1, 2, 3. \tag{17.1}$$

Here the coefficients $\Lambda^{\mu}{}_{\nu}$ form a set of numbers satisfying certain relationships, the x^{μ} are the 'old' coordinates used in $\mathcal{F}$, the x'^{μ} are the 'new' coordinates used in $\mathcal{F}'$, the a^{μ} represent a shift in the origin of spacetime coordinates, and there is an implied summation of the repeated index ν. The less general transformations, which do not involve changes in the origin of coordinates, take the form

$$x'^{\mu} = \Lambda^{\mu}{}_{\nu} x^{\nu}, \quad \mu = 0, 1, 2, 3. \tag{17.2}$$

and are known as *Lorentz transformations*.

A useful simplification is to align the respective spatial axes so that there is no spatial rotation of any of them during the transformation. A final simplification is to take the velocity of the 'new' frame $\mathcal{F}'$ to be along the x-axis, with component v

---

[40] 'Standard' here means that each inertial frame chorus of observers has a copy of the latest version of the International Organization for Standards (ISO) handbook specifying the SI system of units and uses it to follow a universally agreed protocol for setting up their apparatus, system of units, and measurement protocols.

[41] Note that x^2 does not mean x-squared but the second spatial coordinate, y.

Images of Time. First Edition. George Jaroszkiewicz.
© George Jaroszkiewicz 2016. Published in 2016 by Oxford University Press.

in that direction, relative to the 'old' frame $\mathcal{F}$. These simplifications reduce to the standard Lorentz transformation

$$
\begin{bmatrix} x'^0 \\ x'^1 \\ x'^2 \\ x'^3 \end{bmatrix} = \begin{bmatrix} \gamma & -\gamma\beta & 0 & 0 \\ -\gamma\beta & \gamma & 0 & 0 \\ 0 & 0 & 1 & 0 \\ 0 & 0 & 0 & 1 \end{bmatrix} \begin{bmatrix} x^0 \\ x^1 \\ x^2 \\ x^3 \end{bmatrix},
\tag{17.3}
$$

where $\gamma \equiv 1/\sqrt{1-\beta^2}$ is the Lorentz factor and $\beta \equiv v/c$.

A crucial principle used by Einstein in his discussion of SR in 1905 [Einstein, 1905b] was that no inertial frame has a special status above any other inertial frame. Therefore, the inverse transformation taking us back to $\mathcal{F}$ from $\mathcal{F}'$ should take the same form, the only difference being a reversal of the relative velocity, that is, we replace β by $-\beta$. This gives the transformation

$$
\begin{bmatrix} x^0 \\ x^1 \\ x^2 \\ x^3 \end{bmatrix} = \begin{bmatrix} \gamma & \gamma\beta & 0 & 0 \\ \gamma\beta & \gamma & 0 & 0 \\ 0 & 0 & 1 & 0 \\ 0 & 0 & 0 & 1 \end{bmatrix} \begin{bmatrix} x'^0 \\ x'^1 \\ x'^2 \\ x'^3 \end{bmatrix},
\tag{17.4}
$$

which is readily shown to be the inverse transformation of (17.3).

Implicit in this transformation is the assumption that the 'one-way speed' of light is the same in each direction, that is, the speed of light in the positive x-direction is the same as in the opposite direction. Although this is intuitively reasonable, it was noted by Tangherlini that the Michelson–Morley experiment did not prove that this one-way speed assumption is necessary: all that is needed is that the *average* speed around a closed path in a given frame is the speed of light [Tangherlini, 1958]. In Chapter 18 we discuss Generalized Transformations and Tangherlini's version of SR.

Simultaneity in special relativity

SR has a number of features not found in Newtonian CM. Perhaps the most counterintuitive and unsettling is the loss of absolute simultaneity. Tangherlini's work can be seen as an attempt to restore absolute simultaneity.

To help us see how loss of absolute simultaneity comes about in SR, we simplify the discussion by ignoring the transverse y and z coordinates and concentrate only on t and x. In the spacetime diagram in Figure 17.1 we plot the coordinate t down-to-up the page and x left-to-right on the page. Then, given the Lorentz transformation, we can plot the t' and x' axes as indicated.

Events A and B are simultaneous according to the $\mathcal{F}'$ chorus at time $t' = T'$. Suppose they have $\mathcal{F}'$ coordinates $x'_A = (cT', a)'$, $x'_B = (cT', b)'$, where $a \neq b$,

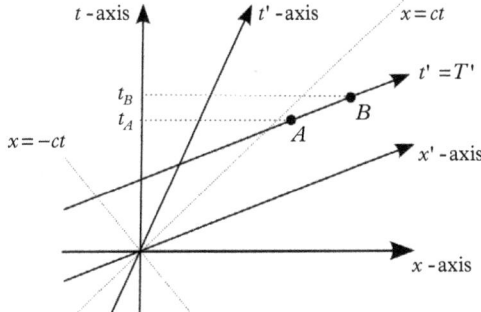

Fig. 17.1 *The loss of absolute simultaneity between inertial frames. Events A and B are simultaneous in frame F′, occurring at time t′ = T′ in that frame, but are not simultaneous according to frame F observers.*

Then using the inverse Lorentz transformation we find the times in frame $\mathcal{F}$ for events A and B to be

$$t_A = \gamma(v)\left[cT' + \frac{va}{c^2}\right], \quad t_B = \gamma(v)\left[cT' + \frac{vb}{c^2}\right]. \tag{17.5}$$

Clearly, $t_A \neq t_B$ if $a \neq b$: events A and B are not simultaneous according to frame $\mathcal{F}$ observers if these events are not spatially coincident in frame $\mathcal{F}'$.

There is complete symmetry between inertial frames in SR. Events that are simultaneous relative to frame $\mathcal{F}$ will not be appear to be simultaneous relative to frame $\mathcal{F}'$.

Time dilation

Lorentz transformations undermine the concept of Absolute Time in two ways: absolute simultaneity does not hold and moving clocks suffer time dilation, that is, they appear to run slow relative to any inertial frame in which they are observed to move. In the previous chapter, we derived time dilation relative to the assumed Aether frame. In fact, this result holds for any inertial frame related to the others by Lorentz transformations, as we show now.

Without loss of generality we can restrict the discussion to one time, one space dimension, and the Lorentz transformation between frames $\mathcal{F}$ and $\mathcal{F}'$, given by

$$t' = \gamma(v)\left(t - \frac{vx}{c^2}\right), \quad x' = \gamma(v)(x - vt), \tag{17.6}$$

where $\gamma(v)$ is the Lorentz factor $\gamma(v) \equiv 1/\sqrt{1 - v^2/c^2}$.

Consider a standard clock at rest at the origin of spatial coordinates $x' = 0$ in frame $\mathcal{F}'$. Then by the second equation in (17.6) any event on its worldline must have coordinates (t, x) in frame $\mathcal{F}$ given by $x = vt$.

Now consider two events, O and A on the worldline of that clock. Event O is the origin of spacetime coordinates for both frames $\mathcal{F}$ and $\mathcal{F}'$, and A is an event at a time T' as measured by the chorus in $\mathcal{F}'$. If the coordinates of A as seen by the $\mathcal{F}$ chorus are (T, X) then using $X = vT$ and the first equation in (17.6) we find $T' = \sqrt{1 - \dfrac{v^2}{c^2}}\, T$, which is the time dilation we found in Chapter 16.

Empirical evidence for time dilation is commonly attributed to the experiments of Ives[42] and Stillwell [Ives & Stilwell, 1938]. They looked at the Doppler effect, an effect seen in *inter-frame* experiments where an extended monochromatic beam of light is prepared in one inertial frame and observed in another, relatively moving inertial frame. We show in the Appendix that if $\mathcal{F}'$ has velocity $\mathbf{v}$ relative to $\mathcal{F}$ and the beam is moving in direction $\mathbf{n}'$, then the frequencies ν, ν' satisfy

$$\nu = \gamma(v) \left\{ 1 + \frac{\mathbf{v} \cdot \mathbf{n}'}{c} \right\} \nu', \tag{17.7}$$

where $\gamma(v)$ is the Lorentz factor as before. In the special case $\mathbf{v} \cdot \mathbf{n}' = 0$, $\mathcal{F}'$ is moving transverse to the beam and so the two frequencies differ by the Lorentz factor only. This is the *transverse Doppler effect*.

A complication is that sources and detectors are generally localized in time and space. An observation by an observer chorus of a beam of light is a mathematical fiction: a localized detector is more like a superobserver, in that retardation effects come into play. There is therefore some need to take this into account when discussing the transverse Doppler effect. The work of Ives needs careful interpretation, because localized moving sources of light do not produce perfect extended monochromatic beams of light and because line of sight between source and detector is constantly changing in a transverse Doppler experiment.

A more convincing manifestation of time dilation is the extension of the mean lifetime of *muons* travelling through the Earth's atmosphere. The muon is an elementary particle much like an overweight electron, about 200 times more massive and unstable, with a mean lifetime of about 2.2 millionths of a second when at rest in the laboratory. High-energy 'secondary' muons are produced in the upper atmosphere by incoming cosmic rays interacting with it. Because of their relatively high speed towards the Earth, time dilation extends the lifetime of such secondary muons so that they can reach the ground and even penetrate significantly into it before they decay. Measurements confirm the SR predictions [Rossi & Hall, 1941]. In this scenario, the concerns of Ives regarding length contraction are not relevant.

---

[42] He was in fact interested in validating the work of Larmor and Lorentz on Fitzgerald length contraction in apparatus.

The clock hypothesis

In the following, the adjective *instantaneous* refers to the numerical value of some quantity being described at a given instant of time as agreed by the observer chorus in some frame. At a different instant of time, the measured instantaneous value might be different.

The clock hypothesis in SR is the assertion that the instantaneous rate of a moving clock, relative to a given inertial frame chorus, depends only on the instantaneous speed v of that clock as measured in that frame, and not on the instantaneous acceleration, if any, of that clock or of any other factors such as the clock's mass.[43]

Explicitly, suppose we use standard Cartesian coordinates (x, y, z) to describe the instantaneous position of a moving clock relative to frame $\mathcal{F}$. Then its instantaneous speed v is given by $v = \sqrt{\left(\dfrac{dx}{dt}\right)^2 + \left(\dfrac{dy}{dt}\right)^2 + \left(\dfrac{dz}{dt}\right)^2}$. If τ is the instantaneous time as registered by that clock, then the clock hypothesis asserts that

$$\frac{d\tau}{dt} = \sqrt{1 - \frac{v^2}{c^2}}, \tag{17.8}$$

where c is the speed of light. This is consistent with the time dilation result of the previous section

There are two important physical predictions that follow from the clock hypothesis:

1. No moving clock could go faster than the speed of light, because then the right-hand side of (17.8) would be imaginary.
2. A moving clock always appears to be going *slow* relative to clocks of the observer chorus at rest in a given inertial frame. There is never any speeding up of a moving clock's rate in SR.

Both points are consistent with all available empirical evidence. No material objects or signals that could be used to convey physical information across space have been observed moving faster than the speed of light, and the predicted time dilation effects have been observed in moving aircraft and in the delay of unstable particles.

This gives rise to an obvious question. Suppose a standard clock moves along some differentiable path Γ in an inertial frame, such that it starts at frame time t_i and ends at frame time t_f. What interval of time does the clock register over this path?

---

[43] We are discussing *special* relativity, which does not deal with gravity. We discuss the clock hypothesis in general relativity in the next chapter.

The answer is to divide up the path Γ into a number of connected pieces, approximate them by straight lines, assume the clock hypothesis along each piece, sum up the separate time dilated time intervals, and then take the limit. If all this is valid then the registered proper time interval $\Delta\tau$ is the line integral

$$\Delta\tau = \int_{t_i}^{t_f} \sqrt{1 - \frac{\mathbf{v}(t)^2}{c^2}}\, dt = \frac{1}{c} \int_{\Gamma} ds. \tag{17.9}$$

Here ds is the infinitesimal Minkowski distance [Minkowski, 1908] along the path Γ, related to infinitesimal coordinate changes (dt, dx, dy, dz) by the so-called *line element*

$$ds^2 = c^2\, dt^2 - dx^2 - dy^2 - dz^2. \tag{17.10}$$

A condition on the path Γ is that it is *timelike*, meaning that ds^2 is positive at every point along Γ. This is a fundamental characteristic of physical trajectories or worldlines in SR.

An important point is that the reality of ds requires $|\mathbf{v}(t)| \leqslant c$, so we deduce $\Delta\tau \leqslant t_f - t_i$.

In 1972, two physicists took caesium regulated clocks around the world using commercial airlines. They took one journey westwards and one journey eastwards and compared the travel times registered by their clocks. Because of the Earth's east-to-west rotation, the speed of transit going eastwards relative to a non-rotating frame is greater than that going westwards. SR predicts different proper times of travel in that case. All the observations confirmed the predictions of SR [Hafele & Keating, 1972a]. There are effects predicted from GR that were factored into their calculations: these concern the effects of spacetime curvature (gravity) on clocks [Hafele & Keating, 1972b]. The Hafele–Keating experiment has been repeated several times with greater accuracy and the combined relativistic predictions confirmed to the general satisfaction of physicists.

The Twin Paradox

It is a fact of human life that we are well conditioned to see the universe in classical terms. Perhaps too well conditioned. In particular, time seems to be absolute and the same for everyone. The SR predictions of time dilation and loss of absolute simultaneity remain contentious amongst a relativity small community of individuals that includes respected scientists and hardened conspiracy theorists. Most physicists have no problem in understanding SR and GR and accepting that the experiments to date fully support those theories. But, as with quantum mechanics and the Bell inequality experiments that come down in its favour, there remain those who adamantly refuse to accept the results of those experiments.

It is, of course, part of the principles of science to be sceptic: the great motto 'Nullius in verba' tells us that. Therefore, there should be a role in science for disagreement, provided that disagreement is logical and well argued. Intuition is *not* enough. And that brings us to the famous (some would say notorious) *Twin Paradox*. This thought experiment appears to undermine SR because of the following argument against time dilation.

Two twins, *A* and *B*, are born on Earth in a future when humans have colonized our galaxy. Twin *B* joins the Space Fleet at age 20 and is sent on a journey to a distant colonized planet by a space vessel that travels close to the speed of light relative to the Earth. On the outwards journey, the space vessel accelerates over a relatively small part of the journey and thereafter coasts most of the way to the new planet at uniform speed relative to the Earth frame. As the vessel approaches the planet, it decelerates and Twin *B* lands safely. After a few days on the planet, Twin *B* returns to Earth in a similar way.

Whilst Twin *B* was away, Twin *A* stayed on Earth. The SR argument is that because Twin *B* was in relative motion almost all the time, then *B*'s clock would be reading a smaller elapsed time when they returned to Earth, compared to the clocks left behind on the Earth. In effect, Twin *A* would appear to have aged normally according to the Earth-bound clocks whilst Twin *B* would appear to have aged less, perhaps significantly less (depending on the details of the journey).

The objection to this conclusion, raised by sceptics of relativity, is usually based on the assertion that each Twin moves relative to the other and that the relative motion is symmetrical: therefore time dilation (if any) should be applied to each Twin in the same way, so there should be no difference in ages when they meet back on Earth. A prominent critic of SR who long argued against its conclusions regarding the Twin Paradox was Herbert Dingle, an established academic [Dingle, 1967].

Although science is not and should not be carried out by opinion and consensus,[44] it is fair and balanced to say that Dingle's objections are incorrect[45] and have generally not been accepted by the scientific community. The standard response to Dingle is that the Twins do *not* follow symmetric worldlines. Twin *A* remains at rest in one inertial frame whilst Twin *B* has to change inertial frames substantially during their journey. One of the puzzling aspects of the Paradox is that the age difference gets larger the longer the journey, but the accelerations required to change frame are always the same in scope. There is no paradox if it is accepted that proper time is a *non-integrable* function, that is, a path-dependent quantity, analogous to work in thermodynamics. This answer is consistent with the position taken throughout this book, that contextuality plays a fundamental role in the study of time.

This issue is significant because it says a great deal about what science is or should be. Dingle regarded the Twin Paradox explanation in SR as not based

---

[44] String theory seems to be an exception to this rule.
[45] A theory must be proved to be inconsistent by its own standards, that is, internally, and not by the standards of another theory.

on empirical evidence. Whilst that would be a legitimate objection if valid, the empirical evidence for time dilation is in fact conclusively in favour of the SR and GR predictions.

Lightcones

The Minkowski spacetime line element (17.10) plays a crucial role in the physics of SR and in unravelling its causality properties. Indeed, line elements in GR serve the same purpose and hold the key to spacetime physics in general. We can use lightcones to study the geometry of black holes, as in Chapter 19, and to discuss time travel, as in Chapter 20.

To see this, consider an infinitesimal displacement (dt, dx, dy, dz) in some inertial frame with zero Minkowski distance, that is, set $ds = 0$ in (17.10). Then we deduce that observers in that frame can interpret this as the displacement of an object moving with the speed of light: $ds = 0$ in (17.10) gives $\dfrac{d\mathbf{x}}{dt} \cdot \dfrac{d\mathbf{x}}{dt} = c^2$.

This brings us to the concept of *lightcone*, a concept of immense importance to physics. Pick any event P in $\mathcal{M}^4$ and set up a standard inertial frame (t, x, y, z) with the origin of coordinates at P. The Minkowski distance s_{PQ} between P and any other event Q in $\mathcal{M}^4$ is given by

$$s_{PQ}^2 = c^2 t_Q^2 - x_Q^2 - y_Q^2 - z_Q^2, \tag{17.11}$$

where (t_Q, x_Q, y_Q, z_Q) are the coordinates of Q. Note carefully the signature of the quadratic form (17.11), which is $(1, -3, 0)$. Signature is discussed in the Appendix.

Now comes the point. We ask the question: *given P*, what is the set of events in $\mathcal{M}^4$ that are at zero Minkowski distance from P? In other words, what is the set of events $\{Q\}$ such that $s_{PQ} = 0$?

It is not hard to see that, because of the signature of the quadratic form (17.11), the complete set contains P itself and what appears to be the two branches of the surface of a cone with its vertex at P and with axis along the time coordinate in the frame, shown in Figure 17.2. The lightcone at P divides spacetime into several disjoint sets. First, there is the vertex P itself. Then there is the surface of the lightcone, with two branches. Events on this surface are *lightlike*, or *null*, relative to P. The *absolute elsewhere* is the region exterior to the lightcone. Event D is in that region and is *spacelike* relative to P. D and P cannot causally influence each other: to do so would mean that physical signals had travelled faster than the speed of light. The interior of the lightcone contains the time coordinate axis (t) as shown. It has a forwards branch known as the *absolute future* (relative to P) and a backwards branch known as the *absolute past* (relative to P). P can causally influence events in its absolute future, such as A, but can only be influenced *by* events in its absolute past, such as B.

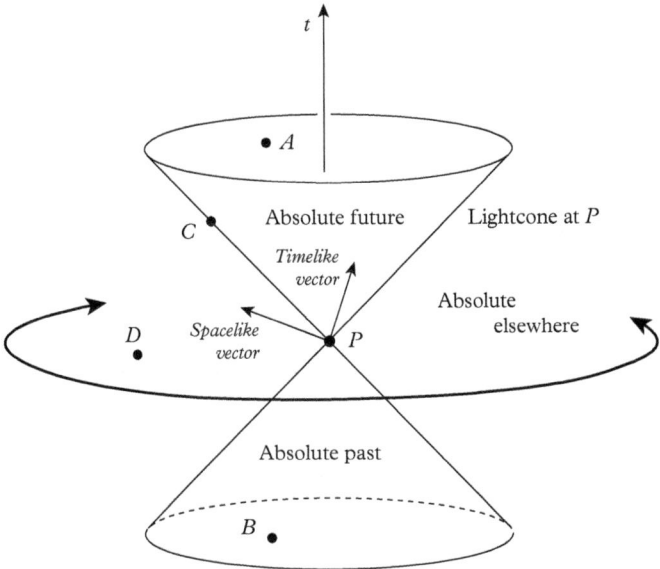

Fig. 17.2 *The lightcone with vertex at P, a typical event in Minkowski spacetime. A and B are timelike relative to P, C is relatively lightlike, and D is relatively spacelike.*

Important points to note about lightcones are:

1. There is a lightcone with the above structure at *every* point in Minkowski spacetime.
2. The existence of lightcones is predicated on having a Lorentzian signature. A Euclidean space does not have lightcones, since the only solution to the equation $c^2t^2 + x^2 + y^2 + z^2 = 0$ for real t, x, y, and z is $t = x = y = z = 0$. Theories that have imaginary time do not have lightcones, which makes their application to physics problematical. In particular, the discussion of signature in Chapter 12 tells us that the partial differential equations encountered in such theories will usually be elliptic rather than hyperbolic, leading to questions about the flow of information.

The Klein–Gordon equation

When Erwin Schrödinger published his famous papers on the non-relativistic quantum wave equation in 1926 [Schrödinger, 1926c] he was taking a step back in one respect: he was ignoring SR. By 1926, SR was over 20 years old and thoroughly familiar to Schrödinger. So the question is, *why did he return to Newtonian space-time?*

History records that in fact, he had already considered a relativistic free particle wave equation, but for the technical reasons that there was no obvious positive probability density associated with the relativistic equation and antiparticles had yet to be thought of or discovered,[46] he conveniently cut his losses and formulated quantum mechanics non-relativistically.

We are interested here in the equation Shrödinger discarded. It is usually called the Klein–Gordon equation (KGE), but also the Schrödinger–Fock–Klein–Gordon equation. We shall discuss some of Fock's thinking on causality towards the end of this chapter.

In $\mathcal{M}^4$ the KGE is given by

$$\hbar^2 \left\{ \frac{\partial^2}{c^2 \partial t^2} - \frac{\partial^2}{\partial x_1^2} - \frac{\partial^2}{\partial x_2^2} - \frac{\partial^2}{\partial x_3^2} \right\} \varphi(x) + m^2 c^2 \varphi(x) = 0, \qquad (17.12)$$

where φ is a real or complex-values function of inertial coordinates $x \equiv (ct, x_1, x_2, x_3)$. This is a hyperbolic partial differential equation, so according to Table 12.1, Cauchy initial data on an initial spacelike hypersurface will propagate into the future without any problems. The question we will answer now is: *what is the causality structure of solutions to this equation?*

The causal propagation of special relativistic fields

To understand the method of solution of the homogeneous equation (17.12), consider an analogous problem, the harmonic oscillator, involving the dynamical variable $q(t)$. The equation for this is

$$\frac{d^2 q}{dt^2}(t) = -\omega^2 q(t), \qquad (17.13)$$

where ω is a non-zero real constant. It is a simplified version of (17.12): essentially we have dropped the spatial derivatives. The general solution to (17.13) is $q(t) = A\cos(\omega t) + B\sin(\omega t)$, where A and B are arbitrary constants. It suits our purpose to write this solution in the equivalent form

$$q(t) = \cos(\omega(t - t_0)) q(t_0) + \omega^{-1} \sin(\omega(t - t_0)) \dot{q}(t_0). \qquad (17.14)$$

This expression is in Newtonian paradigm form, as it shows how the initial data at time t_0 is used to predict the state of the system at any later time $t > t_0$.

A final touch is to write the solution in the form

$$q(t) = \Delta(t - t_0, \omega) \frac{\overleftrightarrow{\partial}}{\partial t_0} q(t_0), \qquad (17.15)$$

---

[46] The indefinite conserved current associated with the complex KGE can be interpreted as a charge density.

where the symbol $\overset{\leftrightarrow}{\dfrac{\partial}{\partial t}}$ is defined by

$$f \overset{\leftrightarrow}{\frac{\partial}{\partial t}} g \equiv f \left(\frac{\partial g}{\partial t} \right) - \left(\frac{\partial f}{\partial t} \right) g \qquad (17.16)$$

for differentiable functions f, g of t and Δ is the *Schwinger function* for this system, defined by

$$\Delta (t, \omega) \equiv \omega^{-1} \sin (\omega t). \qquad (17.17)$$

The Schwinger function Δ has the following properties:

$$
\begin{aligned}
1. & \qquad \lim_{t \to 0} \Delta (t, \omega) = 0 \\[2mm]
2. & \qquad \lim_{t \to 0} \left\{ \frac{\partial}{\partial t} \Delta (t, \omega) \right\} = 1 \\[2mm]
3. & \qquad \left(\frac{\partial^2}{\partial t^2} + \omega^2 \right) \Delta (t, \omega) = 0.
\end{aligned}
\qquad (17.18)
$$

The Schwinger function encodes the essential details of the dynamics, leaving the initial information located explicitly in the boundary conditions represented by $\varphi (t_0)$ and $\dot{\varphi} (t_0)$.

It can be shown that the unequal-time Poisson brackets for the classical theory are given by

$$\{q (t), q (t_0)\}_{PB} = -m^{-1} \Delta (t - t_0, \omega). \qquad (17.19)$$

In the quantum theory this becomes the *unequal-time commutation relation* [Peierls, 1952]

$$\left[\hat{q} (t), \hat{q} (t_0) \right] = -im^{-1} \hbar \Delta (t - t_0, \omega) \qquad (17.20)$$

for the non-relativistic quantum operator $\hat{q}(t)$ of position in the so-called *Heisenberg picture*, discussed in Chapter 24.

Returning now to the full SR equation (17.12), the method of solution is to take the Fourier transform, viz., defining the transformed function $\tilde{\varphi}(t, \mathbf{p})$ by

$$\tilde{\varphi}(t, \mathbf{p}) \equiv \int_{\infty} d^3 \mathbf{x} e^{i \mathbf{p} \cdot \mathbf{x} / \hbar} \varphi (t, \mathbf{x}). \qquad (17.21)$$

Then equation (17.12) becomes

$$\frac{\partial^2}{\partial t^2} \tilde{\varphi} (t, \mathbf{p}) = -\frac{c^2 \left(\mathbf{p} \cdot \mathbf{p} + m^2 c^2 \right)}{\hbar^2} \tilde{\varphi} (t, \mathbf{p}), \qquad (17.22)$$

which now looks just like the harmonic oscillator equation (17.13) with the identification

$$\omega_p = +c\sqrt{\mathbf{p} \cdot \mathbf{p} + m^2 c^2}/\hbar, \quad > 0. \tag{17.23}$$

Hence we may write

$$\tilde{\varphi}(t,\mathbf{p}) = \Delta\left(t - t_0, \omega_p\right) \overset{\leftrightarrow}{\frac{\partial}{\partial t_0}} \tilde{\varphi}(t_0,\mathbf{p}). \tag{17.24}$$

Transforming back is a routine though tedious calculation, giving finally the solution

$$\varphi(x) = \int d^3\mathbf{y}\, \Delta(x-y) \overset{\leftrightarrow}{\frac{\partial}{\partial y_0}} \varphi(y), \tag{17.25}$$

where $x \equiv (x_0 \equiv ct, x_1, x_2)$, $y \equiv (y_0 \equiv ct_0, y_1, y_2, y_3)$, $t > t_0$, and $\Delta(x)$ is the Schwinger/Pauli–Jordan function given explicitly by

$$\Delta(x) = -\frac{1}{2\pi}\varepsilon(x_0)\,\delta(x^2) + \frac{mc}{4\pi\hbar\sqrt{x^2}}J_1\left(\frac{mc\sqrt{x^2}}{\hbar}\right)\theta(x^2)\varepsilon(x_0). \tag{17.26}$$

Here J_1 is one of the Bessel functions [Arfken, 1985], $\theta(x)$ is the Heaviside step function, $\varepsilon(x)$ is the sign function, $\delta(x)$ is the Dirac delta and $x^2 \equiv c^2 t^2 - x_1^2 - x_2^2 - x_3^2$.

The properties of the Heaviside step function, the sign function, and the Dirac delta fully encode the required causality conditions, which are that no effects can propagate faster than the speed of light. In Figure 17.3 we show what the above

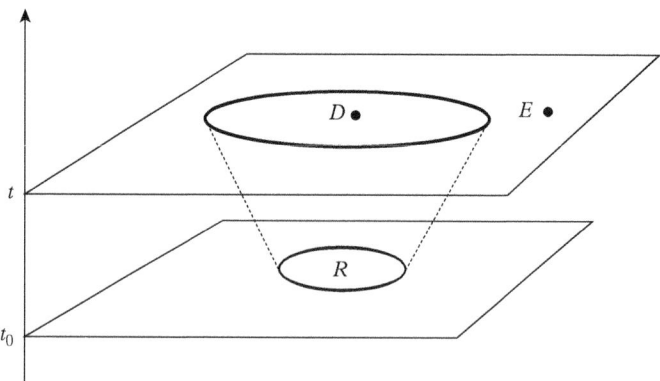

Fig. 17.3 *Propagation from region $\mathcal{R}$ in one spacelike hypersurface at time t_0 to a later spacelike hypersurface at time t.*

solution represents. At initial time t_0, Cauchy initial data is prepared by the observer over region $\mathcal{R}$ over some initial spacelike hypersurface[47] in $\mathcal{M}^4$ at laboratory time y_0. The field then propagates into the observer's future, subject to the light-cone limitation shown. Detector D can pick up a signal whilst detector E will not.

Finally, in the quantum theory, the quantum field operator $\hat{\varphi}(x)$ satisfies the unequal time commutation relation

$$[\hat{\varphi}(x), \hat{\varphi}(y)] = \frac{ic}{\hbar} \Delta(x - y), \qquad (17.27)$$

which encodes all the causality structure of relativistic quantum field theory. In particular, we see that the quantum field operators defined at relatively space-like points of spacetime commute, that is, the right-hand side of (17.27) is then zero, which is equivalent to saying that no physical effects carried by the field can propagate between those events. No quantum event in region $\mathcal{R}$ can influence a detector at E.

Fock–Kemmer front velocity and the memory field

The Russian physicist Vladimir Fock [1898–1974] was one of the pioneers of relativistic quantum mechanics, formulating the relativistic wave equation known as the Schrödinger–Fock–Klein–Gordon equation, commonly referred to as the Klein–Gordon equation (KGE). Relative to a standard inertial frame the free particle KGE is

$$\hbar^2 \ddot{\varphi} - c^2 \hbar^2 \nabla^2 \varphi + m^2 c^4 \varphi = 0, \qquad (17.28)$$

where m is the rest mass of the associated particles and $\ddot{\varphi} \equiv \dfrac{\partial^2}{\partial t^2} \varphi$.

Fock developed an analysis of the causality structure of the KGE that does not require explicit solutions. His method combines the field equation itself and the information available to the observer chorus in an inertial frame in an explicit way. Fock wrote an influential monograph on general relativity [Fock, 1964] and in Appendix E of that book he applied his analysis to the propagation of scalar field wavefront discontinuities in curved spacetimes. Fock's book was subsequently translated by Nicholas Kemmer [1911–98] and he in turn developed a novel and powerful notation for Fock's analysis. He discussed his notation in lectures on electrodynamics that were recorded by this author [Kemmer, 1971]. In this section we review Fock's approach using Kemmer's notation.

In classical field theory, several kinds of velocity concept need to be distinguished. Two of these are well known to quantum physicists: *phase velocity* and

---

[47] This hypersurface is in fact three-dimensional.

group velocity. The relationship between these two velocities came to the attention of physicists when de Broglie introduced the pilot wave concept in his doctoral thesis [de Broglie, 1924]. He suggested that a relativistic particle of non-zero rest mass would be accompanied by a 'pilot wave'. If v is the observed speed of the particle in the laboratory and w the speed of the wave, then de Broglie deduced that $vw = c^2$, where c is the speed of light. Here w is the speed with which a peak in the wave would move.

Three comments are relevant here:

1. The pilot wave concept is paradoxical: a wave is non-local (extends over a region of space) whereas a particle is localized, ideally at a point in space. de Broglie's hypothesis heralded Schrödinger wave mechanics, which was written down two years later and led to the development of modern quantum mechanics.

2. According to special relativity, massive particles travel at subluminal speeds, that is, $v < c$. Hence if de Broglie's relation $vw = c^2$ is correct then we must have $w > c$, that is, pure de Broglie waves travel faster than the speed of light. The speed w here is the phase speed, which is the speed of an individual crest of the wave as the wave moves through space.

   The result $w > c$ does not imply that de Broglie waves can carry observable signals faster than the speed of light. The interpretation of de Broglie waves is subtle and involves statistical *correlations*, which should not be confused with causal signal effects. de Broglie correlation speeds in excess of $10^4 c$ have been reported [Scarani *et al.*, 2000].

3. Particle velocity is often discussed in terms of wave packets built up of a weighted sum of pure de Broglie waves. When such a wave packet has a single peak centred on a specific frequency, the so-called group velocity v_g of the peak corresponds approximately to the classical particle velocity associated with the packet. It is important not to take this literally: it is possible to write down a single-electron wave packet that has two or more peaks. Indeed, this possibility is the basis of attempts by Maris and collaborators to 'split' the electron [Wei *et al.*, 2015].[48]

Although wave velocity and group velocity are the two velocities usually discussed in this context, causality requires the discussion of another velocity known as *front velocity*. A front is a surface of discontinuity moving from a source and spreading out into the wider environment. At a given chorus time, members of the chorus on one side of the front cannot report anything because the wave has not yet propagated to their position, whilst on the other side of the front, every observer has seen the wave.

---

[48] This exciting work has the potential to revolutionize the interpretation of what is meant by *observation*.

To illustrate Fock's method of analysing such fronts, consider a classical field $\varphi(t, \mathbf{x})$ propagating over some region of Minkowski spacetime. It is assumed to satisfy some partial differential equation $\vec{\mathcal{L}}\varphi = 0$, where the differential operator $\vec{\mathcal{L}}$ contains the partial derivatives of time and space, that is $\vec{\mathcal{L}} \equiv \vec{\mathcal{L}}(\partial_t, \nabla; t, \mathbf{x})$.

Solutions of this equation may be assumed to convey energy and momentum throughout spacetime. Imagine that these solutions can be followed, in the sense that a signal is sent at a certain time t_0 and chorus observers throughout spacetime record what happens. After the wavefront has passed through this system of observers, each chorus member has a record of the time and the position when the front reached them. Suppose then that all of this information is passed to a superobserver, who correlates it in the form of a function $f(\mathbf{x})$ of position, such that the value of the function is the time at which the front passed position $\mathbf{x}$:

$$f(\mathbf{x}) = t. \tag{17.29}$$

For example, suppose a spherical pulse of light is sent from the origin of coordinates at time $t = 0$. Then the front condition is $r/c = t$, where $r = |\mathbf{x}|$ is the distance from the spatial origin of coordinates and c is the speed of light.

Classical causality requires that no signal can be received by an observer at position $\mathbf{x}$ for any time $t < f(\mathbf{x})$. This also has to hold for any quantum theoretical prediction.

Given that the field $\varphi(t, \mathbf{x})$ propagates throughout spacetime, define the spatial field

$$[\varphi] \equiv \varphi(f(\mathbf{x}), \mathbf{x}), \tag{17.30}$$

where we call the left-hand side of (17.30) the *Fock–Kemmer bracket* of φ. Clearly this is a field that is distributed over space only and encodes a memory of when the front had reached a given point in space. We can think of it as a *memory front*. It is an extraordinarily subtle concept.

To investigate the information contained in the memory field $[\varphi]$, suppose we differentiate it with respect to the spatial coordinates $\mathbf{x}$. Then we find

$$\nabla[\varphi] = \nabla f[\dot{\varphi}] + [\nabla\varphi], \tag{17.31}$$

from which we deduce

$$\nabla[\dot{\varphi}] = \nabla f[\ddot{\varphi}] + [\nabla\dot{\varphi}]. \tag{17.32}$$

Using (17.31), we take the gradient again, giving

$$\begin{aligned}
\nabla^2[\varphi] &= \nabla^2 f[\dot{\varphi}] + (\nabla f)^2[\ddot{\varphi}] + \nabla f \cdot [\nabla\dot{\varphi}] + \nabla f \cdot [\nabla\dot{\varphi}] + [\nabla^2\varphi] \\
&= \nabla^2 f[\dot{\varphi}] + (\nabla f)^2[\ddot{\varphi}] + 2\nabla f \cdot [\nabla\dot{\varphi}] + [c^{-2}\ddot{\varphi} + m^2 c^2 \hbar^{-2}\varphi].
\end{aligned} \tag{17.33}$$

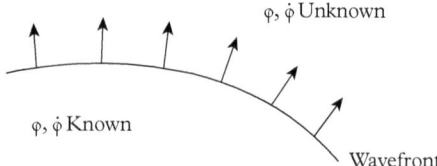

Fig. 17.4 *A Fock surface, where the propagating field φ has a discontinuity, divides physical space at a given time into a region where the field and its conjugate momentum is known, and a region that the field has not yet reached.*

Now from (17.32),

$$\nabla f \cdot \nabla[\dot\varphi] = (\nabla f)^2 [\ddot\varphi] + \nabla f \cdot [\nabla \dot\varphi], \qquad (17.34)$$

so

$$2\nabla f \cdot [\nabla \dot\varphi] = 2\nabla f \cdot \nabla[\dot\varphi] - 2(\nabla f)^2 [\ddot\varphi]. \qquad (17.35)$$

From this we find

$$\nabla^2[\varphi] - \nabla^2 f[\dot\varphi] - 2\nabla f \cdot \nabla[\dot\varphi] - m^2 c^2 \hbar^{-2}[\varphi] = (c^{-2} - (\nabla f)^2)[\ddot\varphi]. \qquad (17.36)$$

The point now is that all terms on the left-hand side of (17.36) represent Cauchy-type information available to the chorus, even on a surface of discontinuity. This sort of information includes $[\varphi]$, $[\dot\varphi]$, $\nabla[\varphi]$, $\nabla[\dot\varphi]$, $\nabla^2[\varphi]$, but not $[\ddot\varphi]$, $[\nabla\dot\varphi]$, and so on. On a surface of discontinuity (Figure 17.4), it should not be possible according to causality to use the physically available information to work out $[\ddot\varphi]$.

Hence we conclude that the right-hand side of (17.36) must vanish, and since we assume $[\ddot\varphi] \neq 0$, we deduce that such a surface satisfies the equation

$$(\nabla f)^2 = c^{-2}. \qquad (17.37)$$

This is essentially the equation of a *characteristic*.

In his book [Fock, 1964], Fock showed that the equivalent calculation for an arbitrary spacetime in general coordinates gives $\partial_\mu f \partial^\mu f = 0$.

The interpretation of these basic results is that not even de Broglie pilot wavefronts can propagate faster than the speed of light. In those discussions where a de Broglie wave accompanies a particle and has a phase velocity faster than the speed of light, the hidden assumption is that the *wavefront* has already passed by.

18

Generalized transformations

Introduction

Despite the continued success of special relativity (SR), there remains a core of
theorists who are dissatisfied with the standard Lorentz transformations discussed
in the previous chapter. This dissatisfaction is motivated in part by the loss of
absolute simultaneity in the transition from Galilean relativity to SR. This has
prompted some theorists to explore the properties of *generalized transformations*,
a class of linear transformations between inertial frames that includes Galilean
transformations (GTs) and Einstein–Lorentz transformations (ELTs) as special
cases. In this chapter we discuss generalized transformations in terms of a toy
model, a simplified version of space-time.

Constraints

Our toy model includes the essential properties of generalized transformations
and excludes properties that are regarded as superfluous. The model is based on
the following assumptions:

1. All aspects of gravitation such as curvature of spacetime are excluded and
   the concept of an infinitely extended inertial frame is valid.
2. There is a preferred inertial frame $\mathcal{F}_0$ in which the speed of light in all spatial
   directions is the same and denoted by c. This frame can be interpreted in
   several ways.
   (a) It is the frame of Newton's Absolute Space and Time.
   (b) It is the rest frame of the *Aether*, the medium that nineteenth-century
       physicists believed was responsible for the transmission of electromag-
       netic waves.
   (c) It is a local frame in which the dipole anisotropy of the cosmic micro-
       wave background radiation (CMBR) field is zero [WMAP, 2013].

Images of Time. First Edition. George Jaroszkiewicz.
© George Jaroszkiewicz 2016. Published in 2016 by Oxford University Press.

3. Each inertial frame consists of a chorus of observers each at rest relative to that frame and each carrying a standard clock with a rate common to that frame and determined by the physics of the *Universal Handbook*.[49]

4. Given two relatively moving inertial frames, there are no length contraction effects transverse to the direction of motion. Therefore, any discussion involving two such frames can be restricted to one time dimension and one spatial dimension in the direction of relative motion.

5. Given point 4, the standard time and space coordinates of an event A in frames $\mathcal{F}_0$ and $\mathcal{F}'$ are denoted

$$A \rightleftharpoons (t_A, x_A) \rightleftharpoons (t'_A, x'_A)' \qquad (18.1)$$

respectively, where the symbol $\rightleftharpoons$ means '*is represented by*'.

6. Ignoring transverse coordinates, a generalized coordinate transformation from $\mathcal{F}_0$ to $\mathcal{F}'$ can always be reduced to the form

$$t' = \alpha t - \frac{\theta}{c}\beta x, \qquad x' = \delta(x - vt), \qquad (18.2)$$

where α, β, θ, and δ are dimensionless parameters and $\beta \equiv v/c$. Without loss of generality we take $\alpha > 0$, $\delta > 0$. Here t and x are time and space coordinates of an event as observed by the $\mathcal{F}_0$ chorus whilst t' and x' are the time and space coordinates of the same event as observed by the $\mathcal{F}'$ chorus. These time and space coordinates are considered *physically observable* (i.e., measurable) coordinate values and not just mathematical coordinate patch artefacts. Therefore, we must take care in defining the various protocols associated with various measurements.

7. It is implicit in all such discussions that *after* all observations and experiments have been concluded, chorus observations are reported back to some primary observer: the information so collected forms the basis of a reconstruction in the mind of that primary observer of past events.

The constant v in (18.2) is by inspection the velocity of the origin of $\mathcal{F}'$ coordinates as seen in the $\mathcal{F}_0$ frame. This is readily seen by setting $x' = 0$ in (18.2).

At this stage, the generally accepted null result of the Michelson–Morley experiment has not been introduced. We shall come to it presently.

General formulae

In a $1 + 1$ dimensional spacetime, consider two arbitrary overlapping coordinate patches $\mathcal{P} \equiv \{(t, x)\}$ and $\mathcal{P}' \equiv \{(t', x')'\}$ related by some coordinate

---

[49] A hypothetical handbook of all standard protocols for setting up physical unit standards and performing experiments.

transformation $(t, x) \rightarrow (t', x')$ written in the form

$$t' = T(t, x), \quad x' = X(t, x), \tag{18.3}$$

where T and X are smooth functions of t and x. Suppose a particle is moving through a region of spacetime covered by both patches, such that relative to patch $\mathcal{P}$, the worldline of the particle is given by

$$x = x(t), \quad \dot{x} \equiv \frac{dx}{dt} = u(t), \quad \ddot{x} \equiv \frac{d^2 x}{dt^2} = a(t). \tag{18.4}$$

Our interest is in the instantaneous velocity $u' \equiv \dfrac{dx'}{dt'}$ and acceleration $a' \equiv \dfrac{d^2 x'}{dt'^2}$ of the particle at a given event on its worldline, as measured by the $\mathcal{P}'$ chorus. If dots denote differentiation with respect to time in $\mathcal{P}$, we readily find the relations

$$u' = \frac{\dot{X}}{\dot{T}}, \quad a' = \frac{\dot{T}\ddot{X} - \dot{X}\ddot{T}}{\dot{T}^3}, \tag{18.5}$$

assuming $\dot{T} \neq 0$.

Application to generalized transformations

Given the generalized transformation (18.2) we readily find

$$\dot{T} = \alpha - \frac{\theta \beta}{c} u, \quad \ddot{T} = -\frac{\theta \beta}{c} a,$$
$$\dot{X} = \delta(u - v), \quad \ddot{X} = \delta a, \tag{18.6}$$

giving for the velocity and acceleration transformations the rules

$$u' = \frac{\delta(u - v)}{\alpha - \dfrac{\theta \beta}{c} u}, \quad a' = \frac{\delta \left(\alpha - \theta \beta^2 \right) a}{\left(\alpha - \dfrac{\theta \beta}{c} u \right)^3}. \tag{18.7}$$

From these results we deduce the following:

1. Setting $u = 0$ gives v', the velocity of $\mathcal{F}_0$ as seen by observers in frame $\mathcal{F}'$, to be

$$v' = -\frac{\delta}{\alpha} v. \tag{18.8}$$

2. Setting $u = +c$ gives c_R, the *one-way speed of light* in the positive direction as measured in the $\mathcal{F}'$ frame:

$$c_R = \frac{\delta(c - v)}{\alpha - \theta \beta}. \tag{18.9}$$

3. Setting $u = -c \equiv -c_L$ gives c_L, the *one-way speed of light* in the negative direction as measured in the $\mathcal{F}'$ frame:

$$c_L = \frac{\delta(c + v)}{\alpha + \theta\beta}. \tag{18.10}$$

Relative scaling

The advent of SR heralded a fundamental reappraisal of the processes of observation. It became clear that *protocol*, or *specific procedure for observation*, plays a central role that must be clearly specified in any discussion of observation. Neglect of this accounts for many of the apparent paradoxes of SR: there are no paradoxes once protocol is clearly specified.

In SR, time dilation and Fitzgerald length contraction are well-known effects that are readily explained in terms of protocol. In this section we establish the generalized transformation protocols for length and time interval measurements as they are defined in the preferred frame $\mathcal{F}_0$. Figure 18.1 shows the spacetime coordinate axes of this frame. Events A and B are on the ends of a rod moving uniformly with velocity v relative to frame $\mathcal{F}_0$, that is, at rest in frame $\mathcal{F}'$. The worldlines of the two ends of the rod are shown running diagonally.

Event B is located on the origin of spatial coordinates in $\mathcal{F}'$, at one end of the moving rod. Hence we may write

$$B \rightleftharpoons (t_B, vt_B) \rightleftharpoons (t_B', 0)'. \tag{18.11}$$

Then the transformation (18.2) gives $t_B' = (\alpha - \theta\beta^2)t_B \equiv \sigma_t t_B$, where $\sigma_t \equiv \alpha - \theta\beta^2$ is the time dilation factor of moving clocks, as seen by the preferred frame observers.

Event A is observed at $\mathcal{F}_0$ time $t_A =$ zero, because the protocol for length measurement is that the length L_A of the rod as seen in frame $\mathcal{F}_0$ is defined as the

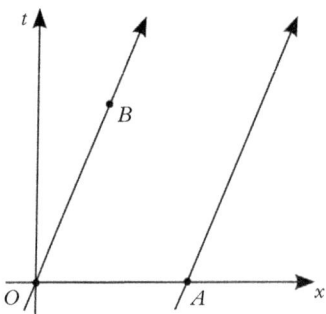

Fig. 18.1 *Moving ruler as seen in the preferred frame $\mathcal{F}_0$.*

coordinate distance from event O to event A, given that O and A are simultaneous in frame $\mathcal{F}_0$. Hence we may write

$$A \rightleftharpoons (0, L_A) \rightleftharpoons (t'_A, L'_A) \tag{18.12}$$

so applying (18.2) gives

$$t'_A = \alpha.0 - \frac{\theta\beta}{c}L_A, \quad L'_A = \delta(L_A - v.0), \tag{18.13}$$

or $L_A = \sigma_x L'_A$, where $\sigma_x \equiv \delta^{-1}$ is the length contraction factor of moving rods, as seen in frame $\mathcal{F}_0$.

Inverse relative scaling

An analogous protocol exists for interval measurements performed by the $\mathcal{F}'$ chorus on clocks and rods at rest in frame $\mathcal{F}_0$. Figure 18.2 shows the events involved.

Event C is along the worldline of a clock at rest in frame $\mathcal{F}_0$ situated at the origin of spatial coordinates, so we have

$$C \rightleftharpoons (t_C, 0) \rightleftharpoons (\alpha t_C, -\delta v t_C)'. \tag{18.14}$$

Hence we may write $t_C = \sigma'_t t'_C$, where $\sigma'_t \equiv \alpha^{-1}$ is the time dilation factor of moving clocks as seen by $\mathcal{F}'$ observers.

Event D is at one end of a ruler of length L_D at rest in frame $\mathcal{F}_0$ and observed at time $t'_D = $ zero in frame $\mathcal{F}'$. Hence we have

$$D \rightleftharpoons (t_D, L_D) \rightleftharpoons (0, L'_D)' \tag{18.15}$$

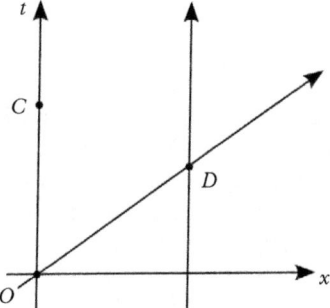

Fig. 18.2 *Scale changes seen in the moving frame $\mathcal{F}'$.*

where L'_D is the length of the ruler as defined by the $\mathcal{F}'$ chorus. The generalized transformation (18.2) gives

$$0 = \alpha t_D - \frac{\theta \beta}{c} L_D$$

$$L'_D = \delta L_D - v \delta t_D = \frac{\delta}{\alpha}(\alpha - \theta \beta^2) L_D = \sigma'_x L_D, \qquad (18.16)$$

so the relevant contraction factor σ'_x is given by $\sigma'_x = \frac{\delta}{\alpha}(\alpha - \theta \beta^2)$.

The Michelson–Morley constraint

There is a hidden assumption in the derivation of the Lorentz transformation discussed in Chapter 17 that need not be made. It was pointed out by Tangherlini that no experiment, including those of Michelson and Morley [Michelson & Morley, 1887], ever measures the *one-way* speed of light but in fact always a *round-trip average* [Tangherlini, 1958]. This means that the postulate that the speed of light is the same in all directions is not forced on us by the Michelson–Morley null result.

We may express this mathematically as follows. Suppose that the speed of light c in a given frame depends on position $\mathbf{x}$ and direction $\mathbf{n}$, where $\mathbf{n}$ is a unit vector in a given direction. Now imagine a pulse of light is sent around a fixed, closed circuit *in that frame*, with points along the circuit parametrized by a real parameter λ. The total length L_{Circuit} and the total time T_{Circuit} around the circuit are given by the closed line integrals

$$L_{\text{Circuit}} = \oint \left| \frac{d\mathbf{x}}{d\lambda} \right| d\lambda, \quad T_{\text{Circuit}} = \oint \frac{\left| \frac{d\mathbf{x}}{d\lambda} \right|}{c\left(\lambda, \frac{d\mathbf{x}}{d\lambda} \right)} d\lambda. \qquad (18.17)$$

Then the Michelson–Morley null result is summarized by the statement $L_{\text{Circuit}} = cT_{\text{Circuit}}$.

We may use this analysis to find a critical relationship between the parameters. Consider an inertial frame in $1 + 1$ dimensions for which $c_R \neq c_L$, but one for which the null result of the Michelson–Morley experiment holds. An experiment is conducted that consists of a pulse of light sent a distance L in the positive x direction, that reflects from a mirror there and returns to the origin, the source of the light. The total time T is given by

$$T = \frac{L}{c_R} + \frac{L}{c_L}. \qquad (18.18)$$

On the other hand, the observers know that the total distance travelled is $2L$. If they believe that light has constant speed c in all directions, then they will assert that

$$T = \frac{2L}{c}. \tag{18.19}$$

Hence for any frame for which $c_L \neq c_R$, the relation

$$\frac{1}{c_L} + \frac{1}{c_R} = \frac{2}{c} \tag{18.20}$$

must hold. Using (18.9) and (18.10) gives the condition $\delta = \gamma^2(\alpha - \theta\beta^2)$, where γ is the standard Lorentz factor $\gamma = 1/\sqrt{1 - \beta^2}$.

Although a path may be closed from the perspective of a given frame, from the perspective of the absolute frame it need not be closed. The question of what constitutes a closed path therefore does not have an absolute answer: the concept of 'returning to the same position' depends on what is mean by 'the same position'.

Some standard transformations

The above results hold for any choice of parameter α, θ, and δ. In Table 18.1. we show the results for three significant choices of parameters, that is, Galilean transformations (GTs), Einstein–Lorentz transformations (ELTs), and Tangherlini–Chang–Rembielinski transformations (TCRTs) [Tangherlini, 1961; Chang, 1979; Rembieliński, 1980]. The latter transformations have the merit of preserving simultaneity as in GTs whilst incorporating time dilation as in ELTs.

The Minkowski metric

In this section we show that the concept of Minkowski metric need not be restricted to SR alone: we can introduce it even into Galilean–Newtonian space-time. However, how this is interpreted is contextually dependent on the physics of the spacetime concerned.

We defined our generalized transformation (18.2) with the following simplification in mind. Because time and space have different physical dimensions, we rescale time coordinates to gave the dimensions of a length by defining $\bar{x} \equiv ct$. Then the generalized transformation (18.2) can be written in the following matrix terms:

$$\begin{bmatrix} \bar{x}' \\ x' \end{bmatrix} = \begin{bmatrix} \alpha & -\theta\beta \\ -\delta\beta & \delta \end{bmatrix} \begin{bmatrix} \bar{x} \\ x \end{bmatrix}, \tag{18.21}$$

or

$$x'^{\mu} = G^{\mu}{}_{\nu} x^{\nu}, \tag{18.22}$$

Table 18.1 *Comparison between Galilean transformations (GTs), Einstein–Lorentz transformations (ELTs), and Tangherlini–Chang–Rembiélinski transformations (TCRTs). MM denotes if the Michelson–Morley experimental outcome holds.*

	GT	ELT	TCRT
α	1	γ	γ^{-1}
θ	0	γ	0
δ	1	γ	γ
v'	$-v$	$-v$	$-\gamma^2 v$
u'	$u-v$	$\dfrac{u-v}{1-\dfrac{uv}{c^2}}$	$\gamma^2(u-v)$
d'	a	$\dfrac{a}{\gamma^3\left(1-\dfrac{uv}{c^2}\right)^3}$	$\gamma^3 a$
c_R	$c-v$	c	$\dfrac{c^2}{c+v}$
c_L	$c+v$	c	$\dfrac{c^2}{c-v}$
MM	no	yes	yes
σ_t	1	γ^{-1}	γ^{-1}
σ_x	1	γ^{-1}	γ^{-1}
σ'_t	1	γ^{-1}	γ
σ'_x	1	γ^{-1}	γ

where the transformation matrix is given by

$$[G^{\mu}{}_{\nu}] = \begin{bmatrix} \alpha & -\theta\beta \\ -\delta\beta & \delta \end{bmatrix}. \tag{18.23}$$

The inverse transformation matrix is therefore

$$[\overline{G}^{\mu}{}_{\nu}] = \frac{1}{\delta(\alpha-\theta\beta^2)}\begin{bmatrix} \delta & \theta\beta \\ \delta\beta & \alpha \end{bmatrix}, \tag{18.24}$$

which requires $\alpha \neq \theta\beta^2$.

Now consider the preferred frame $\mathcal{F}_0$. We define the line element

$$ds^2 = c^2 dt^2 - dx^2 = x^{\mu}\eta_{\mu\nu}dx^{\nu} = dx'^{\alpha}\overline{G}^{\mu}{}_{\alpha}\eta_{\mu\nu}\overline{G}^{\nu}{}_{\beta}dx'^{\beta} \tag{18.25}$$

where

$$[\eta_{\mu\nu}] \equiv \begin{bmatrix} 1 & 0 \\ 0 & -1 \end{bmatrix} \tag{18.26}$$

are the components of the Minkowski metric tensor in standard inertial coordinates.

Now consider a generalized transformation. The matrix of components in the new frame is given by

$$[\eta'] = \overline{G}^T \eta \overline{G}, \tag{18.27}$$

that is

$$[\eta'] = \frac{1}{\delta^2(\alpha - \theta\beta^2)^2} \begin{bmatrix} \delta^2\gamma^{-2} & \beta\delta(\theta - \alpha) \\ \beta\delta(\theta - \alpha) & -\alpha^2 + \theta^2\beta^2 \end{bmatrix} \tag{18.28}$$

We now discuss the line element as seen in the three types of frame discussed in Table 18.1.

Galilean transformations

For GTs we have $\alpha = \delta = 1, \theta = 0$, which gives the line element

$$ds^2 = c^2 dt'^2 - (dx' + \beta c dt')^2. \tag{18.29}$$

The value of the line element is that it gives a condition for light (null) geodesics, that is, we set $ds = 0$. Then we recover the Galilean relativity rule for the speed of light, that is, $c_R = c - v$, $c_L = c + v$, as expected.

Einstein–Lorentz transformations

For ELTs we have $\alpha = \theta = \delta = \gamma$, which gives the standard line element

$$ds^2 = c^2 dt'^2 - (dx')^2, \tag{18.30}$$

so setting $ds = 0$ leads to $c_R = c_L = c$ as expected.

Tangherlini–Chang–Rembielinski transformations

For TCRTs we have $\alpha = \gamma^{-1}, \theta = 0, \delta = \gamma$, which gives the line element

$$ds^2 = (cdt' - \beta dx')^2 - dx'^2. \tag{18.31}$$

Setting $ds = 0$ gives

$$c_R = \frac{c^2}{c^2 + v}, \quad c_L = \frac{c^2}{c^2 - v}, \tag{18.32}$$

as stated in Table 18.1.

The splitting of causality

Special relativity in the form developed by Einstein makes a specific assumption about the one-way speed of light that impacts on the transformation rules between inertial frames. In this section we discuss a simple thought experiment that could in principle provide a test of Einstein–Lorentz synchronization (ELS). If ELS synchronization is upheld, then classical causality should appear to be split in an observable way.

The experiment involves signalling between two inertial frames in relative motion. Whenever signalling apparatus and detecting apparatus are not at rest in a common inertial frame of reference (or related by a simple translation or spatial rotation), we shall refer to such an experiment as an *interframe experiment*. Examples of interframe experiments are those involving Doppler shifts, such as in observational cosmology.

Consider an interframe experiment conducted over a finite interval of time, such that signals sent from apparatus QP at rest in frame $\mathcal{F}$ at time $t = 0$ are observed by apparatus $Q'P'$ at rest in frame $\mathcal{F}'$ at time T', as measured in that frame. Figure 18.3 shows the essential details.

In Figure 18.3, event O is the common origin of spacetime coordinates, events P and Q are simultaneous in $\mathcal{F}$ at time $t = 0$ whilst events P' and Q' are simultaneous in $\mathcal{F}'$ at time $t' = T' > 0$. We shall use the convention that an

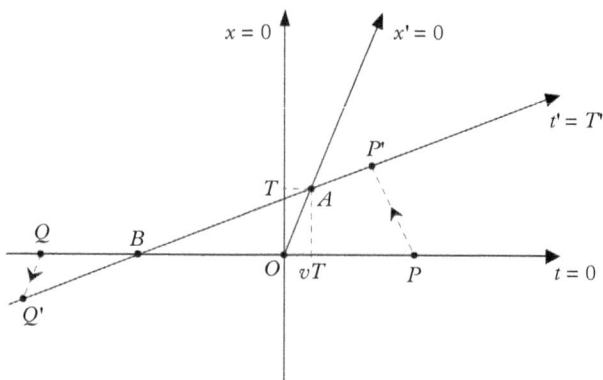

Fig. 18.3 *Event B is a quantum (or information) horizon.*

event P has coordinates (t_P, x_p) in $\mathcal{F}$, coordinates $[t'_p, x'_p]$ in $\mathcal{F}'$, and we write $P \sim (t_P, x_P) \sim [t'_p, x'_p]$. From Figure 18.3, it will be seen that a critical feature is event B, where the lines of simultaneity $t = 0$ and $t' = T'$ intersect. Assuming the velocity component v is positive, then for B we have

$$B \sim (0, -\frac{c^2 T'}{\gamma(v)v}) \sim \left[T', -\frac{c^2 T'}{v}\right]. \tag{18.33}$$

Event B is the focus of attention in this experiment. A novel and interesting interpretation of the significance of event B can be found from basic quantum mechanics. First, we recall that in de Broglie wave mechanics, the speed w of a pilot wave associated with any material physical particle moving with subluminal speed v satisfies the relation $vw = c^2$ [de Broglie, 1924]. This suggests that such a pilot wave cannot be used to convey physical signals, because it travels at super-luminal speed, given $v < c$. According to these ideas, event B may be regarded by observers in $\mathcal{F}'$ as the crest at time T' of such a pilot wave associated with a particle at rest in frame $\mathcal{F}$, if it were sent out from O in the same direction as $\mathcal{F}$ appears to move. Note that this interpretation of event B should be regarded as no more than a mathematical curiosity, because any genuine de Broglie wave would not be localized at a single point. Moreover, the significance of event O as the common origin of coordinates is an artefact to do with the choice of coordinates. Nevertheless, we shall argue below that B does have something critical to do with quantum processes.

Several circumstances conspire to mask events such as B in conventional phys-ics: (i) the speed of light c is large on ordinary laboratory scales; (ii) the relative speed $|v|$ is usually small compared to c, (iii) signal detection is limited to a lo-calized region, such as around event P'; and the time T' of observation of a signal at P' is usually relatively large. In consequence, event B will normally be beyond the limits of observation in typical experiments. In the case of the experiment un-der consideration, however, the architecture is different: non-localized detection is arranged in such a way that B is within the space-time region involved in the experiment.

For this experiment, we imagine that a quantum state has been prepared by ap-paratus $\mathcal{A}_{QP}$, at rest in frame $\mathcal{F}$, and a contingent quantum outcome subsequently detected by apparatus $\mathcal{A}'_{QP'}$, at rest in frame $\mathcal{F}'$.

The critical word here is '*subsequently*'. Quantum physics, as it is performed in real laboratories, can discuss only the possibility of information travelling forwards in time. Both signal emitter and signal detector in any quantum experiment must agree that the former acts before the latter. Otherwise, the physical significance of the Born probability rule would be completely undermined. In quantum theory and in the real world, we cannot know the outcome of an experiment before it is performed. We shall call the requirement that P is earlier than P' in both frames of reference *quantum causality*.

From Figure 18.3, it is clear that there is no problem with quantum causality as far as events P and P' are concerned. But consider events Q and Q' on the other side of B. If quantum causality is valid, then signals prepared at Q cannot be received by Q'. In essence, event B acts a barrier to quantum causality, and on this account we shall refer to B as a *quantum horizon*.

Ordinarily, such horizons are ignored in conventional physics because, under most circumstances, B appears to be very far from events such as P and P'. For example, in experiments looking at the transmission of quantum information, speeds in excess of 10^5 c have been reported [Scarani *et al.*, 2000]. In practice, high-energy particle theory conventionally takes the scattering limit $T' \rightarrow \infty$, $v = 0$ in the calculation of scattering matrix elements. Finite-time processes and inter-frame experiments of the sort discussed by us here are generally avoided, presumably because it is assumed that there is no significant novel physics involved. A consequence of this assumption is that this scattering limit simplifies the calculations, because all quantum horizons are at spatial infinity.

We now consider the implications of the relativity principle and ask the following question: if according to the relativity principle frames $\mathcal{F}$ and $\mathcal{F}'$ are 'just as good as each other', why does the quantum horizon B appear to distinguish between the two?

A little thought soon resolves the question. If the relativity principle is valid, then there must be a symmetry between the two frames. There is no doubt that a quantum signal can be prepared at P and received at P', if P' is in or on the forwards lightcone with vertex P. Quantum causality rules out the transmission of a quantum signal from P' to P, and the transmission of a signal from Q to Q'. But nothing currently known in physics forbids the possibility of a physical signal being sent from Q' to Q, if Q is in the forwards lightcone with vertex Q'. Indeed, symmetry demands such a possibility. This is the essence of the split causality experiment proposed here.

Based on the above considerations, we propose the following experimental test of the principle of special relativity. It will undoubtedly be difficult to perform, but would test the principle of relativity in a spectacular and convincing way.

We envisage the use of four spacecraft P, Q, P', and Q', sufficiently far from gravitating bodies to justify the use of the SR transformation rule. P and Q are in the same rest frame $\mathcal{F}$ and situated at some distance from each other. By prior signalling arrangement, clocks on P and Q craft have been synchronized. Likewise, P' and Q' are in their own rest frame $\mathcal{F}'$ and all their clocks have been synchronized.

With reference to Figure 18.3, spacetime homogeneity means that we may always transfer the origin of spacetime coordinates in both frames $\mathcal{F}, \mathcal{F}'$ to the quantum horizon B. This means that the hyperplanes of simultaneity involved in the experiment are now at times $t = 0$ in $\mathcal{F}$ and $t' = 0$ in $\mathcal{F}'$, as shown in Figure 18.4.

The experiment consists of P sending a brief light pulse signal towards P' at time $t = 0$, whilst simultaneously in $\mathcal{F}$, Q opens a detector in order to receive

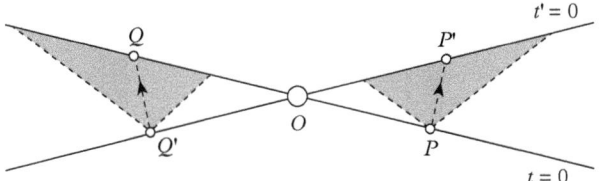

Fig. 18.4 *O is a quantum horizon. Shaded regions are forwards lightcones. Arrowed signals shown with subluminal transmission speeds, which do not alter the overall conclusions.*

light from Q' for a similar brief period. In addition, the same protocol is carried out in frame $\mathcal{F}'$ at time $t' = 0$: Q' sends a brief light pulse towards Q whilst simultaneously in $\mathcal{F}'$, P' opens a detector to receive a signal from P. The whole experiment is illustrated in Figure 18.4.

After the signals have been sent and received, observers from all spacecraft can meet at leisure and compare results. If it turns out that P sent a signal at the same time $t = 0$ that Q received a signal, and that P' received a signal at the same time $t' = 0$ that Q' sent a signal, then ELS would be upheld. If not, then Tangherlini synchronization with all its implications would be preferred. If ELS holds, then quantum causality would have been respected but classical causality would appear to be split in a remarkable and counterintuitive fashion. On the other hand, if no such result was ever detected despite repeated attempts, this would rule out ELS, with implications for physics.

Empirical evidence for a preferred frame

Ever since Maxwell's theory gave a theoretical basis for the speed for light, theorists have been fixated on the question of an absolute frame of reference. Is there such a thing or not? This preoccupied the minds of Larmor, Lorentz, Tangherlini, and many others. In 1905, Einstein bypassed this issue by asserting the principle of special relativity: that the laws of physics apart from gravitation are the same in form in all standard inertial frames. The outstanding empirical success of the consequences of this principle led to a general loss of interest in any conjectured Aether frame.

However, interest in a preferred frame has not waned completely, for several reasons. First, it is not true that we cannot identify any special inertial frame: if we ignore gravitation, any localized laboratory defines a reasonable approximation to a 'preferred' frame, that is, itself. Of course, such a frame is not absolute. Second, the discovery of the CMBR field and the dipole anisotropy gives a specific empirical method of defining a local absolute inertial frame (modulo rotations). This should be possible everywhere in the universe, by standard cosmological principles, so in effect there *is* a preferred frame of reference permeating the cosmos. Third, it is an imperative in science to test every principle, every conjecture to the limit and to keep doing so. We are not dealing with absolute truths here but empirical, contextual truths.

In this regard, experimentalists continue to test in different ways the standard SR paradigm that there is no preferred frame of reference over and above our local matter frame of reference and the local CMBR field: these latter frames are usually regarded as secondary frames of reference resulting from some initial unspecified conditions located in the past.[50] Experimentalists such as Hughes [Hughes *et al.*, 1960], Drever [Drever, 1961], and many more recently [Allmendinger *et al.*, 2014] have tested for the isotropy of mass and space and deviations from the predictions of SR that do not focus on light propagation in the manner of the Michelson–Morley experiments. The results to date are all consistent with the null results of those historic experiments: the local behaviour of the laws of non-gravitational physics is consistent with there being no preferred frame. But of course, that is also consistent with what Fitzgerald, Larmor, and Lorentz were trying to prove, using a preferred (Aether) frame.

---

[50] In this respect, the CMBR field has a dual nature: it is a universal field, an essential aspect of the expansion of the universe, and also it seems accidental.

19

General relativistic time

Space-time versus spacetime

Over the millennia, humanity's view of time has changed in a series of steps, each stimulated by a paradigm shift, or way of looking at the universe. Aristotelian space-time $\mathcal{A}^4$ was followed by Galilean–Newtonian space-time $\mathcal{N}^4$ and then by Minkowski spacetime $\mathcal{M}^4$. We discussed these in earlier chapters. In this chapter we discuss the step after $\mathcal{M}^4$, which we will refer to as general relativistic (GR) spacetime, denoted by $\mathcal{G}^4$.

An important difference between $\mathcal{G}^4$ and the other models is that they are specific in their mathematical structures, whereas $\mathcal{G}^4$ is really a class of spacetimes, each with its characteristic metric or distance structure. There is, unfortunately, no unique GR spacetime that we can say for sure models the universe. There are two reasons for this. First, each GR spacetime is dynamically coupled to the local distribution of energy and mass via Einstein's field equations [Einstein, 1915a], so each different distribution of energy and mass leads to a different $\mathcal{G}^4$ spacetime. Second, Einstein's field equations, being local, do not say anything about the topology, or large-scale structure, of spacetime: we don't know whether it is spherical, flat, or hyperbolic at very large distances, or even like a vast four-dimensional doughnut with many holes. Recent astrophysical data suggests that at the largest observable spatial distances it is flat [WMAP, 2013], but we can have no idea what it is like beyond those distances.

Another, more disturbing, point is that whilst it could be said that the spacetime we are in now is well-defined locally, that is no more than a classical approximation to what is surely a much different reality [Donoghue, 1994], in the same way that local temperature is a simple indicator of thermal equilibrium in our unimaginably complex local environment. It surely would be hubris of the worse kind to image that we know what 'reality' is really like. In Chapter 27 we discuss Snyder's quantized spacetime theory, a theory that suggests that a geometrical model is not the last word on the nature of time and space.

To understand $\mathcal{G}^4$, it helps to understand $\mathcal{M}^4$. There is good evidence [Petkov, 2012] that Hermann Minkowski [1864–1909] was thinking about a unified, *absolute* four-dimensional *spacetime* structure well before Einstein came around to

Images of Time. First Edition. George Jaroszkiewicz.
© George Jaroszkiewicz 2016. Published in 2016 by Oxford University Press.

a geometric perspective. For some reason, Arnold Sommerfeld in 1907 '. . . *was unable to resist rewriting Minkowski's judgement of Einstein's formulation of the principle of relativity.*'[51] Sommerfeld '. . . *also suppressed Minkowski's conclusion, where Einstein was portrayed as the clarifier, but by no means as the principal expositor, of the principle of relativity.*' [Petkov, 2012]. The evidence is that Minkowski alone understood the significance of the spacetime (no hyphen) perspective when he spoke at a scientific meeting in 1908 and made his ideas on spacetime public. At that talk, he famously declared:

> The views of space and time which I wish to lay before you have sprung from the soil of experimental physics, and therein lies their strength. They are radical. Henceforth space by itself, and time by itself, are doomed to fade away into mere shadows, and only a kind of union of the two will preserve an independent reality.
>
> [Minkowski, 1908]

He went on to propose that, not only should time and space be thought of as a single four-dimensional manifold, but that it should have a distance structure now known as the Minkowski metric. If Δt, Δx, Δy, and Δz are the coordinate intervals between any two events in Minkowski spacetime, relative to coordinates in a given standard inertial frame, then Minkowski's distance Δs between those events is given by a '*line element*' Δs written in the form

$$\Delta s^2 = c^2 \Delta t^2 - \Delta x^2 - \Delta y^2 - \Delta z^2, \qquad (19.1)$$

where c is the speed of light. After 1908, the four-dimensional spacetime continuum of SR with the above line element was referred to by relativists as *Minkowski spacetime* (with no hyphen) and denoted $\mathcal{M}^4$.

A refinement of this concept is to consider infinitesimal coordinate displacements $dx^\mu = (dx^0, dx^1, dx^2, dx^3)$, where $x^0 \equiv ct$, $x^1 \equiv x$, $x^2 \equiv y$, and $x^3 \equiv z$.[52] The choice of *upper index* is conventional. Then the $\mathcal{M}^4$ line element is given by (17.10).

Lorentzian signature metrics

Minkowski's line element (17.10) does not correspond to a metric according to the standard mathematical definition of a metric space: it does not satisfy the non-negativity axiom given in the Appendix. The relative sign change between the $c^2 dt^2$ terms and the three other terms makes this distance rule fundamentally different to that for a four-dimensional Euclidean space: it is possible to find events

---

[51] The principle that the laws of physics apart from gravitation take the same form in every inertial frame.

[52] Psychology is important. We choose to denote time by x^0, not x^4, to remind us that time is *not* equivalent to space. The factor of c (speed of light) gives x^0 the physical dimensions of a length, to match the other coordinates.

P and Q in Minkowski spacetime such that $ds^2 < 0$, something that is inadmissible for a genuine metric space. For this reason mathematicians often refer to Minkowski's line element as defining a *pseudo-metric*. The lightcone structure generated by this pseudometric is discussed in Chapter 17. Lightcones play a crucial role in all GR spacetimes.

A useful way of discussing Minkowski's line element is to write it out in matrix terms, that is,

$$[ds^2] = [dx^0, dx^1, dx^2, dx^3] \begin{bmatrix} 1 & 0 & 0 & 0 \\ 0 & -1 & 0 & 0 \\ 0 & 0 & -1 & 0 \\ 0 & 0 & 0 & -1 \end{bmatrix} \begin{bmatrix} dx^0 \\ dx^1 \\ dx^2 \\ dx^3 \end{bmatrix}. \tag{19.2}$$

Now all the essential information about the sign changes is contained in the central 4×4 matrix, referred to as the *matrix of components of the metric tensor relative to the given frame*. A more compact notation is to write

$$ds^2 = dx^\mu \eta_{\mu\nu} dx^\nu, \tag{19.3}$$

where we use the summation convention.[53] Here the 16 values $\eta_{\mu\nu}$ are the components of the metric tensor relative to the coordinate frame being used.

By inspection, the diagonal components of the metric tensor in (19.2) are $(+1,-1,-1,-1)$, in contrast to $(+1,+1,+1,+1)$ we would have for four-dimensional Euclidean space.[54]

Minkowski spacetime $\mathcal{M}^4$ can be regarded in two different ways mathematically. On the one hand it can be modelled as an *affine vector space* with a Lorentzian metric, discussed in the Appendix. In that case any inertial frame gives a global coordinate covering of $\mathcal{M}^4$. On the other hand, $\mathcal{M}^4$ can also be regarded as a particularly bland example of a *four-dimensional pseudo-Riemannian manifold*, discussed next.

Pseudo-Riemannian manifolds

Einstein went far beyond SR when he developed his general relativity (GR). But, as with the development of SR, he was not alone. He was not the only one at that time (c. 1910–15) who was playing with curved spacetime concepts: the Finnish theorist Nordström had published ideas in it by 1913 [Nordström, 1913]. Ultimately, empirical evidence and sound thought experiment analysis pointed towards the work of Einstein: by 1915, he had published what can be considered the definite version of GR that we use to this day [Einstein, 1915a].

---

[53] In this convention, a repeated index in an expression means to sum over the appropriate range of the index.
[54] Only *relative* sign changes are significant here.

The basic structure in GR is the *spacetime manifold*, which is a four-dimensional manifold over which a Lorentz-signature metric tensor is defined. Manifolds and Lorentz-signature metric tensors are discussed in the Appendix.

Manifolds come in many forms. Some such as $\mathbb{R}^n$ can be assigned a global cover, a single set of coordinates called a patch that can assign numerical coordinates to any point of the manifold. Other manifolds such as S^2, the two-sphere, cannot be covered by a single coordinate patch without some point in the manifold being problematical. A well-known example of this is the failure of Mercator's projection to provide a reasonable projection of the world's surface onto a single page of an atlas. In that projection, the points corresponding to the North and South poles are mapped onto lines at the top and bottom of the page respectively.

Borrowing notation from Riemannian geometry, the line element in a coordinate patch is given by

$$ds^2 = dx^\mu g_{\mu\nu}(x)dx^\nu, \tag{19.4}$$

where the $\{dx^\mu\}$ are four infinitesimal coordinate differences and the $g_{\mu\nu}(x)$ are the possibly spacetime dependent components of the metric tensor. In (19.4), the repeated indices μ, ν tell us to sum over each from 0 to 3. This summation convention makes the formalism more attractive and is worth getting used to.

To understand how GR deals with space and time, we need to develop some more geometrical concepts. It is worth reviewing the concept of an affine space, discussed in the Appendix. There, a set $\mathcal{A}$ is associated with a *single vector space* V that relates any pairs of points in $\mathcal{A}$. When we move to spacetime manifolds, we have to drop that single vector space concept and attach a separate four-dimensional vector space at each and every point in a spacetime manifold. Given an event P in a spacetime, the associated vector space T_P is called the *tangent space at P*.

This plethora of tangent spaces makes the structure of GR spacetime much more complicated than that of a mere affine vector space. There are as many tangent spaces as there are points in a manifold. We pointed out above that $\mathcal{M}^4$ can be viewed either as an affine vector space or as a manifold. Given a choice, the affine vector space model for $\mathcal{M}^4$ is the one to use in SR, because inertial frames are affine coordinate frames for $\mathcal{M}^4$ and so cover the whole of that spacetime. We have no such luck in general for arbitrary manifolds. This is one of the factors that makes GR intrinsically more complicated than SR.

We have already encountered something analogous to the tangent space concept in our discussion of Galilean–Newtonian spacetime $\mathcal{N}^4$ in Chapter 14. There Absolute Time is modelled by $\mathbb{R}$, which is the simplest manifold, and a copy of three-dimensional Euclidean space is attached to each point. A difference is that the tangent space T_P at each point in a manifold is a vector space with the same dimensions of the manifold, in our case four, whereas in the case of $\mathcal{N}^4$, the attached spaces are three-dimensional affine spaces. Nevertheless, the issue of *parallel transport* is much the same.

Parallel transport concerns the relationship of direction in different copies of the same vector space. We saw in Chapter 14 that Newton's first law of motion can be thought of as a way of defining (or identifying) 'straight' lines in inertial frames. But with tangent spaces we have no such law. Given two events P and Q in a GR spacetime, how are directions in their respective tangent spaces related?

Einstein's theory of general relativity showed how to do that, via the metric tensor $g_{\nu\mu}(x)$ that sits at the heart of the line element (19.4). Knowing the metric tensor, we can calculate a set of quantities known as the *metric connection*. The details are complicated but can be found in any book on GR. Suffice it to say that the metric tensor is like the key to the structure of space and time: given the line element, we can immediately read off the metric tensor components. With the metric tensor we can then calculate the metric connection. This connection tells us how to relate direction in one tangent space to direction in another.

There is more, much more. Knowing the metric connection, we can work out a fundamental quantity called the *Riemann curvature tensor*. This is a formidable beast. Suffice it to say that it encodes all the geometrical concepts familiar to students of spherical geometry (geometry on a sphere). Knowing the Riemann curvature tensor, we can work out how space and time intervals are distorted or curved over the spacetime manifold.

We do not need to work out the Riemann curvature tensor, or even the metric connections, to unravel the important geometrical structure of a GR spacetime. We can do a great deal of investigation using the line element alone, looking at lightcones. Recall that in Chapter 17, setting $ds^2 = 0$ gave us the possible trajectories (worldlines) of particles moving at the speed of light, that is, the lightcones in $\mathcal{M}^4$. We can do the same in GR spacetimes. We shall use this approach in the next section to investigate the spacetime geometry of the famous Schwarzschild line element that is used to discuss black hole physics.

The Schwarzschild metric

We mentioned above the work of Minkowski and Nordström. In a real sense they were competitors of Einstein. Other equally brilliant theorists lurked in the background. One such was Karl Schwarzschild [1873–1916]. Within a month of Einstein publishing his finalized theory of GR in 1915, Schwarzschild had found a solution to Einstein's field equations for a point mass, a solution that has become known as a *black hole* [Schwarzschild, 1916]. Droste independently also found a solution and worked on the anomalous precession of the perihelion of Mercury that Newtonian gravitation could not explain [Droste, 1917]. In a strange time-reversal of history, Einstein had worked out the same precession problem *before* Schwarzschild had found his line element [Einstein, 1915b], a veritable triumph of intuition over precision.

The Schwarzschild solution continues to fascinate theorists and experimentalists alike for a very good reason: it looks as if at the centre of most large galaxies lurks a supermassive black hole, sucking in energy and matter in the form of gas and, ultimately, stars. There is now evidence that at the centre of our galaxy there is such a monster, with a mass just over four million times the mass of the Sun [Ghez *et al.*, 2008].

To discuss the Schwarzschild line element, we need first to describe our chosen coordinate patch, that is, our four coordinates. One of these, denoted t, appears at first sight to play the role of time. The other three are the familiar spherical polar coordinates r, θ, and ϕ centred on the point mass concerned.[55] In such a coordinate patch, the Schwarzschild metric takes the form [Rindler, 1969]

$$ds^2 = \left(1 - \frac{r_s}{r}\right)c^2\,dt^2 - \frac{1}{\left(1 - \frac{r_s}{r}\right)}dr^2 - r^2\,d\theta^2 - r^2\sin^2\theta\,d\phi^2, \qquad (19.5)$$

where r_s is known as the *Schwarzschild radius*. Important points to note are:

1. At large finite values of r, this line element represents a curvature of space-time that, when coupled to Einstein's principle of equivalence,[56] has the effective, physical space-time interpretation that there is an inverse square Newtonian gravitational force field due to a particle of Newtonian mass $M = r_s c^2 / 2G$ at the origin $r = 0$ of coordinates.

2. In the limit $r \to \infty$, the metric reverts to the flat space (Minkowski) metric.

3. There seems to be a singularity at the *Schwarzschild radius* r_s as well as at $r = 0$.

4. There is a strange interchange between the timelike/spacelike characteristics of the t and r coordinates for $r < r_s$. We shall explore this phenomenon presently.

Timelike, lightlike, and spacelike coordinates

To pin down the nature of the coordinates in a given patch in a GR spacetime, consider an infinitesimal displacement such that $dx^\mu = 0$ for all coordinates except for a given one, x^α. Then for such a displacement, (19.4) gives $ds^2_\alpha = g_{\alpha\alpha}\,(dx^\alpha)^2$, with no sum over α.

Because of the signature of GR metrics, ds^2_α can be positive, zero, or negative. We say x^α is a *timelike* coordinate if ds^2_α is positive, x^α is a *lightlike* coordinate if ds^2_α is zero, and x^α is a *spacelike* coordinate if ds^2_α is negative. Finding suitable coordinate patches with various timelike, lightlike, or spacelike coordinates is a powerful tool in the analysis of Schwarzschild's line element, and other line elements such as Gödel's, discussed in Chapter 20, that admits time travel.

---

[55] The supermassive black hole at the centre of our galaxy is not a point: Schwarzschild's line element is an idealization.

[56] This principle asserts that the laws of physics in a freely falling, non-rotating localized laboratory are those of special relativity.

Looking at the Schwarzschild line element (19.5), we see that the coordinate t is timelike for $r > r_s$ and spacelike for $r < r_s$, whereas it is the other way around for the radial coordinate r. This shows that in GR, coordinates do not necessarily represent what we may think they are physically. Merely labelling a coordinate with a t does not make it necessarily a time.

Timelike, lightlike, and spacelike vector fields

GR spacetimes are manifolds and these have tangent vector spaces. A *vector field* **v** is a rule that picks out one particular vector $\mathbf{v}_P$ in the tangent space T_P at every point P in the manifold. Vector fields are important in GR in many ways. A particularly important one is that they can be used to describe *frame fields*. These are local laboratory frames of reference needed to give us a physical handle on time. The technical formalism can be formidable depending on the details of the metric, but suffice it to say that when in doubt, just think of what a human endophysical observer would look for if they were in a localized[57] laboratory: they would want clocks to tick and rulers to measure spatial intervals in their laboratory. By Wheeler's participatory principle, GR does not make empirical sense if such a perspective cannot be found at a given point in a GR spacetime.

Given a vector field **v**, we can pick a vector $\mathbf{v}_P$ in the tangent space T_P at P and use the metric tensor to classify the vector field at that point P. Essentially, the metric tensor gives us a definition of 'length squared', or dot product, of the vector, denoted $\mathbf{g}(\mathbf{v}, \mathbf{v})$. We say $\mathbf{v}_P$ is *timelike* at P if $\mathbf{g}(\mathbf{v}, \mathbf{v})_P > 0$,[58] *null* or *lightlike* at P if $\mathbf{g}(\mathbf{v}, \mathbf{v})_P = 0$, or *spacelike* at P if $\mathbf{g}(\mathbf{v}, \mathbf{v})_P < 0$. For want of a better term, we shall refer to the sign of $\mathbf{g}(\mathbf{v}, \mathbf{v})$ as the *signature* of **v** *relative to* **g**, or just the *signature* if **g** is understood.

Note that the signature of a vector field at a point in a manifold is *intrinsic*, that is, does not depend on choice of coordinate patch. Note also that a vector field may change its signature from point to point if the metric is of Lorentzian signature.

A *frame field* $\{\mathbf{e}_\mu : \mu = 0, 1, 2, 3\}$ is a set of four vector fields that can be thought of giving a local set of inertial frame axes. By convention, $g(\mathbf{e}_0, \mathbf{e}_0) = +1$, $g(\mathbf{e}_1, \mathbf{e}_1) = g(\mathbf{e}_2, \mathbf{e}_2) = g(\mathbf{e}_3, \mathbf{e}_3) = -1$. The vector $\mathbf{e}_0$ is like a local time coordinate axis and the other three are like standard Euclidean space axes. Note that (i) we could change these axes locally by a local Lorentz transformation without any change in physical content, and (ii) if the GR spacetime is curved, then the frame field will twist and turn accordingly as we move from point to point. The vector field $\mathbf{e}_0$ part of a frame field gives us an idea of the 'flow' of time in a GR spacetime. Part of the problem with GR is that some GR spacetimes can have frame fields such that the 'flow' of $\mathbf{e}_0$ loops back upon itself, which is really time travel. This is discussed in Chapter 20.

---

[57] That is, sufficiently small-scaled compared to the wider universe outside their laboratory.
[58] Here the subscript P means 'evaluated at P'.

Gravitational time dilation

We may use the Schwarzschild metric (19.5) to determine how clocks are affected by gravitation. Consider two standard clocks at rest[59] in a spacetime with this line element, at radial coordinates r_1 and r_2 respectively, such that each radius is much greater than the Schwarzschild radius r_s. Now consider the clock at r_1 moving along a worldline of coordinate interval $\Delta t = T$, the changes in the other coordinates being zero. Note that such a worldline is constrained, that is, forced. The natural, 'freefall' movement of an unconstrained object would be to follow what is known as a *geodesic* towards the origin of the spatial coordinates. That is not what is happening here, which is just like us sitting on the surface of the Earth, restricted by the ground from falling into the centre of the planet.

Given this, the change Δs_1 in s on the left-hand side of the line element is given by $\Delta s_1 = \sqrt{1 - r_s/r_1 c}\, T$. Now by the clock hypothesis, the interpretation of Δs for a timelike worldline is that it is equal to $c\Delta\tau$, where $\Delta\tau$ is the proper time, or elapsed time of a standard clock on that worldline. Hence if $\Delta\tau_1$ and $\Delta\tau_2$ are the respective time intervals registered by the clocks for the same coordinate interval T, we deduce

$$\Delta\tau_2 - \Delta\tau_1 = \frac{GMT}{c}\left[\sqrt{1 - \frac{r_s}{r_2}} - \sqrt{1 - \frac{r_s}{r_1}}\right]. \tag{19.6}$$

If r_2 is greater than r_1, then this result predicts that $\Delta\tau_2$ is greater than $\Delta\tau_1$. This is easy to see by taking r_2 to spatial infinity.

The conclusion is that clocks lowered down a gravitational potential well will run slower than clocks higher up the gravitational potential. This is gravitational time dilation, a phenomenon that has been empirically confirmed [Chou *et al.*, 2010].

Black hole geometry

In the following, we shall assume that the source of the Schwarzschild metric is some massive object well within its Schwarzschild radius r_s. This is an extreme assumption for normal matter. Even for protons, which have an effective radius of about a Fermi (10^{-13} cm), the associated Schwarzschild radius turns out to be about 10^{-50} cm, which is very much smaller than even the so-called Planck length $L_P \equiv \sqrt{\hbar G/c^3} \simeq 1.62 \times 10^{-35}$ metres, widely believed to represent the shortest distance scale which is meaningful in physics.

For the Sun, the associated Schwarzschild radius is 2.95 km, which would place it deep within the interior of the Sun, if it were a valid concept. We note,

---

[59] 'At rest' means relative to a coordinate patch such that along a worldline of such an object, the spatial coordinates such as r do not change.

however, that the Schwarzschild metric was derived for regions where the energy–momentum tensor is zero, so it is not sensible to discuss the metric at values of r smaller than the physical radius of an object.

For the present, assume that the source of the metric is interior to its Schwarzschild radius. Given the Schwarzschild line element (19.5), we consider light (photons) moving radially, which means setting $ds = d\theta = d\phi = 0$. Then we deduce that lightcones have to satisfy the differential equation

$$c\frac{dt}{dr} = \pm \frac{r}{(r - r_s)}. \tag{19.7}$$

This integrates to $ct = -r - r_s \ln |r - r_s| + const$ for *ingoing* photons and $ct = r + r_s \ln |r - r_s| + const$ for *outgoing* photons.[60]

It is useful to plot these functions over a range of different constants. In Figure 19.1, region *I* corresponds to $r > r_s$ whilst region *II* corresponds to $r < r_s$.

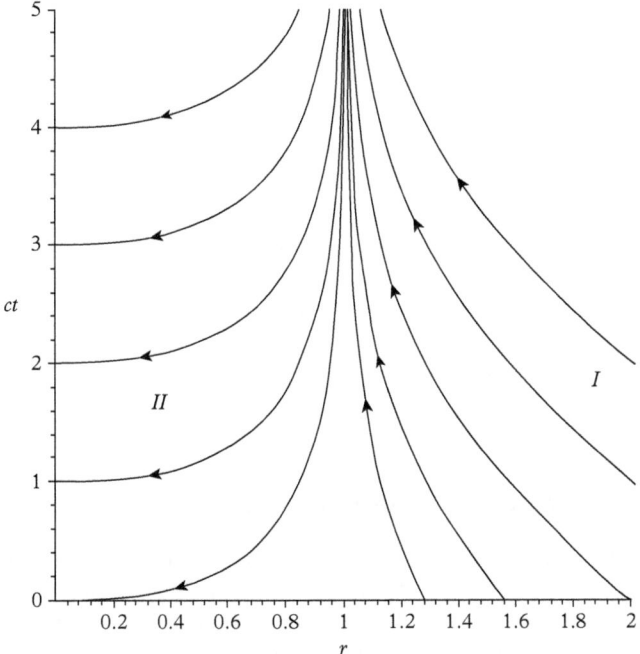

Fig. 19.1 *Radially infalling light for $r_s = 1$.*

---

[60] Note that technically the argument of a logarithm should be dimensionless. An arbitrary constant term ensuring this is implicit.

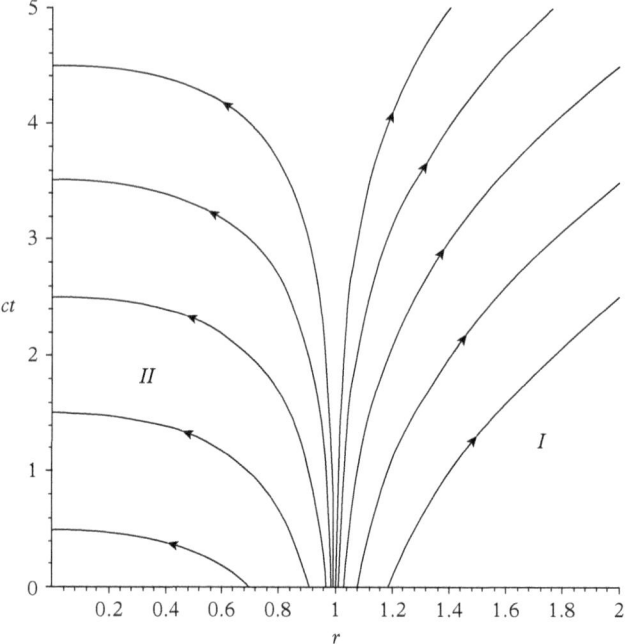

Fig. 19.2 *Radially outgoing light for $r_s = 1$.*

A similar diagram, Figure 19.2, shows outgoing photons. From these diagrams we see that all photons in region *II* must end up at $r = 0$. This point corresponds to a real singularity where the curvature of the Schwarzschild metric diverges. We can now understand the lightcone structure in the $t - r$ plane by combining these two diagrams into Figure 19.3. From this diagram we deduce that in region *II*, all timelike worldlines (i.e., those associated with massive particles) must eventually end up at the origin. There is no escape from the singularity.

Eddington–Finkelstein coordinates

The above spacetime diagrams show that the worldlines of radially moving photons and massive particles cross the Schwarzschild radius $r = r_s$ only at $t = \pm\infty$. This suggests that the 'line' $r = r_s$, $-\infty < t < \infty$ might really not be a line but a point, analogous to what happens with Mercator's projection at the North and South poles on maps of the Earth.

To get a better description of what happens at $r = r_s$, we shall look at ingoing particles. We redefine the coordinate t by

$$ct \equiv c\tilde{t} - r_s \ln |r - r_s| \tag{19.8}$$

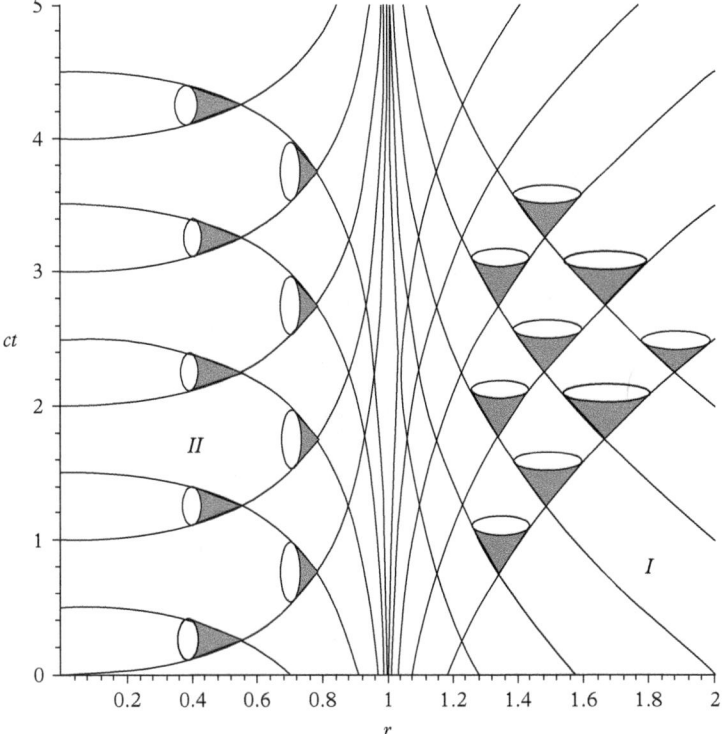

Fig. 19.3 *Schwarzschild spacetime forward lightcone structure.*

and then the line element (19.5) takes the form

$$ds^2 = \left(1 - \frac{r_s}{r}\right) c^2 d\tilde{t}^2 - \frac{2r_s c}{r} d\tilde{t} dr - \left(1 + \frac{r_s}{r}\right) dr^2 - r^2 d\Omega^2, \qquad (19.9)$$

which is regular for $0 < r < \infty$. The coordinates $(\tilde{t}, r, \theta, \phi)$ are called *advanced Eddington–Finkelstein coordinates*.

In these coordinates, incoming and outgoing photon worldlines are given by

$$c\tilde{t} = -r + const, \qquad c\tilde{t} = r + 2r_s \ln |r - r_s| + const, \qquad (19.10)$$

giving the spacetime diagram Figure 19.4.

In these coordinates, we see that ingoing photons cross the Schwarzschild radius at finite values of $\tilde{t}$. These lines are continuous at that radius. A massive particle, which follows a timelike trajectory, can approach and cross the Schwarzschild radius at finite $\tilde{t}$. However, any particle or photon which is inside the region $r < r_s$ can never escape that region. The Schwarzschild radius defines what is known as an *event horizon*, a boundary of no return.

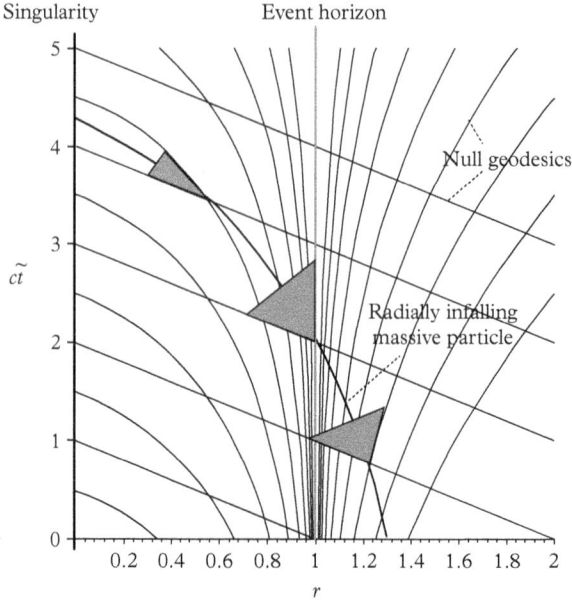

Fig. 19.4 *Forwards lightcones in Eddington–Finkelstein coordinates.*

From this diagram, we see that although objects can cross the event horizon into region *II* from region *I*, there is no escape. Therefore, to an observer in region *I*, no signals can ever reach them from region *II*: to all intents and purposes, region *II* is invisible to such an observer. A compact object with such an event horizon is called a *black hole*, a term attributed to Wheeler.

White holes

It is possible also to look at outgoing photons and construct so-called *retarded Eddington–Finkelstein* coordinates. This time we redefine the time coordinate *t* by $ct = ct^* + r_s \ln |r - r_s|$. In these coordinates the metric looks like

$$ds^2 = \left(1 - \frac{r_s}{r}\right) c^2 \, dt^{*2} + \frac{2r_s c}{r} dt^* dr - \left(1 + \frac{r_s}{r}\right) dr^2 - r^2 d\Omega^2 \qquad (19.11)$$

Now radial null geodesics are given by

$$ct^* = r + const, \qquad ct^* = -r - 2r_s \ln |r - r_s| + const, \qquad (19.12)$$

In this coordinate patch, outgoing photons now cross the Schwarzschild radius at finite t^*, but ingoing photons take forever. Now particles move *away* from the singularity and are forcibly expelled from the region $r > r_s$.

To understand what is going on, we have to realize that the original Schwarz-schild coordinates cover only part of the full geometry. This more complete geometry contains both a black hole and a white hole singularity.

The above analysis shows the importance when discussing GR spacetimes of taking into account how they may have evolved into the configurations stated. There is no statement in any of the above, and indeed in most line element calculations, of exactly where those spacetimes came from. The black hole and white hole solutions are different in the way that an exophysical observer relates to them, in terms of boundary and initial conditions.

The spinning disc

The spinning disc provides an interesting scenario that seems halfway between SR and GR. It concerns a 'frame' of reference consisting of a chorus of observers, each member of the chorus rotating uniformly around a fixed axis in three-dimensional space at constant angular velocity ω. Both axis and angular velocity are specified relative to an inertial frame $\mathcal{F}$, in which the axis is fixed. Without loss of content, we can ignore the spatial dimension in the direction of the axis.

There are at least two problems with this scenario.

First, relative to $\mathcal{F}$, a chorus member situated a radial distance r from the axis will have instantaneous speed $v = r\omega$. Therefore, there will be a problem for observers at or beyond the critical radial distance $r_c = c/\omega$, because such an observer will be travelling faster than the speed of light, according to frame $\mathcal{F}$. A naive application of SR seems incorrect for such observers.

The second problem is synchronization. The standard method of discussing this in SR is to use light signals sent back and forth between chorus members. Such an approach will work in the case of uniform motion, even if we wish to discuss generalized transformations, as we did in Chapter 18. For the spinning disc, however, the observers are actually non-inertial, assuming the frame $\mathcal{F}$ is an inertial frame. There are, therefore, problems in defining hyperplanes of simultaneity for such a disc chorus.

This scenario is called *Ehrenfest's paradox* and has been the source of debate for a hundred years. Einstein's thoughts about this problem were significant during his development of GR, as it became clear that conventional ideas about the spinning disc, such as assuming rigidity and Euclidean distance relationships, would not be applicable in such a case. The general conclusions are (i) there are no rigid bodies in SR; (ii) it is important to distinguish those cases where an object is initially at rest and then accelerated to a final state and those cases where the final state is just assumed; (iii) the spatial geometry relevant to a chorus such as that on a spinning disc need not be Euclidean; and (iv) synchronization may be impossible [Grøn, 2004].

20
Time travel

Introduction

No book on time would be complete without a discussion of time travel (TT). In this chapter we review selected aspects of that concept in some detail: there are many discussions in the literature on TT, exploring the paradoxes and the constraints. Ours is not a review of those discussions simply because many of them are contextually incomplete. Our interest is in those aspects that may have some empirical validity.

Although TT is a common motif in science fiction, it has been and remains the focus of many serious studies in general relativity (GR). The fact is, TT is mathematically admissible in classical GR, as we shall show, but seems logically inadmissible in any context involving irreversibility, such as quantum mechanical (QM) or thermodynamics. The current failure of the quantum gravity programme to reach any conclusions may be related to this fact.

Information flow

We do not need to discuss GR and QM to see what problems could arise were TT possible: temporal paradoxes arise in all approaches to the subject. Some reflection on TT soon leads to the conclusion that these paradoxes are all to do with *information*.

To see this, suppose we wanted to win a lot of money on the National Lottery. Do we actually need to construct a time machine at almost certainly enormous cost, time travel physically into the day after tomorrow, read in that day's news-papers the numbers that have/will have been selected 'at random' in the previous day's National Lottery draw, travel back to today, and then place an enormous bet on that outcome, before tomorrow's Lottery Draw? Not really. All we would need to achieve the same result would be to obtain that information from the future. But of course, that is the rub: we can no more get this information by wishful thinking than we can actually travel in time and get it.

The flow of information is therefore the central issue in TT. This can be illustrated by *differential equations* (DEs), of the sort discussed in Chapter 12.

Images of Time. First Edition. George Jaroszkiewicz.
© George Jaroszkiewicz 2016. Published in 2016 by Oxford University Press.

Conventionally, these are *local in time*, meaning that they involve functions and their derivatives, all of which are evaluated at the same instant of the observer's time. Consider the simple first order DE

$$\frac{dx(t)}{dt} = x(t). \tag{20.1}$$

The solution is easily found: $x(t) = e^t x(0)$. Suppose now that the equation was modified 'very slightly' to

$$\frac{dx(t)}{dt} = x(t - T), \tag{20.2}$$

where T is a positive constant representing a *delay* by a time T in the information about the current position of the system under observation (SUO) reaching the observer/mathematician. Although physically realistic, this equation is surprisingly subtle. To see this, suppose that the solution exists and is an analytic function of time. This means that we may make a Taylor expansion of x about t. Then (20.2) becomes

$$\frac{dx(t)}{dt} = x(t) - \frac{dx(t)}{dt}T + \frac{1}{2}\frac{d^2x(t)}{dt^2}T^2 - \dots. \tag{20.3}$$

This is an *infinite*-order differential equation, a very different proposition to the original first-order differential equation (20.1). Equation (20.2) is an example of a *difference-differential* equation, higher order versions of which require sophisticated mathematical techniques [Pinney, 1958]. Such equations are usually not studied by physicists, who tend to represent delays (or retardations) via additional dynamical variables such as electromagnetic fields.

Equation (20.2) is physically realistic in that it reflects the fact that, in the real world, many processes are affected by delays. Indeed, before our age of near instant communications, significant delays (retardations) of information were a constant fact of life: letters would arrive days or weeks after posting. Even a Roman emperor had to wait many days before news from the frontier could reach him in Rome: the average courier speed over land at that time was about 50 miles a day [Ramsey, 1925].

Naval architects ran head-on into difference-differential equations when they started to use Newtonian mechanics to study the stability of ships. Consider a traditional masted sailing boat with its mast making an instantaneous angle $\theta(t)$ with the vertical, as in Figure 20.1. The forces destabilizing the ship depend not only on the instantaneous position of the centre of mass, but also on the amount of water in the bilges swirling around in the lower decks. Forces due to such water would depend on the orientation of the boat a few seconds earlier, due to retardation effects. Much the same effect is experienced when a large ship passes a small boat: the boat starts to feel the passage of the ship *after* the ship has passed by.

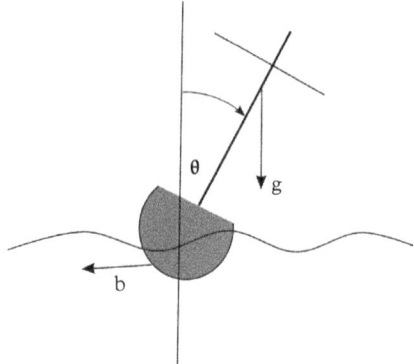

Fig. 20.1 *Boat oscillating at sea: g represents gravitational forces, b represents bilge water forces.*

Prediction and retrodiction

To appreciate the effects of retardation, we shall explore equation (20.2) for the case $T > 0$ (delay). To 'solve' this equation we have to establish what we mean, in physical terms rather than mathematical terms. Time is, after all, a physical phenomenon, not a mathematical one. Our interpretation is that we are a primary observer working in process time and the moment of the present is actually $t = 0$. We have information from the past up to the present, and we want to use that information to predict the future values of x, knowing that the variable $x(t)$ *will* behave according to the rule (20.1) for times after zero.

Given that we know[61] that this SUO obeys (20.2) we are entitled to integrate the equation from the present moment ($t = 0$) to a future time just less than T. Assuming continuity of solution, this gives

$$x(t) = x(0^+) + \int_{-T}^{t-T} m(u)\,du \quad 0 < t \leqslant T, \tag{20.4}$$

where $x(0^+)$ is the limit of $x(t)$ as t tends to zero from the future direction. Here $m(u)$ is the value of x at time u, for $-T < u \leqslant 0$, which we are entitled to have: this value is data held in memory at time $t = 0$, having been put there in the past at time $u \leqslant 0$[62] and is available to us at time zero to feed into the above solution.

Now integrating further, from T to $t < 2T$, we find

$$x(t) = x(T^+) + (t - T)x(0^+) + \int_0^{t-T}\int_{-T}^{v-T} m(u)\,du\,dv \quad T < t \leqslant 2T, \tag{20.5}$$

---

[61] This knowledge is based on all our prior experience of this particular SUO. Of course, if the SUO spontaneously changed so that (20.2) was no longer valid, then all bets would be off.
[62] This is an idealization that assumes measurement and memory storage takes zero time.

where we have used (20.2) and (20.4). If we insist on continuity of solution at $t = 0, T$, and so on, we find

$$x(0^+) = m(0^-), \quad x(T^+) = m(0^-) + \int_{-T}^{0} m(u)\,du, \quad (20.6)$$

and so on. It is clear that, in principle, we could use the initial data (the values of m over the interval $[-T, 0]$) at time zero to predict the value of x at any future time.

Suppose now we consider the difference-differential equation.

$$\frac{dx}{dt}(t) = x(t + T), \quad t > 0, T > 0. \quad (20.7)$$

Our first thoughts might be to rewrite this equation as

$$x(t) = \frac{dx}{dt}(t - T). \quad (20.8)$$

This would suggest that the problem has been trivially solved: if we know $m(t)$ for $t < 0$ then we just differentiate it and we have an immediate solution at a later time. In fact, this is precisely where physics and mathematics have fundamentally different perspectives on time. The point is that we have to take a *process* time perspective, not a manifold one.

To illustrate this, let us really simplify the issue enormously. Consider two investigations:

Scenario 1: investigate $x(t) = m(t - T)$,
Scenario 2: investigate $x(t) = m(t + T)$.

The interpretation is: we have information in memory obtained about m and we want to know how the variable x depends on that information. We will illustrate the issue with a diagram involving **two** times: one of these is t_x, the time for which we want to have a prediction (or retrodiction) of the value of x and the other time t_m is the time stamp of data held in memory. In Figure 20.2, t_m is plotted left–right and t_x is plotted down–up.

Scenario 1 is consistent with causality and *prediction*. Event A represents the observation at time t of the variable m that is registered at that time in memory. Assuming the equation of motion $x(t + T) = m(t)$ for Scenario 1 holds at time t, then the observer at time t is entitled to *predict*, at their time t, that the value of x at time $t + T$ will be $m(t)$.

On the other hand, Scenario 2 is not consistent with prediction but with *retrodiction*. Now Event A cannot be used to predict the future states of x but only to say what x must have been at time $t - T$, that is, retrodict $x(t - T)$, an event that has already happened by time t.

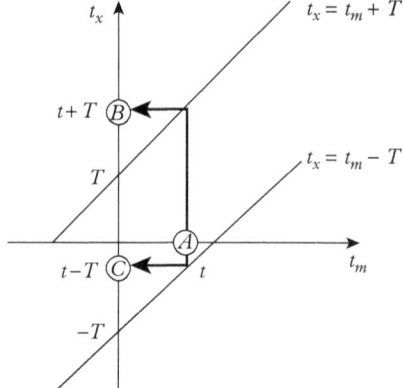

Fig. 20.2 *Scenarios 1 and 2. Event A is data capture at time t, from which Scenario 1 gives us a prediction of x at event B, but Scenario 2 gives a retrodiction at event C.*

Our analysis of these scenarios is based on a contextual interpretation of equality. Despite appearances, 'equation' (20.7) is not equivalent to (20.8). (20.7) is really not an equality in the mathematical sense of the word but an *assignment*, where the position on the left is not equivalent to that on the right. The interpretation of (20.7) is that the left-hand side is calculated from a knowledge of the right-hand side. This is like the assignment expression $x := 2$ in a computer algebra program such as Maple. There is no meaning to rewriting this assignment the other way around, viz., $2 := a$ is inadmissible. Therefore, we cannot argue that (20.8) is equivalent to (20.7). In any case, the status of the variable m is different to that of x: $m(t)$ is empirical data acquired and held in memory whilst $x(t)$ is the observer's prediction or retrodiction as to what might be found if looked for at time t. There still remains the issue of actually measuring $x(t)$ and comparing it to the prediction. Prediction and retrodiction are *not* observations.

As to empirically validating the observer's expectation of x, the physics of validating prediction is different to that involved in validating retrodiction. Suppose we solve Scenario 1 at time t. This means that at that time, we know $m(t)$ and therefore predict that at time $t + T$, x will have the value $m(t)$. We can do nothing until our clocks have reached at least time $t + T$, and only then can we measure what x actually is at that time. If there is no delay in the measurement, we should get a pristine measurement and, hopefully, confirmation of our prediction.

On the other hand, if Scenario 2 is in operation, then knowing $m(t)$ we make the retrodiction that at the earlier time $t - T$, x had the value $m(t)$. But we can confirm it only at time t, a time T after it had actually happened. The possibility exists therefore that there could be corruption of evidence, possibly leaving no real evidence left by time t that x indeed had the value $m(t)$ at time $t - T$. This erasure of past signals is the problem all archaeologists, geologists, and early universe investigators face. The past and the future are not symmetric at this level of observation: they have different architectures.

Tachyons

Hypothetical particles that travel faster than the speed of light are called *tachyons*. There is no confirmed evidence for such objects. In SR, such particles are regarded as unphysical from several perspectives. First, relative to a standard inertial frame, the Lorentz factor $\gamma(v) \equiv \dfrac{1}{\sqrt{1 - v^2/c^2}}$ associated with such a particle would be imaginary: the speed v of an observed particle has to be less than c the speed of light for $\gamma(v)$ to be real. Second, no one has ever observed any physical particle or signal propagating at any speed greater than the speed of light. This does not invalidate correlations, such as those reported by [Scarani *et al.*, 2000], as no physical, causal effects can be actioned by such correlations. Third, if there were such a particle in the usual sense of the word, then we could expect there to be rest frames of reference for such particles, implying observers moving faster than c relative to any normal standard inertial frame. Apparatus at rest in such superluminal frames of reference would require new, contextually complete descriptions.

One way of explaining away the non-existence of tachyons is that 'particles' such as photons and electrons should not be thought of as little lumpy objects moving around spacetime but as signals in apparatus. If preparation and detection apparatus can only be set up via relative subluminal speeds, then correlations could be misinterpreted as superluminal signals. Essentially, tachyons as currently understood are outside the conventional framework of modern physics.

Spreadsheet time travel

A spreadsheet such as Excel can be used to illustrate significant features of differential/difference equations, causality, lightcones, and time travel, as we now illustrate. In Figure 20.3, we show a simulation of a discrete one-dimensional space and a discrete time: time runs left to right, starting with the column labelled A, then B, and so forth, whilst space coordinates run top to bottom, starting with 1, 2, 3, etc. The dynamics is encoded as follows. In every cell # in every column apart from the one labelled A (initial time), the program instruction looks at data in the earlier column, calculates a value $V[\#]$ from it, and then deposits it in that cell. For instance, the cell instruction for cell $D5$ is $V[D5] = \max(V[E4] + V[E5[+V[E6], 10)$, the maximum of 10 being there purely to prevent overflow. This particular algorithm is a crude but effective simulation of a first-order difference equation.

	A	B	C	D	E	F	G	H	I	J	K	L
1	0	0	0	0	0	0	0	0	0	0	0	0
2	0	0	0	0	0	0	0	0	0	0	0	1
3	0	0	0	0	0	0	0	0	0	0	1	10
4	0	0	0	0	0	0	0	0	0	1	10	10
5	0	0	0	0	0	0	0	0	1	9	10	10
6	0	0	0	0	0	0	0	1	8	10	10	10
7	0	0	0	0	0	0	1	7	10	10	10	10
8	0	0	0	0	0	1	6	10	10	10	10	10
9	0	0	0	0	1	5	10	10	10	10	10	10
10	0	0	0	1	4	10	10	10	10	10	10	10
11	0	0	1	3	10	10	10	10	10	10	10	10
12	0	1	2	6	10	10	10	10	10	10	10	10
13	1	1	3	7	10	0	10	10	10	10	10	10
14	0	1	2	6	10	10	10	10	10	10	10	10
15	0	0	1	3	10	10	10	10	10	10	10	10
16	0	0	0	1	4	10	10	10	10	10	10	10
17	0	0	0	0	1	5	10	10	10	10	10	10
18	0	0	0	0	0	1	6	10	10	10	10	10
19	0	0	0	0	0	0	1	7	10	10	10	10
20	0	0	0	0	0	0	0	1	8	10	10	10
21	0	0	0	0	0	0	0	0	1	9	10	10
22	0	0	0	0	0	0	0	0	0	1	10	10

Fig. 20.3 *A spreadsheet simulation of lightcone effects in a* 1 + 1 *spacetime. Time runs left–right and space runs up–down. Data is inserted in cell* $A13$ *and a 'lightcone' appears to the right. The arrowed triangle indicates that information from the 'future' of an event cannot be used consistently.*

Initial data is placed in column A, and consists of zeros except in one cell, $A13$, where the value is one as indicated. This value is responsible for a lightcone-like structure of non-zero-values appearing towards the right (the direction of time), with its vertex at $A13$. This lightcone appears to disappear instantly if $V[A13]$ is suddenly set to zero rather than one.[63]

---

[63] In Asimov's *End of Eternity* [Asimov, 1950], the Eternals stand outside spacetime and watch a reality change, which does not take place instantly as far as they are concerned. Likewise, if a value is changed in a given cell in our simulation, there may be a noticeable time delay before it is seen on screen, simply because the central processor needs its own time to complete calculations and update displays. This delay will depend on the scale of the change.

To illustrate the significance of the 'lightcone' and the flow of information, consider cell $F13$, as indicated. The algorithm in that cell was changed according to the rule

$$V[F13] \equiv \max(V[E12] + V[\mathbf{E13}] + V[E14], 10)$$
$$\rightarrow \max(V[E12] + V[\mathbf{J11}] + V[E14], 10), \qquad (20.9)$$

that is, an attempt is made to take data from a cell, $J11$, that is inside the lightcone centred on $F13$, that is, in the absolute future of $F13$. The response of the Excel programme is practically instantaneous, with the appearance of the arrowed triangle as shown and with a warning that an inconsistent calculation is being attempted. In essence, the Excel program recognizes a causal paradox, or the equivalent of what could happen if information from the future is brought to the present and then used to alter that future.

No such loop appears if data is brought from outside the lightcone centred on $F13$, even if that data is in the 'future' of $F13$, as long as the two cells are relatively 'spacelike' For example, if instead of (20.9) we use

$$V[F13] \equiv \max(V[E12] + V[\mathbf{E13}] + V[E14], 10)$$
$$\rightarrow \max(V[E12] + V[\mathbf{J5}] + V[E14], 10), \qquad (20.10)$$

the calculation goes ahead without any problem. This example illustrates a breakdown of 'absolute simultaneity'.

What is remarkable about such simulations is that a Minkowski spacetime-like metric emerges not from being imposed onto the cell structure *per se*, but simply because of the dynamical relationship between the cell content values and the local architecture of their functional relationships. Other concepts such as relativistic time dilation can be investigated [Jaroszkiewicz, 2014].

Despite appearances, such simulations are not simply 'Block Universe' models. The diagrams obtained do look like Block Universe accounts of events and their relationships, but in fact there is a process time action underpinning these calculations. That process time is the clock time of the central microprocessor that is performing the calculations, deciding whether there are inconsistencies or not, displaying results, and constantly going back and forth to each cell, including those in the 'past' in the simulation. Essentially, the computer is playing the role of an absolute primary observer, with its own absolute sense of time. That time is not on show in the diagram, but its presence can be deduced for example by the inconsistency warning triangle shown in Figure 20.3.

The Gödel metric

In 1949, Kurt Gödel published a paper discussing a four-dimensional GR spacetime with closed timelike curves (CTCs) [Gödel, 1949]. He started by writing down the line element

$$ds^2 = a^2 \left(dt^2 - dx^2 + \tfrac{1}{2}e^{2x}dy^2 - dz^2 + 2e^x dt dy \right), \tag{20.11}$$

where a is a positive constant with the dimensions of a length and t, x, y, and z are taken dimensionless. He explored the consequences physically and geometrically.

Physics of Gödel's spacetime

The physics of this spacetime is explored using Einstein's field equations for GR (q.v. Appendix):

$$R_{\mu\nu} - \tfrac{1}{2}Rg_{\mu\nu} + \Lambda g_{\mu\nu} = -\kappa T_{\mu\nu}, \tag{20.12}$$

where the $R_{\mu\nu}$ are the components of the Ricci tensor, R is the Ricci scalar field, $g_{\mu\nu}$ are the components of the metric tensor that can be read off from the line element (20.11), Λ is the cosmological constant, $T_{\mu\nu}$ are the components of the stress–energy–momentum tensor, and $\kappa \equiv 8\pi G/c^4$, where G is the Newtonian constant of gravitation.

The Ricci tensor and the Ricci scalar can be readily calculated knowing the line element. The important physical properties of this spacetime are these:

1. The Ricci scalar R is found to be $R = -a^{-2}$, which means that this spacetime is homogeneous: there are no special points where there is a singularity.

2. Assuming this spacetime is that of the universe, we can model the stress–energy–momentum tensor as that due to a *dustlike* distribution of matter. This means that galaxies are regarded as individual grains of dust in a cloud of dust. Now one of the characteristics of a dustlike system is that there is no *pressure* in the system coming from the particles themselves: the grains of dust are thinly spread out and essentially do not interact with each other directly. In such a case, the stress–energy–momentum tensor for a dust cloud is assumed to have the form $T^{\mu\nu} \equiv \rho u^\mu u^\nu$, where ρ is the intrinsic mass density and u^μ are the components of the flow four-vector field associated with the dust cloud, satisfying the constraint $u_\mu u^\mu = c^2$.

3. Assuming the Ricci tensor can be written in the form

$$R_{\mu\nu} = \alpha g_{\mu\nu} + \beta u_\mu u_\nu \tag{20.13}$$

where α and β are to be determined, a simple calculation gives two solutions: (i) $\alpha_1 = 0, \beta_1 = 1/a^2 c^2$ and (ii) $\alpha_2 = 0, \beta_2 = -1/a^2 c^2$. The first solution then gives $\rho_1 = -1/a^2 c^2 \kappa$, which is negative and therefore rejected.[64]

[64] Gödel's spacetime has positive energy density everywhere. Negative energy density is a common problem with worm hole solutions in GR.

Hence solution (ii) applies and we find

$$\Lambda = -\frac{1}{2a^2}, \qquad \rho = \frac{c^2}{8\pi Ga^2}. \tag{20.14}$$

That the cosmological constant is negative here is of interest. When Einstein introduced such a term in his field equations, it was intended to provide a repulsive effect to counteract the inherent attractive nature of gravity, so that his model universe would be static. At that time, Einstein was unaware of the Lemaître–Hubble expansion of the universe. In the weak field, Newtonian space-time limit, the gravitational field intensity for a mass M located at the origin of coordinates is given by

$$\mathbf{g}(\mathbf{r}) = -\frac{GM}{r^3}\mathbf{r} + \frac{c^2}{3}\Lambda\mathbf{r}, \tag{20.15}$$

so we see that a positive Λ represents a weak, long-range repulsion. Gödel's negative Λ therefore acts in the opposite way, acting as a form of *pressure* serving to stabilize his spacetime. This pressure has to be regarded as an inherent property of empty space, if the stress–energy–momentum tensor is purely dust-like.

4. To understand the role of Gödel's negative cosmological constant, we need to look at the geodesics, or free particle motion, in his spacetime. Having established that his spacetime was acceptable in terms of energy density, Gödel went on to discuss the *compass of inertia*, the behaviour of free particle worldlines in his spacetime. He showed that the off-diagonal terms in the metric tensor acted to create a 'swirling' effect, such that freely moving objects do not lock onto the distant stars but their trajectories start to veer away. The effect is unlike that found with some other spacetimes with intrinsic rotation, such as van Stockum's rotating infinite cylinder metric [van Stockum, 1937], because Gödel's spacetime is homogeneous. Gödel's negative cosmological constant acts as a form of centripetal force, stabilizing his model universe.

Geometry of Gödel's spacetime

By all accounts, Gödel was preoccupied with time and with the inevitability of death. His interest in his spacetime was driven by more than academic curiosity: his spacetime has closed timelike curves (CTSs), as we shall discuss presently. His spacetime has an inherently different temporal architecture to conventional spacetimes used previously to model the universe, such as the generic Friemann–Lemaître–Robertson–Walker (FLRW) line element [Hobson *et al.*, 2006]

$$ds^2 = c^2\,dt^2 - a(t)^2 g_{ij}(\mathbf{x})dx^i dx^j, \tag{20.16}$$

where $g_{ij}(\mathbf{x})$ represents the local spatial geometry and the t-coordinate plays the role of an absolute cosmic time. Gödel's paper shattered the veneer of conventionality that GR had built up. Previously, it had been believed that the universe ran on conventional lines, with a time that was locally somewhat flexible, but overall pressed forwards with the expansion of the universe.

Halfway through his paper, Gödel gave a coordinate transformation from the 'old' coordinates (t, x, y, z) to 'new' coordinates $(\tilde{t}, r, \phi, u)$, such that in the new coordinates, his line element (20.11) takes the form

$$ds^2 = 4a^2(d\tilde{t}^2 - dr^2 + f(r)d\phi^2 + g(r)d\phi dt - du^2), \qquad (20.17)$$

where $f(r) \equiv \sinh^4(r) - \sinh^2(r)$ and $g(r) \equiv 2\sqrt{2}\sinh(r)$. It is this form of the line element that creates all the fuss and interest. The exploration of the complete architecture of Gödel's spacetime has been covered elsewhere [Hawking & Ellis, 1973]: we are interested only in the CTCs. The essential trick that gives CTCs is the behaviour of the function $f(r)$ in (20.17). Consider a trajectory such that the coordinates $\tilde{t}$, r, and u are held constant. Then $d\tilde{t} = dr = du = 0$ and so $ds^2 = 4a^2 f(r)d\phi^2 = 4a^2 \left[\sinh^4(r) - \sinh^2(r)\right] d\phi^2$. By our discussion of timelike, null, and spacelike intervals in earlier chapters, we can see that the coordinate ϕ is timelike if $f(r)$ is positive, null if $f(r)$ is zero, and spacelike if $f(r)$ is negative. It is easy to see that if $r > r_G \equiv \ln(1 + \sqrt{2})$ then ϕ is a timelike coordinate. This means that, in this coordinate patch, for values of r greater than the Gödel radius r_G, the lightcones have 'tipped over', as shown in Figure 20.4. Then a circuit along such a trajectory with ϕ running from 0 to 2π takes us back to our starting point. But, at each point of such a circuit, we have always been moving into the forwards lightcone at that point, which is physically admissible. Therefore, we have travelled in time.

Or have we? This question arises because we should ask: are we quite sure the coordinate ϕ is periodic? Merely using what looks like an angular coordinate

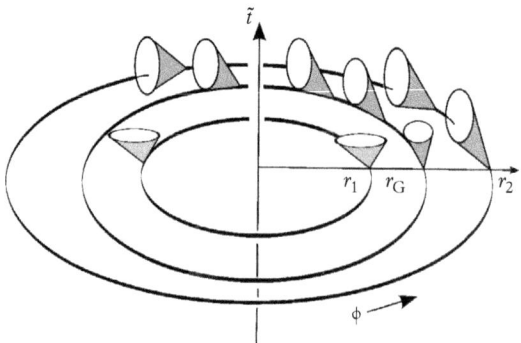

Fig. 20.4 *Lightcones in Gödel spacetime, with* $r_1 < r_G < r_2$.

does not thereby endow that coordinate with periodicity. We saw in our discussion of the Schwarzschild geometry in the previous chapter that the coordinate t was not timelike everywhere. Another way of looking at this question is to replace the symbol ϕ with say x and *start* from 20.17. Would we be justified in asserting that x was periodic? What would induce us to assume it was?[65]

Timelike geodesics

An interesting question is whether the CTCs we have just discussed are geodesics: can a free particle move around in time in a natural way? An analysis readily shows that the paths discussed above are *not* geodesics: we would have to employ a rocket to travel along such a worldline.

A number of theorists have constructed spacetimes with geodesic CTCs: Som and Raychaudhuri's spacetime has a geodesic CTC at a critical radius [Som & Raychaudhuri, 1968], Calvão, Soares, and Tiomno's spacetime contains geodesic CTCs for a range of radii [Calvao *et al.*, 1990], and Grøn has found a line element that permits geodesic CTCs everywhere [Grøn & Johannesen, 2010].

---

[65] Suppose we were given the line element $ds^2 = d\phi^2 - dx^2 - dy^2 - dz^2$. Would we be entitled to assume ϕ was periodic? Hopefully not, since this is really the Minkowski line element (17.10).

21

Imaginary time

Introduction

The representation of time by real numbers has been successful, particularly in classical mechanics (CM). Another possibility, *imaginary time*, is the analytic continuation of real time from the real numbers $\mathbb{R}$ into the complex number plane $\mathbb{C}$. This is the focus of this chapter. Imaginary time has found significant application in fundamental areas of quantum physics such as high energy particle physics and quantum cosmology.

We need to distinguish between *imaginary* time and *complex* time. A complex number z can be written in the form $z = u + iv$, where u and v are real and i is the square root of -1. Therefore $\mathbb{C}$, the set of complex numbers, requires two real parameters to label all of its elements. When time is asserted to be complex, that then implies that time is being parametrized by *two* real dimensions. We now refer back to Chapter 12, where we discussed the consequences of time having two or more dimensions. The general conclusion there was that such a possibility leads to unstable ultrahyperbolic solutions to differential equations [Tegmark, 1997]. Another problem with complex numbers is that they are not an ordered set, so they are not consistent with the ordering property of time that we would naturally expect.

Clearly, complex time is problematical and we are going to avoid it. When we refer to *imaginary* time we mean a complex number of the form $z = i\tau$, where τ is real. This form of complex number has a single real parameter, τ, that has the required ordering we want for time. We shall use the term *complex time* for a point in the complex plane that is parametrized by two independent real variables. Imaginary time is one dimensional, complex time is two dimensional. If we need to think of a parametrized curve (a function of a single real variable) in the complex plane not necessarily along the imaginary axis, we shall refer to it as *complex path time*. The set of complex numbers on a complex path is ordered by the values of the real parameter involved.

Images of Time. First Edition. George Jaroszkiewicz.
© George Jaroszkiewicz 2016. Published in 2016 by Oxford University Press.

Minkowski's imaginary time

Time is not space. We can move in all three spatial dimensions, that is, left–right, up–down, forwards–backwards, but we cannot move back in time. A great para-dox of twentieth-century physics is that both special relativity (SR) and general relativity (GR) became very successful ignoring this basic fact. Neither SR nor GR has anything in them that gives an obvious arrow of time: the forwards and backward branches of lightcones have to be specified externally[66] and white holes are 'as good' solutions to the point mass field equations of Einstein as black holes.

An equally great paradox is that going into imaginary time has given major insights into the structure of the universe. The first significant instance of this was in 1908, when Minkowski [1908] reformulated Einstein's 1905 paper on SR in geometrical terms. He first proposed the spacetime line element

$$ds^2 = c^2 dt^2 - x^2 - dy^2 - dz^2, \qquad (21.1)$$

where c is the speed of light. After discussing lightcones and worldlines, he went further and suggested the formal replacement $w \equiv ict$, where i is the square root of minus one, giving the line element

$$ds^2 = -dw^2 - x^2 - dy^2 - dz^2. \qquad (21.2)$$

In this imaginary time formulation, the Minkowski distance rule looks just like the Euclidean distance rule, except now it is applied to a four-dimensional Euclidean space, if we take w real. This step by Minkowski emphasized the role of geometry in physics, an emphasis which led eventually to Einstein's *GR* and which has had a profound influence on mathematical physics ever since.

There is a fundamental problem however: the lightcones have disappeared. If we set $ds = 0$ then we readily conclude from (21.2) that $dw = dx = dy = dz = 0$, if all coordinates are real. In other words, *there are no lightcones in a Euclidean space*. The transition from real to imaginary time not only abolishes all the normal properties associated with time, such as causality, but changes the mathematical structures involved. For example, the invariances of (21.2) contain the compact rotation group $O(4)$, whereas the invariances of (21.1) contain the non-compact group $O(3,1)$,[67] which is quite a different group.

Application to wave mechanics

Minkowski's use of imaginary time has little value when restricted to classical mechanics but is useful in quantum mechanics (QM). Consider the real-time

---

[66] By reference to the observer's arrow of time or some relevant arrow of time, such as Hubble expansion or the laws of thermodynamics.
[67] The group $O(p, q)$ leaves the quadratic form (A.10) invariant.

Schrödinger equation for a particle in a time-independent potential,

$$i\hbar\frac{\partial}{\partial t}\Psi\left(t,\mathbf{x}\right)=\vec{H}\Psi\left(t,\mathbf{x}\right), \tag{21.3}$$

where $\vec{H}$ is the Hamiltonian operator, assumed positive. This equation has general solution

$$\Psi\left(t,\mathbf{x}\right)=\sum_{n}\Psi_{n}\Phi_{n}\left(\mathbf{x}\right)e^{-iE_{n}t/\hbar}, \tag{21.4}$$

where the sum is over all energy eigenstates Φ_n and the Ψ_n are complex coefficients. Now perform a *Wick rotation*, which is an analytic continuation of the time parameter from the real line into the complex plane. This is formally equivalent to Minkowski's replacement $ct \to -iw$. Writing $t = -i\tau$, where τ is real, and defining $\Psi\left(-i\tau,\mathbf{x}\right)\equiv\tilde{\Psi}(\tau,\mathbf{x})$, Schrödinger's equation (21.4) takes the form

$$-\hbar\frac{\partial}{\partial\tau}\tilde{\Psi}\left(\tau,\mathbf{x}\right)=\vec{H}\tilde{\Psi}\left(\tau,\mathbf{x}\right), \tag{21.5}$$

with formal solution

$$\tilde{\Psi}\left(\tau,\mathbf{x}\right)=\sum_{n}\Psi_{n}\Phi_{n}\left(\mathbf{x}\right)e^{-E_{n}\tau/\hbar}. \tag{21.6}$$

Assuming this series converges for large and positive values of τ, the dominant contribution in the sum (21.6) comes from the term involving the lowest energy E_0, that of the ground state Φ_0. Making the standard assumption that the ground state Φ_0 is unique, we deduce $\lim_{\tau\to\infty}\tilde{\Psi}\left(\tau,\mathbf{x}\right)e^{E_0\tau/\hbar}=\Psi_0\Phi_0\left(\mathbf{x}\right)$, which gives us a way of finding approximations to ground-state wavefunctions.

Imaginary time gives another insight into the Schrödinger equation. Consider the free particle in three dimensions. With the same Wick rotation, the Schrödinger equation becomes

$$\frac{\partial}{\partial\tau}\tilde{\Psi}\left(\tau,\mathbf{x}\right)=(\hbar/2m)\nabla^{2}\tilde{\Psi}\left(\tau,\mathbf{x}\right), \tag{21.7}$$

recognized as the diffusion equation for a space-time-dependent density $\rho\left(\tau,\mathbf{x}\right)$ with diffusion coefficient $D = \hbar/2m$. In other words, the Schrödinger equation in imaginary time looks like a diffusion process in real time.

Propagators and Green's functions

Given the classical oscillator Lagrangian

$$L=\frac{1}{2}\dot{x}^{2}-\frac{1}{2}\omega^{2}x^{2} \tag{21.8}$$

for a unit mass particle, where ω is real, a fundamental quantity in quantum theory is the ground-state expectation value

$$\langle 0 | T\hat{x}(t)\,\hat{x}(0) | 0 \rangle = i\hbar\Delta_F(t), \tag{21.9}$$

where $|0\rangle$ is the ground state, $\hat{x}(t)$ is the Heisenberg picture position operator at time t, and T is the time ordering operator defined by

$$T\hat{x}(t)\,\hat{x}(0) = \Theta(t)\,\hat{x}(t)\,\hat{x}(t) + \Theta(-t)\hat{x}(0)\hat{x}(t), \tag{21.10}$$

Θ being the Heaviside or step function. In (21.9), $\Delta_F(t)$ is the *Feynman propagator*, one representation of which is the contour integral

$$\Delta_F(t) = \frac{1}{2\pi}\int_\Gamma dv e^{-ivt}\frac{1}{v^2 - \omega^2 + i\varepsilon}, \tag{21.11}$$

the contour of integration Γ being along the real axis in the complex v plane. Here the infinitesimal parameter ε has been introduced via Feynman's $+i\varepsilon$ prescription in order to recover the desired causality structure. In the complex-v plane, the singularities of the integrand in (21.11) occur at $v = \pm(\omega - i\varepsilon)$. Contour integration then gives the closed form

$$\Delta_F(t) = \frac{-i}{2\omega}\left\{\Theta(t)\,e^{-i\omega t} + \Theta(-t)\,e^{i\omega t}\right\}, \tag{21.12}$$

which shows that positive energy waves propagate forwards in time and negative energy waves propagate backwards in time. This inspired the Feynman–Stückelberg interpretation of positrons as electrons moving backwards in time [Stueckelberg, 1941; Feynman, 1949; Bjorken & Drell, 1964].

Related to the Feynman propagator is the Dyson anticausal propagator: it satisfies the same differential equation as the Feynman propagator but propagates positive energy waves *backwards* in time and negative energy waves *forwards* in time. This illustrates one of the problems with the Block Universe model of space-time: there is no obvious reason, no context, for choosing the Feynman propagator over the Dyson propagator, other than bringing in primary observers and their sense of time.

Performing a Wick rotation of the integration variable v in (21.11) so as to avoid the singularities, we define the *Euclidean Green's*[68] function $\Delta_E(\tau) \equiv -i\Delta_F(-i\tau)$ with boundary conditions $\lim_{\tau\to\pm\infty}\Delta_E(\tau) = 0$, giving

$$\Delta_E(\tau) = -\int\frac{dv}{2\pi}\frac{e^{-iv\tau}}{v^2 + \omega^2} = \frac{-1}{2\omega}\{e^{-\omega\tau}\Theta(\tau) + e^{\omega\tau}\Theta(-\tau)\}. \tag{21.13}$$

---

[68] Or Green functions; named after George Green [1793–1841].

The Euclidean Green's function has a limit as τ tends to infinity whereas the Feynman propagator in its closed form (21.12) does not. One of the issues in modern physics is that the real-time formulation of QM frequently has such problems but the imaginary-time formulation does not. Our view is that this is related to fact that the primary observer is not factored into the theory at all except in an ad-hoc way. The Feynman $+i\varepsilon$ prescription does the job required, but nevertheless some degree of dissatisfaction remains, because it has to be put in 'by hand'.

Path integrals

Another application of imaginary time is in Feynman's path integral approach to QM [Feynman, 1948]. This was originally formulated in real time and is used for non-relativistic Schrödinger wave mechanics and relativistic quantum field theory. For example, in Schrödinger mechanics, our interest is often the amplitude $\langle \mathbf{y}, t_f \,|\, \mathbf{x}, t_i \rangle$ for the particle to go from $(t_i, \mathbf{x})$ to $(t_f, \mathbf{y})$. The path integral formulation expresses this as a sum

$$\langle \mathbf{y}, t_f \,|\, \mathbf{x}, t_i \rangle \sim \exp\{iS_{\Gamma_1}/\hbar\} + \exp\{iS_{\Gamma_2}/\hbar\} + \ldots \tag{21.14}$$

of sub-amplitudes, where the sum is taken over all possible continuous paths $\Gamma_1, \Gamma_2, \ldots$, each of which runs from the initial spacetime event $(t_i, \mathbf{x})$ to the final spacetime event $(t_f, \mathbf{y})$. For each path Γ, the corresponding S_Γ is calculated by the time integral of the Lagrangian over that path, that is,

$$S_\Gamma \simeq \int_{t_i}^{t_f} dt\, L\,(\dot{\mathbf{x}}, \mathbf{x})\bigg|_\Gamma. \tag{21.15}$$

Formally, the infinite sum (21.14) is written [Feynman & Hibbs, 1965]

$$\langle \mathbf{y}, t_f \,|\, \mathbf{x}, t_i \rangle \sim \int [d\mathbf{z}]\, \exp\{iS(\mathbf{y}, t_f; \mathbf{x}, t_i)/\hbar\} \tag{21.16}$$

and interpreted as a functional integral.

The path integral does not specify which path the particle actually takes: according to Feynman's prescription, all possible paths have to contribute to the amplitude. This is a direct realization of the principle of contextuality in QM that we have encountered before in this book: *if we are not monitoring an SUO, we cannot say what it is doing, or exclude any accessible possibility.*

In general, Feynman's path integrals are not well-defined mathematically. This is related to the fact that, for real time, the arguments of the exponentials being integrated over in (21.16) are pure imaginary. One remedy has been the imaginary-time approach. In the 1920s, Norbert Wiener developed the theory of integration in function spaces, the so-called Wiener integral, in his study of

Brownian motion,[69] giving the technology to solve diffusion equations. In 1947, Kac realized that going to imaginary time in the Feynman path integral led to a Wiener integral, equivalent to turning the free particle real-time Schrödinger equation into the parabolic diffusion equation (21.7). This approach gives Euclidean (imaginary time) Green's functions, discussed above for the harmonic oscillator.

The general strategy is to formulate a path integral in real time, rewrite it in imaginary time by performing a Wick rotation, calculate the Euclidean Green's functions, and then perform a reverse Wick rotation back to real time. The real problem with this strategy is that it is not clear why this rigmarole should be necessary. Clearly, we do not really understand time.

Quantum gravity

We have seen that in classical GR, the important feature of the metric tensor is its signature, which for our universe is $(1, 3, 0)$, or equivalently, $(1, -1, -1, -1)$. This has a serious impact on *quantum gravity*, the attempt to unify QM and GR.

It is not clear at this time what the 'quantization' of gravity means.[70] It has generally not been a successful programme. Originally the approach was to regard the components $g_{\mu\nu}$ of the metric tensor as operators, that is, make the substitution $g_{\mu\nu} \to \hat{g}_{\mu\nu}$, the idea being that this would correspond in some sense to a spacetime metric which could fluctuate. Successful quantum theories such as quantum electrodynamics seem to work with such a strategy applied to the relevant fields. With the rise of quark–gluon theories, it was appreciated that perhaps the metric *connection* rather than the metric *tensor* should be the object to quantize. This makes sense to us: connections involve observers and context.

Some schemes attempting to quantize gravity are:

1. *Lorentzian quantum gravity* attempts to quantize gravity in real time, using a Lorentz signature metric in four-dimensional spacetime. No satisfactory version exists.

2. *Euclidean quantum gravity* uses the substitution $t = i\tau$ everywhere, turning Lorentzian signature metrics into Riemannian (curved version of Euclidean) metrics. In the path integral approach, weight factors of the form e^{-S} are used instead of e^{iS}.

3. *Riemannian quantum gravity* uses the substitution $t = i\tau$ everywhere, turns Lorentzian signature metrics into Riemannian signature metrics, but uses real-time factors e^{iS} in path integrals.

---

[69] The diffusion of particles suspended in a liquid.
[70] The author once attended a seminar given by C. Isham where *thirteen* different approaches to quantum gravity were reviewed.

The physical significance of any of these models is not clear, because they are all contextually incomplete.[71] Our view is that *ad hoc* modifications to real time physics is misguided: even if one of these approaches were to work (which none do), we would not know why.

Quantum thermodynamics

In applications to scattering problems and bound-state calculations, quantum field theory is usually done in real time. Continuation to imaginary time gives *quantum statistical mechanics*, that is, thermodynamical quantum field theory. If we define the operator $\hat{K} \equiv \hat{H} - \mu \hat{N}$, where μ is the chemical potential, $\hat{N}$ is the particle number operator, and assume that the Hamiltonian $\hat{H}$ is time independent, then the Grand Partition function Z_G for a quantum system in thermodynamic equilibrium is given by $Z_G \equiv Tr\{e^{-\beta\hat{K}}\}$. Here $\beta \equiv 1/k_B T$, where k_B is the Boltzmann constant, T is the absolute temperature of the system, and Tr is the *trace* operation. Now in the real-time Heisenberg picture, quantum field operators at different times are related by the rule

$$\hat{\psi}_H^\alpha (t, \mathbf{x}) \equiv e^{i\hat{H}t/\hbar} \hat{\psi}_S^\alpha (\mathbf{x}) e^{-i\hat{H}t/\hbar}, \tag{21.17}$$

where $\hat{\psi}_S^\alpha (\mathbf{x})$ is a Schrödinger picture field operator (at time $t = 0$) indexed by α. In the thermal Heisenberg picture, field operators are defined by the rule

$$\hat{\psi}_K^\alpha (\tau, \mathbf{x}) \equiv e^{\hat{K}\tau/\hbar} \hat{\psi}_s^\alpha (\mathbf{x}) e^{-\hat{K}\tau/\hbar}, \quad 0 \leqslant \tau < \beta\hbar, \tag{21.18}$$

where $\tau \equiv it$ corresponds to imaginary time.

The single particle temperature Green's functions are defined by

$$G_{\alpha\beta} (\tau, \mathbf{x}; \tau', \mathbf{x}') \equiv -Tr\left\{ \hat{\rho}_G T_\tau \hat{\psi}_K^\alpha (\tau, \mathbf{x}) \hat{\psi}_K^{\beta+} (\tau', \mathbf{x}') \right\}, \quad 0 \leqslant \tau, \tau' < \beta\hbar, \tag{21.19}$$

where T_τ now denotes the τ-ordering operator and $\hat{\rho}_G \equiv Z_G^{-1} e^{-\beta\hat{K}}$ is the Grand Canonical density operator. We note that the real-time operator $\hat{\psi}_S^{\alpha+} (\mathbf{x})$ is the adjoint of $\hat{\psi}_S^\alpha (\mathbf{x})$ but $\hat{\psi}_K^{\alpha+} (\tau, \mathbf{x})$ is not the adjoint of $\hat{\psi}_K^\alpha (\tau, \mathbf{x})$ as long as τ is real.

A fundamental property of the temperature Green's functions is *periodicity* in imaginary time. Specifically, we find

$$G_{\alpha\beta} (0, \mathbf{x}; \tau', \mathbf{x}') = \pm G_{\alpha\beta} (\beta\hbar, \mathbf{x}; \tau', \mathbf{x}'), \tag{21.20}$$

where the plus sign applies for bosonic (commuting) fields and the minus sign applies for fermionic (anticommuting) fields. We conclude that temperature Green's functions are periodic in the τ argument of the fields, with period $\beta\hbar$, a phenomenon alien to real-time physics. This has implications for the domain of integration in partition functions, as we shall see in the next section.

---

[71] At best we would classify them as mathematical metaphysics.

Black hole thermodynamics

Periodicity in imaginary time is used in black hole physics and quantum cosmology. Consider the real-time Schwarzschild line element metric (19.5). Transforming to imaginary time $t = -i\tau$ and rescaling the radial coordinate by the rule $R \equiv 2r_s\sqrt{1 - r_s/r}$ in the region $r > r_s$, where $r_s \equiv 2GM/c^2$ is the Schwarzschild radius, gives

$$d\sigma^2 = \left(\frac{Rc}{2r_s}\right)^2 d\tau^2 + \frac{r^4}{r_s^4} dR^2 \tag{21.21}$$

along the radius, that is, for $d\Omega = 0$. Here we define $d\sigma^2 \equiv -ds^2$: it gives us the Euclidean signature metric for the regions of integration in the relevant partition function. Unlike real-time path integrals, the domain of τ is compact, because τ is periodic.

We stated above that imaginary time τ is related to temperature in quantum statistical mechanics and that the temperature Green's functions were periodic. This leads to the definition of the dimensionless variable $\theta \equiv c\tau/2r_s$ and then (21.21) becomes, for R close to zero,

$$d\sigma^2 = R^2 d\theta^2 + dR^2 + \ldots, \qquad 0 \leqslant R < 2r_s. \tag{21.22}$$

The first two terms on the right-hand side look like the metric on the surface of a cylinder, provided θ is an angular coordinate with period 2π. Imposing this condition means that τ has period $P = 8\pi GM/c^3$. If we go further and regard the system as in thermodynamic equilibrium at some temperature T then we find $P = \beta\hbar = \hbar/k_B T$, from which we deduce $T = c^3\hbar/8\pi Gmk_B$. This is precisely the temperature of the black body spectrum of emitted particles near a black hole of mass m, first predicted by Bekenstein using thermodynamics [Bekenstein, 1973] and Hawking using field theory [Hawking, 1976].

Quantum cosmology

Another application of imaginary time has been to quantum cosmology. With Hubble's discovery of the red shift, Einstein's GR gave a paradigm for classical evolutionary models of an expanding universe. A conceptual problem occurred when these models were used to retrodict the origin of the universe in a hypothesized Big Bang. Theorems by Hawking and Penrose predict a classical spacetime singularity at the origin of time, which is usually regarded as physically incorrect or inadmissible. A possible way out is to invoke QM and imaginary time. In the path integral approach of Hawking, Hartle, and others, the ill-defined QM vacuum functional $Z \sim \int [dg] \exp\{iS_G/\hbar\}$ is rewritten in imaginary time, that is $\tilde{Z} \sim \int [dg] \exp\{-\tilde{S}_G/\hbar\}$, where S_G is the real-time Einstein–Hilbert action and $\tilde{S}_G$ its imaginary time equivalent. Here, $[dg]$ refers to a functional integral over the relevant gravitational degrees of freedom. A specific proposal discussed by several theorists is that in the part of the integral involving the early universe, time is imaginary and therefore periodic. This is the Hartle–Hawking 'no-boundary' proposal [Hartle & Hawking, 1983], giving a picture of the early universe where the classical singularity is avoided by the instanton (imaginary time) properties of a curved Euclidean signature spacetime. There should be concerns however about the interpretation of this approach. It is not obvious how much is put in 'by hand' and how much occurs naturally, and there are at least three variants of this scenario.

Conclusions

The imaginary-time formulation is a useful extension of continuous real time in various branches of SR, GR, and QM, but there are deep uncertainties as to the physical meaning of what is involved. Of course, QM itself suggests that the universe might not be explicable in every sense of the word, but we should certainly worry about the operational meaning of imaginary time. If the only way the early universe could be understood is in terms of what amounts to a 'black box' approach, then the objective of obtaining a fundamental theory seems to have been defeated. Imaginary time touches on deep issues involving thermodynamics and the relationship of mixed states versus pure states in QM. If the view is taken that the *only* thing of physical significance is the irreversible acquisition of information, which is in fact the only thing laboratory physics can deal with, then physics should be described by asymmetric process time rather than the symmetric manifold time of a geometric, Block Universe. Imaginary time lacks any asymmetry or arrow of time, which is the perhaps the real cause for concern.

22

Irreversible time

Introduction

One of the paradoxes of modern physics is that whilst the universe appears to run in an irreversible way when viewed on macroscopic scales, many of the fundamental laws governing elementary particle processes appear to be reversible. How can this conundrum be resolved?

To understand what is meant by irreversibility, we need to ensure that our discussion is contextually complete: it is no good talking about a reversible system under observation (SUO) if we make no reference to the measurement processes that can establish that reversibility. That is where our resolution of the conundrum is found. Observation is an irreversible process by definition: it is a comparison by an observer of information from different times. We cannot establish that something has changed without comparing a memory of what it was like before with what it is like now. That requires a contextually complete description, but unfortunately all too often in such discussions we are given propositions that are contextually incomplete.

Glauber's correlations

In relativistic quantum field theory, Feynman diagrams are useful pictorial methods to aid in the calculation of scattering amplitudes. Typically, such amplitudes are vacuum expectation values of time-ordered products of field operators. For example, a two-particle-to-two-particle scattering amplitude involving spinless fields would require knowledge of a four-point function, such as

$$G(x_1, x_2, x_3, x_4) \equiv (0 \,|\, T\hat{\varphi}(x_1)\hat{\varphi}(x_2)\hat{\varphi}(x_3)\hat{\varphi}(x_4)\,|\,0), \qquad (22.1)$$

where $|0)$ is the vacuum (empty space) state vector, T is the time-ordering operator, and $\hat{\varphi}(x)$ is a field operator at the spacetime event x. What is relevant to us here is that Feynman diagrams are perturbative expansions[72] involving relativistic

---

[72] Or asymptotic expansions, according to Dyson.

Images of Time. First Edition. George Jaroszkiewicz.
© George Jaroszkiewicz 2016. Published in 2016 by Oxford University Press.

field operators with positive energy parts and negative energy parts. The interpretation of these parts is that positive energy terms propagate forwards in time whilst negative energy terms propagate backwards. For charged particles such as electrons, the negative energy backwards propagation is interpreted as forwards propagation of antiparticles (positrons in the case of electrons).

In conventional quantum field theory, irreversible information extraction does not take place in the 'information void', the region between the preparation devices and the outcome detectors, simply because there *are* no detectors there. In the Lehmann, Symanzik, and Zimmerman (LSZ) approach to scattering theory [Lehmann *et al.*, 1955], all apparatus is assumed to be at virtually infinitely large negative and positive times respectively. In the void (or empty space region), there are no detectors, and therefore no information extraction. The calculations reflect this fact, with particles and antiparticles allowed to come and go, pop in and out of the vacuum, as long as necessary conservation laws are upheld, such as total charge and total energy conservation, and so on.

The situation changes when information is extracted locally. Now irreversibility comes into play. The physicist Glauber discussed this in an influential paper looking at photon correlation experiments, where the focus is on the probability of catching one photon at one place and another photon at another [Glauber, 1963b]. We shall follow Glauber's ideas that he developed for photons.

In photon detection experiments, the important operator is the electric field operator $\hat{\mathbf{E}}(\mathbf{x}, t)$, given in principle by a positive energy part $\hat{\mathbf{E}}^{(+)}(\mathbf{x}, t)$ and a negative energy part $\hat{\mathbf{E}}^{(-)}(\mathbf{x}, t)$, such that

$$\hat{\mathbf{E}}(\mathbf{x}, t) = \hat{\mathbf{E}}^{(+)}(\mathbf{x}, t) + \hat{\mathbf{E}}^{(-)}(\mathbf{x}, t). \tag{22.2}$$

The positive energy part has the property that it annihilates the vacuum (the state representing empty space), viz., $\hat{\mathbf{E}}^{(+)}(\mathbf{x}, t)|0\rangle = 0$. Likewise, $\langle 0|\hat{\mathbf{E}}^{(-)}(\mathbf{x}, t) = 0$. On the other hand, $\hat{\mathbf{E}}^{(-)}(\mathbf{x}, t)|0\rangle$ is a one-photon state.

Glauber looked at the way classical electromagnetic theory was discussed. He noted that it was customary to think of the classical electric field $\mathbf{E}(\mathbf{x}, t)$ as the object to study, because it is real, whereas the classical fields $\mathbf{E}^{(+)}(\mathbf{x}, t)$ and $\mathbf{E}^{(-)}(\mathbf{x}, t)$, corresponding to the quantum field operators $\hat{\mathbf{E}}^{(+)}(\mathbf{x}, t)$ and $\hat{\mathbf{E}}^{(-)}(\mathbf{x}, t)$, are complex. He noted that this would be valid only in the situation where the frequency ω of the classical field is so low that the energy $\hbar\omega$ of a photon of that frequency would be negligible. In such a regime, the observer could not tell whether photons were being emitted or absorbed by the electrical charges in the system, that is, there would be no irreversibility in the system as far as the observer was concerned.

Turning to the quantum theory, Glauber looked at the matrix element

$$\langle f|\hat{E}^{(+)}(\mathbf{x}, t)|i\rangle, \tag{22.3}$$

where $|i\rangle$ is an initial state with n photons, $|f\rangle$ is a final state with $n-1$ photons, and $\hat{E}^{(+)}(\mathbf{x},t)$ is the field operator associated with an absorbed polarized photon. The total rate of absorption ρ is given by the square modulus of this matrix element summed over all possible final states. This is found to be given by $\rho = \langle i|\hat{E}^{(-)}(\mathbf{x},t)\hat{E}^{(+)}(\mathbf{x},t)|i\rangle$.

An important observation here is that ρ is zero when the initial state is the vacuum state. This makes sense: a photon detector should register nothing in empty space. On the other hand, Glauber noted that if the operator $\hat{E}(\mathbf{x},t) \equiv \hat{E}^{(+)}(\mathbf{x},t) + \hat{E}^{(-)}(\mathbf{x},t)$ was used instead, then the vacuum expectation value $\langle 0|\hat{E}(\mathbf{x},t)\hat{E}(\mathbf{x},t)|0\rangle$ does *not* vanish. The interpretation of this is that the electric field in the vacuum state undergoes so-called zero-point oscillations, which Glauber notes have nothing to do with the detection of photons.

The critical importance of these comments to us is that they support our view that conventional relativistic quantum field theory is an excellent theory but it needs careful interfacing with the realities of observation. These are that observation is invariably local in time and space, and involves irreversible processes.

It is metaphysical and physically inconsistent to discuss 'truly' reversible processes: how could one ever know that such a phenomenon was going on? This returns us to Wheeler's participatory principle discussed in Chapter 2. We require observers because without them we cannot talk about the universe in any meaningful way. But observations necessarily involve irreversibility. Therefore, the universe that we observe *must* be be irreversible, as far as we are concerned.

That is a contextual conclusion, relevant only to us. This raises the bizarre notion that the universe could be reversible, relative to some exophysical superobserver, but as far as we are concerned, it is not.

Probability

Probability is all about time and information: if we know that the probability of something happening tomorrow is 1,[73] then we are certain that it will happen. Our certainty, unfortunately, is not a guarantee of the event actually happening: we may be deluding ourselves or our information about what is going to happen may be faulty. On the other hand, if we have incomplete information, then the probability we estimate may be less than 1.

Probability has always been a problematic concept because it involves two parties: the system under observation (SUO) to which the laws of probability are being applied and the observer. It is a fundamental mistake to exclude the observer and think of probability as an objective property of an SUO. Once an observer is included, irreversible time is brought in explicitly, because probability makes no sense without it.

---

[73] Mathematicians generally assign the value 1 to something that will certainly happen and a probability value of 0 to something that will not happen, rather than 100% and 0 respectively.

The expansion of the universe

It has been known since the late 1920s that light received from very distant galaxies is red-shifted from its expected wavelengths. This has generally been interpreted as an expansion of the universe. Since the red shift increases the more distant the galaxy is, the conclusion is that there was a period in the past when the universe was highly compressed compared to its condition now. Therefore, the universe appears to be running on irreversible lines.

However, some theorists have speculated that the gravitational attraction of matter will eventually reverse the expansion, resulting in a collapse known as a Big Crunch. Einstein considered a series of expansions and contractions, but a potential problem with that was pointed out by Tolman, who argued that whilst the laws of mechanics might be time reversible, the laws of thermodynamics are not, and they say that the entropy of the universe can only increase.

Our view is that thermodynamics has to be discussed contextually. All the evidence points to that. Statistical mechanics requires a definition of microcanonical ensemble, coarse graining, and such like concepts that depend on choices made by observers. In the same way that quantum cosmology can be criticized for being too loose with the concept of ensemble and observer [Fink & Leschke, 2000], it seems to us to be equally inadmissible to discuss the thermodynamics of the universe in a contextually incomplete way.

Poincaré recurrence

The mathematician Poincaré tackled a difficult problem involving the behaviour of mechanical systems. In the phase-space approach to mechanics, discussed in Chapter 14, suppose the 'volume' of phase space is finite. Such a supposition may appear bizarre, considering that linear momentum in Newtonian mechanics is generally unbounded, that is, can take on any value. However, there are mechanical systems where phase space is of finite volume.

Now consider a cloud of system points, initially filling a small region $\delta V(0)$ of phase space at time $t = 0$. Liouville's theorem asserts that for a system of such points that obey Hamilton's equations of motion (14.7), their volume is conserved, that is, at time $t > 0$, $\delta V(t) = \delta V(0)$. The shape of this volume need not be preserved, but its magnitude will be constant. Therefore, we may imagine this cloud to sweep out a higher dimensional form of 'tube' in phase space, as time progresses. It seems plausible that such a tube will eventually have to return 'close' to the original volume $\delta V(0)$, given that the volume of this phase space is finite. The time to return in some sense to be specified is known as the Poincaré recurrence time.

Recurrence times will vary greatly depending on the conditions of the SUO, its starting positions, and so on. One heuristic estimate for the Poincaré recurrence time for a black hole of solar mass is about $10^{10^{10^{76.66}}}$ Planck times[74] [Page, 1995].

---

[74] Or millenia: it does not matter on this scale.

23
Discrete time

Introduction

> By building our study of the physical world ultimately upon the assumption of the ubiquity of continuous changes at the most macroscopic level, we may be making not merely a gross simplification but an infinite simplification.   [Barrow, 1992]

In Chapter 7 we looked at the mathematics of time, focusing principally on the real numbers. These have the ordering property generally considered essential to any image of time. The reals also have continuity, generally considered to be an essential property of time: continuity is a necessary ingredient in the calculus definition of velocity and acceleration, fundamental concepts in Newtonian classical mechanics (CM). But some reflection should make us wary of assuming too much here. The existence of atoms, the modern view of temperature as a statistical parameter, and the ubiquitous success of quantum mechanics (QM), all provide counter examples to the view that the universe is inherently based on principles of continuity.

We should be prepared, therefore, to view the continuity of time as a contextual concept. When used in some contexts, time can be modelled as a continuous parameter, whilst in other contexts a discrete-time (DT) model is more appropriate:

> After three hundred years during which mathematical physics has been dominated by the continuous geometrical time of Galileo, Barrow, and Newton, the idea that time is atomic, or not infinitely divisible, has only recently come to the fore as a daring and sophisticated hypothetical concomitant of recent investigations in the physics of atoms and elementary particles.   [Whitrow, 1980]

DT explores the notion that time itself has fundamental units or building blocks, rather than consisting of a continuum of points. DT has a long history but only recently has it begun to be explored in depth. The Jewish philosopher Maimonides wrote:

> Time is composed of time atoms, i.e. of many parts, which on account of their short duration cannot be divided ... An hour is. e.g. divided into sixty minutes,

Images of Time. First Edition. George Jaroszkiewicz.
© George Jaroszkiewicz 2016. Published in 2016 by Oxford University Press.

the minutes into sixty seconds, the second into sixty parts, and so on; at last after ten or more successive divisions by sixty, time-elements are obtained which are not subjected to division, and in fact are indivisible . . .   [Maimionides, 1190]

He paid no particular penalty for such thinking. Nicholas d'Autrecourt was not so fortunate: he had to renounce his ideas and burn his books.

Difference equations

Many of the great laws of physics are given in the form of differential equations, which presupposes that dynamical variables are differentiable functions of time. This in turn presupposes that those variables are also continuous functions of time. But what if that were not the case? What if time was not continuous, but some discrete set? Such a possibility has been investigated by several theorists such as Caldirola and the Italian School [Farias & Recami, 2010], and Sorkin and co-workers [Bombelli & Sorkin, 1987].

There are two sorts of reason for thinking of time as discrete. We may well believe that time is continuous, but have reasons for discretizing. For example, our differential equations may be impossible to solve so we want to simulate them on a computer. That requires discretization of time. Every computer game will involve such a process. On the other hand, we may believe, like Caldirola, that time is discontinuous in the first place [Caldirola, 1978]. A popular term for a hypothetical smallest increment of time is *chronon*.

Regardless of the motivation, the question arises: *what replaces differential equations?* A common answer is: *difference equations*. To understand these, let us look at the standard definition of the derivative of a differentiable function of time. Suppose $x(t)$ is such a function. Then its derivative $\dot{x}(t)$ (in Newton's fluxion notation) is defined by the limit

$$\dot{x}(t) \equiv \lim_{h \to 0} \left\{ \frac{x(t+h) - x(t)}{h} \right\}. \tag{23.1}$$

Now we may approximate this derivative by *not taking the limit*, viz., everywhere $\dot{x}(t)$ occurs we replace it with the differential quotient, the term in brackets on the right-hand side of equation (23.1). To see what we mean, suppose we had the differential equation

$$\dot{x}(t) = f(x(t)), \tag{23.2}$$

where f is some given function. Then a suitable discretization of (23.2) might be the difference equation

$$x(t+h) - x(t) = hf(x(t)). \tag{23.3}$$

This difference equation is based on taking the *forwards difference* in the differential quotient. But we know from standard analysis that we could just as easily have written

$$\dot{x}(t) \equiv \lim_{h \to 0} \left\{ \frac{x(t) - x(t-h)}{h} \right\} \tag{23.4}$$

instead of (23.1). Then our discretized equation would have been

$$x(t) = x(t-h) + hf(x(t)). \tag{23.5}$$

Which one of these discretizations is right?

The answer is that there is no 'right' discretization. Differentiability imposes a strong condition on a function. When we undo that condition and replace it with a difference, we are losing a constraint, which means we are getting more general. As a rule, generalizations become less specific than the equations they are generalizing. If that were not the case, we should be able to derive discrete equations from differential equations, and that is absurd.

Mathematicians have in fact long worked with difference equations: we shall show below how they can serve to model discrete time mechanics. To understand the architecture of what is involved, consider the following scenario. We are given some non-empty set A. Now imagine picking out of A a number r of elements, denoted $x_1, x_2, \ldots, x_r$, where r is some positive integer. Suppose now that, given such a subset of A, there is some rule[75] F that points out some element of A, which we denote by x_{r+1}. Then we can write

$$x_{r+1} = F(x_1, x_2, \ldots, x_r). \tag{23.6}$$

There is nothing in this that prevents x_{r+1} being one of the initial elements $x_1, x_2, \ldots, x_r$. Suppose now that we take the initial ordered set $(x_1, x_2, \ldots, x_r)$, replace it with $(x_2, x_3, \ldots, x_{r+1})$, and now use that new set with the rule F to determine another element, denoted x_{r+2}, viz.,

$$x_{r+2} = F(x_2, x_2, \ldots, x_{r+1}), \tag{23.7}$$

and so on. Then we see that given the initial set $\{x_1, x_2, \ldots, x_r\}$ we can generate a sequence $x_{r+1}, x_{r+2}, x_{r+3}, \ldots$ ad infinitum. This is just like a dynamical system evolving in time from a given starting configuration and according to some specific law of motion. We shall say that this is an rth-order discrete dynamical system. Note that this assumes that there is an exophysical observer with a sense of time, an observer that has an internal clock ticking off successive calculations. This is just like a *Turing machine* [Hodges, 2014].

---

[75] Or law of physics if we are dealing with the real world.

The relationship with difference equations is readily understood: we can always write (23.6) in the form $x_{r+1} - x_r = G(x_1, x_2, \ldots, x_r)$, where $G(x_1, x_2, \ldots, x_r) \equiv F(x_1, x_2, \ldots, x_r) - x_r$.

The above example can be generalized to arbitrary sets A and to different integer values of r. For example, a first-order discrete dynamical system has a rule of the form $x_{n+1} = F(x_n), n = 0, 1, 2, \ldots$, a second-order discrete dynamical system has a rule of the form $x_{n+1} = F(x_{n-1}, x_n)$, and so on. It turns out that second-order differential equations, such as those typically found in Newtonian mechanics, are usually replaced by second-order discrete dynamical equations.

The action sum

We outline here how a CM SUO can be related to a DT SUO [Jaroszkiewicz, 2014]. As we saw in earlier chapters, the variational approach to continuous time (CT) Lagrangian dynamics is formulated via an action principle based on the action integral

$$A_{if}[\Gamma] = \int_{t_i}^{t_f} dt L(\mathbf{q}, \dot{\mathbf{q}}, t), \tag{23.8}$$

where t_i and t_f are the initial and final times respectively along some given differentiable classical path Γ in configuration space. In our version of discrete-time mechanics we postulate that the dynamical variables $\mathbf{q}(t)$ are observed or sampled at a finite number of times $t_n, n = 0, 1, \ldots, N$, where $t_0 = t_i$ and $t_N = t_f$, such that the intervals $t_{n+1} - t_n$ are all equal to some fundamental interval T, the chronon. For convenience we will write $\mathbf{q}_n \equiv \mathbf{q}(t_n)$. It is possible to develop a theory where the time intervals vary dynamically along the path, such a mechanics being considered by [Lee, 1983]. In our formulation of DT mechanics we replace the action integral (23.8) by an *action sum* of the form

$$A^N[\Gamma] = \sum_{n=0}^{N-1} F^n, \tag{23.9}$$

where $F^n \equiv F(\mathbf{q}_n, \mathbf{q}_{n+1}, n)$ is the *system function*. The system function has the same central role in DT mechanics as the Lagrangian has in CT mechanics. With it we may construct the equations of motion, define conjugate momenta, construct constants of the motion, and attempt to quantize the system. In principle we could consider higher order system functions which depended on (say) $\mathbf{q}_n, \mathbf{q}_{n+1}, \ldots,$ $\mathbf{q}_{n+r}, r \geq 2$, but the case $r = 1$ represents the simplest possibility which could give rise to non-trivial dynamics and will be considered exclusively from now on. Such system functions are the DT analogues of CT Lagrangians of the canonical form $L = L(\mathbf{q}, \dot{\mathbf{q}}, t)$.

Another reason for considering only a second-order formulation ($r = 1$) is its direct relationship to Hamilton's principal function, discussed presently. Cadzow [Cadzow, 1970] applied a variational principle to an action sum such as (23.9) and derived the equation of motion

$$\frac{\partial}{\partial \mathbf{q}_n} \left\{ F^{n-1} + F^n \right\} \underset{c}{=} \mathbf{0}, \quad 0 < n < N, \tag{23.10}$$

where the symbol $\underset{c}{=}$ denotes an equality holding over a true or dynamical trajectory. We shall refer to (23.10) as a *Cadzow's equation of motion* for the system.

We now discuss the interpretation of equation (23.10). Suppose we have a continuous-time action integral of the form (23.8). First, partition the time interval $[t_0, t_N]$ into N equal subintervals. Then the action integral may be written as a sum of sub-integrals, that is,

$$A_{if}[\Gamma] = \sum_{n=0}^{N-1} \int_{t_n}^{t_{n+1}} dt\, L(\mathbf{q}(t), \dot{\mathbf{q}}(t), t). \tag{23.11}$$

Now suppose that we fixed the coordinates $\mathbf{q}_n$ at the various times $t_0, t_1, \ldots, t_N$ and then chose the path connecting each pair of points $(\mathbf{q}_n, \mathbf{q}_{n+1})$ to be the true or dynamical path, that is, a solution to the Euler–Lagrange equations of motion for those boundary conditions. If this partially extremized path is denoted by $\tilde{\Gamma}_c$ then we may write

$$A_{if}[\tilde{\Gamma}_c] = \sum_{n=0}^{N-1} S^n, \tag{23.12}$$

where $S^n \equiv S(\mathbf{q}_{n+1}, t_{n+1}; \mathbf{q}_n, t_n)$ is known as Hamilton's principal function, being just the integral of the Lagrangian along the true path from $\mathbf{q}_n$ at time t_n to $\mathbf{q}_{n+1}$ at time t_{n+1}.

With reference to the transformation theory discussed in Chapter 14, the canonical momenta $\mathbf{p}_n^{(+)}$, $\mathbf{p}_{n+1}^{(-)}$ at the end points $\mathbf{q}_n, \mathbf{q}_{n+1}$ may be obtained from Hamilton's principal function via the rule

$$\mathbf{p}_{n+1}^{(-)} \equiv \frac{\partial}{\partial \mathbf{q}_{n+1}} S^n, \quad \mathbf{p}_n^{(+)} \equiv -\frac{\partial}{\partial \mathbf{q}_n} S^n, \tag{23.13}$$

where the superscript (+) denotes that the momentum at the initial time t_i carries information *forwards in time,* whereas the superscript (−) denotes that the momentum at the final time t_f is influenced by earlier dynamics with respect to the temporal interval concerned. At this stage the action sum (23.12) has not been extremized fully, as the intermediate points $\mathbf{q}_n, 0 < n < N$ have been held fixed. Now suppose we went further and extremized (23.12) fully by variation of the previously fixed intermediate coordinates $\mathbf{q}_n, n = 1, 2, \ldots, N - 1$. Then we would find that

$$\frac{\partial}{\partial \mathbf{q}_n} \left\{ S^{n-1} + S^n \right\} \underset{c}{=} \mathbf{0}, \quad 0 < n < N, \tag{23.14}$$

which should be compared with Cadzow's equation (23.10). We notice immediately that (23.14) has the same formal structure as Cadzow's equation (23.10) provided we make the identification $F^n \leftrightarrow S^n$.

From this, we see that Cadzow's equation may be understood as the condition that the canonical momentum along the true path from $\mathbf{q}_i$ to $\mathbf{q}_f$ is continuous, that is, $\mathbf{p}_n^{(+)} \underset{c}{=} \mathbf{p}_n^{(-)}$. Another interpretation of Cadzow's equation is that it endows the action sum with the additivity property of action integrals: these satisfy the relations

$$\int\limits_{t_0}^{t_1} dt L + \int\limits_{t_1}^{t_2} dt L = \int\limits_{t_0}^{t_2} dt L, \quad t_0 < t_1 < t_2. \tag{23.15}$$

This property holds for all trajectories in continuous-time mechanics, not just for the true or classical trajectory. In the case of system functions we may write

$$F(\mathbf{q}_{n-1}, \mathbf{q}_n) + F(\mathbf{q}_n, \mathbf{q}_{n+1}) \underset{c}{=} f(\mathbf{q}_{n-1}, \mathbf{q}_{n+1}) \tag{23.16}$$

for some function f of $\mathbf{q}_{n-1}$ and $\mathbf{q}_{n+1}$, because Cadzow's equation (23.10) is equivalent to the statement that $F^n + F^{n-1}$ is independent of $\mathbf{q}_n$ along dynamical trajectories. However, unlike action integrals, this property will not hold off the true or classical trajectory in general.

Caldirola's proper time chronon

In a series of articles [Caldirola, 1978, 1979], Caldirola investigated the notion of a *proper time chronon*, which is given by the quantum mechanical relationship $\tau_1 \equiv \hbar/mc^2$, where m is the rest mass of the particle concerned. This may also be called the *Compton time*, and is a purely quantum concept. We note the absence of G, the Newtonian constant of gravitation, in this expression. For the electron we find $\tau_1 \approx 1.3 \times 10^{-21}$ seconds.

Another model discussed by Caldirola that does not involve Planck's constant encounters a chronon $\tau_2 \equiv e^2/(6\pi\varepsilon_0 mc^2) \approx 6.3 \times 10^{-24}$ seconds, where e is the electric charge and ε_0 is the permittivity of free space. It is clear that there is a great difference between the magnitude of any of these chronons and the Planck time, discussed in Chapter 10, by an enormous factor of about 10^{20} or more.

Caldirola's microverse model

Following his work on proper time chronons, Caldirola and Benza [Benza & Caldirola, 1981] discussed a model which attempted to give a dynamical origin for the existence of proper time chronons. In this model, a particle such as an electron sits in ordinary Minkowski spacetime $\mathcal{M}^4$, which retains all its usual attributes such as providing the arena for physical measurements on the electron system by an external observer. Over and above any normal electron system attributes, Caldirola and Benza proposed that the particle would also carry with it an internal four-dimensional spacetime, V^4. This spacetime was assumed to be a de Sitter spacetime, a special type of general relativistic spacetime. Four-dimensional de Sitter spacetime can be regarded as the Cartesian product manifold $\mathbb{R} \times \mathbb{S}^3$, where $\mathbb{R}$ is time and $\mathbb{S}^3$ is the three sphere embedded in four-dimensional Euclidean space $\mathbb{E}^4$.

Caldirola and Benza wrote down some equations for the dynamical evolution of the curvature of this V^4 with respect to external proper time (i.e. the time associated with some observer O sitting in the rest frame of the electron in $\mathcal{M}^4$), and suggested that the radius of V^4 would undergo periodic oscillation. According to their ideas, there would be critical times when V^4 had collapsed to a point. Subsequently, V^4 would undergo a period of expansion, reach a maximum, and then contract back down again to a point. The whole process would then be repeated indefinitely. The (proper) time measured by O between successive collapsed states of V^4 thereby provide a motivation for the notion of a proper time chronon.

This is not the only place where de Sitter spacetimes are relevant to us in this book. In 1947, Hartland Snyder postulated an algebra for quantized spacetime operators, which we discuss in Chapter 27. In his first paper [Snyder, 1947b], Snyder acknowledges Wolfgang Pauli's comment that Snyder's variables η^μ, $\mu = 0, 1, 2, 3, 4$, 'may be regarded as the homogeneous (projective) coordinates of a real four-dimensional space of constant curvature (a de Sitter space)'.

Discrete-time classical electrodynamics

In this section we show that the discretization of time applied to Maxwell's equations of electrodynamics gives difference equations that encode a discrete-time version of gauge invariance. To see how this works out, we review briefly the continuous-time Maxwell's equations. We use the convention $c = \mu_0 = \varepsilon_0 = 1$.

The continuous-time theory

There are four physical fields and an auxiliary electromagnetic potential field. The physical fields are the electric field $\mathbf{E}$, the magnetic field $\mathbf{B}$, the electric charge density ρ, and the charge current $\mathbf{j}$. These satisfy Maxwell's equations:

$$\nabla \cdot \mathbf{B} = 0, \quad \nabla \times \mathbf{E} + \partial_t \mathbf{B} = \mathbf{0},$$
$$\nabla \cdot \mathbf{E} = \rho \quad \nabla \times \mathbf{B} - \partial_t \mathbf{E} = \mathbf{j}. \tag{23.17}$$

These equations give the continuity equation $\partial_t \rho + \nabla \cdot \mathbf{j} = 0$, which is consistent with charge conservation.

The electric and magnetic fields $\mathbf{E}$, $\mathbf{B}$ can be written as $\mathbf{E} = -\nabla A^0 - \partial_t \mathbf{A}$, $\mathbf{B} = \nabla \times \mathbf{A}$, where $A^0, \mathbf{A}$ are the components of the electromagnetic four-vector potential $A^\mu \equiv (A^0, \mathbf{A})$. The potentials are not physical fields, in that they can be replaced by the gauge transformed fields $A'^\mu = A^\mu + \partial^\mu \chi$, where χ is an arbitrary gauge field. The four-vector potential A^μ satisfies the Lorentz covariant equation

$$\Box A^\nu - \partial^\nu (\partial_\mu A^\mu) = \underset{c}{j^\mu}, \qquad (23.18)$$

where $j^\mu \equiv (\rho, \mathbf{j})$ is the four-vector charge current. The physical fields are gauge invariant.

The discrete-time theory

We now show how we can rewrite the above CT equations in DT terms. All these DT equations can be derived from first principles, that is, from the DT calculus of variations [Jaroszkiewicz, 2014].

The important DT operator is the temporal displacement operator $\mathbb{U}_n$: for any function f_n indexed by discrete time n, we have $\mathbb{U}_n f_n = f_{n+1}$. The inverse temporal operator is denoted $\overline{\mathbb{U}}_n$, which has action $\overline{\mathbb{U}}_n f_n = f_{n-1}$. With these operators we define the operators $\mathbb{T}_n \equiv T^{-1}(\mathbb{U}_n - 1)$, $\overline{\mathbb{T}}_n \equiv T^{-1}(1 - \overline{\mathbb{U}}_n)$, $\mathbb{S}_n \equiv 6^{-1}(\mathbb{U}_n + 4 + \overline{\mathbb{U}}_n)$, and $\Box_n \equiv \mathbb{T}_n \overline{\mathbb{T}}_n - \mathbb{S}_n \nabla^2$. Here T is the chronon, or shortest interval of discrete time. In the limit $T \to 0$, we have the following effective limits

$$\lim_{T \to 0} \left\{ \mathbb{T}_n, \overline{\mathbb{T}}_n, \mathbb{S}_n, \Box_n \right\} \to \left\{ \frac{\partial}{\partial t}, \frac{\partial}{\partial t}, 1, \Box \equiv \frac{\partial^2}{\partial t^2} - \nabla^2 \right\}. \qquad (23.19)$$

Temporal discretization naturally involves *nodes* and *links*. The former are temporal points of zero duration whilst the latter are the temporal intervals between those points. In our discretization scheme, the links have temporal duration T. Here we show what happens for fixed chronon duration: it is possible to develop DT mechanics based on variable chronon duration, something considered by Lee in his DT path integral approach [Lee, 1983].

Some DT fields are associated with temporal nodes whilst others are naturally associated with the temporal links. Table 23.1 shows the associations we have used in our approach to DT Maxwell electrodynamics [Jaroszkiewicz, 2014].

Table 23.1 *The association of the various fields with nodes or links.*

	Node n	Link n
Potential fields	$\mathbf{A}_n$	A^0
Physical fields	$\mathbf{B}_n \equiv \nabla \times \mathbf{A}_n$	$\mathbf{E}_n \equiv -\nabla A_n^0 - \mathbb{T}_n \mathbf{A}_n$
Charge fields	$\mathbf{j}_n$	$j_n^0 \equiv \rho_n$
Gauge fields	χ_n	

Note that the operator $\mathbb{T}_n$ converts a node field into a link field and vice-versa, whilst ∇ leaves the association unchanged.

The physical fields $\mathbf{E}_n$, $\mathbf{B}_n$, $j_n^\mu \equiv (\rho_n, \mathbf{j}_n)$ are invariant to the DT gauge transformation

$$A_n^0 \to A_n'^0 = A_n^0 + \mathbb{T}_n \chi_n, \qquad \mathbf{A}_n \to \mathbf{A}_n' = \mathbf{A}_n - \nabla \chi_n. \tag{23.20}$$

The DT equivalent of Maxwell's equations are then

$$\begin{aligned} \nabla \cdot \mathbf{B}_n = 0, \qquad &\nabla \times \mathbf{E}_n + \mathbb{T}_n \mathbf{B}_n = 0 \\ \nabla \cdot \mathbf{E}_n = \rho_n, \quad &\nabla \times \mathbb{S}_n \mathbf{B}_n - \overline{\mathbb{T}}_n \mathbf{E}_n = \mathbf{j}_n, \end{aligned} \tag{23.21}$$

which give the DT continuity equation $\overline{\mathbb{T}}_n \rho_n + \nabla \cdot \mathbf{j}_n = 0$.

The fields satisfy the second-order DT equations

$$\begin{aligned} \Box_n \mathbf{B}_n = \nabla \times \mathbf{j}_n \qquad &\Box_n \mathbf{A}_n + \nabla \Lambda_n = \mathbf{j}_n, \\ \Box_n \mathbf{E}_n = -\mathbb{S}_n \nabla \rho_n - \mathbb{T}_n \mathbf{j}_n, \quad &\Box_n A_n^0 - \mathbb{T}_n \Lambda_n = -\mathbb{S}_n \rho_n. \end{aligned} \tag{23.22}$$

where $\Lambda_n \equiv \overline{\mathbb{T}}_n A_n^0 + \mathbb{S}_n \nabla \cdot \mathbf{A}_n$. A DT Lorentz gauge is one where the gauge function χ_n satisfies the relation $\Box_n \chi_n = -\Lambda_n$. Then in such a gauge, $\Lambda_n' = 0$ and we have the equations for the transformed potentials:

$$\Box_n \mathbf{A}_n' = \mathbf{j}_n, \qquad \Box_n A_n'^0 = \mathbb{S}_n j_n^0. \tag{23.23}$$

All of these DT equations collapse back to the CT Maxwell's equations in the limit of the chronon T going to zero.

24
Time and quanta

Introduction

Time plays a curious role in quantum mechanics (QM). In the Schrödinger picture, its primary function is to be a label for successive states of a system under observation (SUO) within a formalism that is inextricably linked with processes of measurement, observability, and information extraction. Yet time seems to play no role in those processes per se, standing outside of those processes, preserving its essential classical attribute as a parameter.

Time in QM is logically interpreted as the time in the observer's laboratory, not as an objective attribute of an SUO. We say this because of our particular interpretation of the Schrödinger wavefunction. Not every theorist will agree with this, because some see the wavefunction in objective terms, something that is 'there' even if no one is looking at it. Schrödinger thought in those terms when he originally wrote down his famous equation in 1926 [Schrödinger, 1926c].

Our interpretation of quantum states is motivated partly by the fact that we can choose to describe experiments in the so-called Heisenberg picture (HP) rather than the Schrödinger picture (SP).[76] In the HP, quantum states are 'frozen' relative to the laboratory clock, all time evolution being associated with the observables, the operators representing the processes of observation. The fact that this is consistent with the SP, in which quantum states do change in time, underlies the essential point we want to reinforce in this book, that quantum mechanics is all about the processes of observation.

For convenience, we will use a hybrid notation for the basic quantum state equation of motion, halfway between Schrödinger's original wavefunction representation and Dirac's ket notation for the quantum states [Dirac, 1958]:

$$i\hbar \frac{d}{dt}\Psi_t = \hat{H}_t\Psi_t, \tag{24.1}$$

---

[76] There are other reasons, such as contextual incompleteness, if we do not mention observers or apparatus.

Images of Time. First Edition. George Jaroszkiewicz.
© George Jaroszkiewicz 2016. Published in 2016 by Oxford University Press.

where Ψ_t is a 'pure' state vector for some SUO at laboratory time t and $\hat{H}_t$ is the Hamiltonian operator, which might have some explicit time dependence. Three comments are relevant:

1. The Hamiltonian operator $\hat{H}_t$ plays the role of a *generator of translations in time*, in line with our discussion in Chapter 14.

2. The state vector Ψ_t is an element of a Hilbert space. These abstract spaces are discussed in the Appendix. They are nothing like the space we observe around us in the real world, reinforcing the message that QM is a theory of observation, not of *things*. Not all theorists agree [Bohm, 1952].

3. The Schrödinger–Dirac equation (24.1) is a first-order in time differential equation. All our experience with first-order in time differential equations in classical mechanics (CM) is that solutions are irreversible in time. But remarkably, the Schrödinger–Dirac equation is not dissipative if the Hamiltonian operator is a self-adjoint operator (self-adjoint operators are discussed in the Appendix).

The status of time in QM is at first sight quite different to that of dynamical variables such as particle position coordinates and linear momentum components, which in QM are represented by non-commuting Hermitian operators. Were it not for special relativity (SR), there would perhaps have been no particular concerns about this difference. But SR was built on principles that led to the paradigm shift from Galileo–Newtonian *space-time* to Minkowski *spacetime*. And that generates a real conceptual problem. In SR, time and space are treated in a much more symmetric way, as if there was no inherent difference between the two concepts apart from a sign change in a line element.

This raises a curious problem. Because time and space are treated as democratic partners in Minkowski spacetime, and because spatial coordinates are frequently 'quantized', meaning that they are regarded as quantum operators, this has led theorists to ask the question: should time too be represented by some operator, $\hat{t}$?

This question has attracted the attention of theorists such as Pauli, Dirac, and Snyder. We shall discuss this further on in this chapter.

Schrödinger versus Heisenberg

When Schrödinger formulated his famous quantum wave equation in 1926, he was using the standard temporal architecture of the Galileo–Newtonian space-time paradigm in which observers are exophysical and quantum states are described in the SP. This picture is an intuitive description of quantum states of SUOs, in which a given state vector Ψ_t at time t obeys the Schrödinger–Dirac equation (24.1).

Observables in QM are operators in $\mathcal{H}$ that correspond reasonably closely (but not precisely) to classically observable properties of SUOs. In general, a SP operator $\hat{A}_t$ may have an explicit time dependence, in which case we have the rule

$$\frac{d}{dt}\hat{A}_t = \frac{\partial}{\partial t}\hat{A}_t, \qquad (24.2)$$

where the partial time derivative involves only the explicit dependence in the operator, if any. Dynamical degrees of freedom such as the $\hat{p}^i$ and the $\hat{q}^i$ that correspond to momentum and position coordinates in phase space do *not* have any explicit t-dependence. This reflects the architecture of the SP: the laboratory and all the apparatus in it is fixed.

One of the features of Hilbert space, discussed in the Appendix, is that given any two vectors Φ, Ψ, there is an 'inner product' denoted (Φ, Ψ) that maps the pair of vectors into the complex numbers. One of the deep mysteries surrounding QM is the Born rule [Born, 1926], the proposition that the square modulus $|(\Phi, \Psi)|^2$ gives the conditional probability $P(\Phi|\Psi)$ that, given the SUO was prepared in state Ψ, it would be 'found' in state Ψ.[77] Of course, things are never as simple as that, but essentially, that is it: one of the QM mysteries in a nutshell (there are more).

Feynman rephrased this rule in terms of the inner product (Φ, Ψ) itself: in his way of describing things he would say that '(Φ, Ψ) is the amplitude $\mathcal{A}(\Phi|\Psi)$ for Ψ to go to Φ, and the probability for this to happen is the square modulus of this amplitude, i.e., $P(\Phi|\Psi) = |\mathcal{A}(\Phi|\Psi)|^2$.'

No one understands the origin of the Born rule, although some theorists claim to be able to derive it. What is bizarre is that this rule is all about time: *first* we have to prepare Ψ, and only *then* can we see if we have got Φ. Contextuality is inherent in this description, because probabilities in QM are always *conditional* probabilities. Likewise, amplitudes are always *conditional* amplitudes. This contextuality reinforces our assertion that QM is not about absolute truths but the correct way of dealing with contextual truths in physics.

Amazingly, the Born rule works even when there is a time interval between state preparation of Ψ and outcome detection of Φ. In this case, though, in the intervening time between preparation and detection, the observer must make absolutely no attempt to interfere or intervene with the evolution of Ψ. Provided that there is no intervention between initial and final times then the amplitude $\mathcal{A}(\Phi|\Psi_{t_0})$ for the state Ψ_{t_0} prepared at time t_0 to be found in state Φ_{t_1} at time $t_1 > t_0$ is given by the rule

$$\mathcal{A}(\Phi|\Psi_{t_0}) = (\Phi, \Psi_{t_1}), \qquad (24.3)$$

that is, the inner product at time t_1. Essentially, this is the amplitude for success in an attempt at time t_1 to see if, at *that* time, the prepared state is actually Φ.

---

[77] We ignore any complicating normalization factors here.

To calculate the right-hand side of (24.3) we need to integrate the Schrödinger–Dirac equation (24.1). The result can be represented in the form

$$\Psi_{t_1} = \hat{U}_{t_1,t_0}\Psi_{t_0},\tag{24.4}$$

where U_{t_1,t_0} is the *temporal evolution operator* evolving quantum states from initial time t_0 to final time t_1. A fundamental feature of Schrödinger mechanics is that if the Hamiltonian operator is self-adjoint[78] then the evolution operator is unitary, meaning that for any state, $(\Psi_{t_1},\Psi_{t_1}) = (\Psi_{t_0},\Psi_{t_0})$, corresponding to *conservation of total probability*. Assuming unitary evolution, then $\mathcal{A}(\Phi\,|\,\Psi_{t_0}) = (\Phi, \hat{U}_{t_1,t_0}\Psi_{t_0})$, a representation that shows clearly the temporal architecture of the SP.

If the Hamiltonian is explicitly independent of time, then we can show that

$$\hat{U}_{t_1,t_0} = \exp\left(-i\hat{H}(t_1 - t_0)/\hbar\right).\tag{24.5}$$

In principle the observer is able to reverse the temporal evolution of a state under such circumstances.

The Heisenberg picture

The Heisenberg picture (HP) is a formal trick, shifting the dynamical time-dependence of the Schrödinger–Dirac equation (24.1) from the state vectors onto the operators. The following analogy may help explain what is going on. Consider yourself standing on the ground in a fairground observing a moving merry-go-round carrying a friend. In this analogy, your friend is the analogue of the Schrödinger wavefunction Ψ_t: they are moving whilst you, the observer, is at rest in the laboratory. This scenario is analogous to the SP.

Suppose now you jumped onto the moving merry-go-round and sat next to your friend. From this new perspective the merry-go-round and your friend would appear to be at rest relative to you, whilst the rest of the universe would appear to be moving in a direction opposite to the one your friend appeared to have before you joined them. This scenario is the analogue of the HP. The point is, both pictures describe the *same* reality.

To see how the HP is obtained from the SP, define the HP state $\Psi_H \equiv \Psi_{t_0}$. Then the SP state Ψ_t is related to Ψ_H by

$$\Psi_t \equiv U_{t,t_0}\Psi_H,\tag{24.6}$$

where the subscript H refers to the Heisenberg picture.

To understand the effect of the transition from the SP to the HP on the observables of the theory, that is, the operators we use to squeeze out physical predictions

---

[78] Corresponding to a real Hamiltonian in CM.

from the states, consider some observable $\hat{O}_t$ in the SP, an observable that may have some explicit time dependence. The standard rule in QM is that the expectation value $\langle O_t \rangle$, or average of many separate, ensemble measurements of the observable $\hat{O}$ relative to the state Ψ_t at time t, is given by[79]

$$\langle O_t \rangle \equiv \frac{(\Psi_t, \hat{O}_t \Psi_t)}{(\Psi_t, \Psi_t)}. \tag{24.7}$$

Using (24.6), we readily find it to be given by

$$\langle O_t \rangle = \frac{(\Psi_H, \hat{O}_{H,t} \Psi_H)}{(\Psi_H, \Psi_H)}, \tag{24.8}$$

where $\hat{O}_{H,t}$ is the HP operator defined by $\hat{O}_{H,t} \equiv U^+_{t,t_0} \hat{O}_t U_{t,t_0}$, where U^+_{t,t_0} is the adjoint operator of U_{t,t_0} (adjoint operators are discussed in the Appendix).

The fact that we can use the HP should give anyone who believes in the reality or objectivity of wavefunctions serious food for thought. Clearly, in the HP, all the dynamical evolutionary processes are now associated with the measuring apparatus and quantum states are frozen in time. Therefore, time is not an intrinsic attribute of quantum states: it depends how you choose to look at it.

de Broglie waves

One of the most intruiging concepts in early twentieth-century physics was de Broglie's conjecture in 1924 that particles should be accompanied by waves [de Broglie, 1924]. This conjecture was based on a thorough understanding of special relativity (SR), Einstein's photon hypothesis, and Bohr's theory of the atom. de Broglie was able to 'explain' Bohr's quantization condition for the energy levels of electrons in the hydrogen atom by imagining the wave associated with an orbiting electron 'bending' around the orbit and closing up perfectly.

The Bohr atom is a non-relativistic model, so de Broglie made a non-relativistic approximation to the more correct relativistic theory we discuss now for a uniformly moving free particle of rest mass m_0. The energy E_p and linear momentum p of such a particle with speed v is given by

$$E_p = \frac{m_0 c^2}{\sqrt{1 - v^2/c^2}}, \quad p = \frac{m_0 v}{\sqrt{1 - v^2/c^2}}, \tag{24.9}$$

so we deduce $E_p v = c^2 p$. According to de Broglie, the associated pilot wave has frequency ν and wavelength λ given by $E_p = h\nu$ and $p\lambda = h$, where h is Planck's

---

[79] This assumes Ψ_t is not the zero vector in the Hilbert space. The denominator is there to 'normalize' the probabilities.

constant. Now the phase speed w^{80} of a monochromatic wave of frequency ν and wavelength λ is given by $w = \nu\lambda$, so we deduce

$$c^2 = w\nu. \tag{24.10}$$

This is a most disturbing result, because special relativity requires $v < c$ for particles with non-zero rest mass, and so we deduce $w > c$, that is, the phase speed is greater than the speed of light.

This is the first indication that in quantum mechanics, issues of causality may cause some problems. It is a hint of the strange 'spooky' flavour of quantum processes that so unsettled Einstein that he became an ardent opponent of the conventional probability interpretation of QM.

An interpretation of de Broglie's waves is given by Rindler [Rindler, 1969]. Imagine a swarm of particles travelling along parallel lines with the same velocity. Without loss of generality, restrict the discussion to the x-direction. Suppose in their rest frame $\mathcal{F}'$, they each emit a flash at the same time $t' = 0$ (in that frame). Then in a given inertial frame $\mathcal{F}$ these events are seen at times

$$t = \gamma\,(v)\,\left(0 + vx/c^2\right) = \gamma\,(v)\,vx/c^2 \tag{24.11}$$

and positions

$$x = \gamma\,(v)\,(x + 0) = \gamma\,(v)\,x, \tag{24.12}$$

where $\gamma\,(v) \equiv 1/\sqrt{1 - v^2/c^2}$ is the Lorentz factor. This is a set of events which appears to move with speed $w \equiv x/t = c^2/v$, which is de Broglie's phase velocity. Hence we may think of de Broglie waves as 'waves of simultaneity'.

We can show that de Broglie waves are Lorentz scalar fields, that is, independent of any inertial frame of reference, as follows. For a free point particle satisfying Einstein's relations, the associated pilot wave is given by $\psi_{\mathbf{p}}\,(\mathbf{x},t) \sim \exp\{ip_\mu x^\mu/\hbar\}$, where $\hbar$ is the reduced Planck's constant and $p_\mu x^\mu \equiv E_p t - \mathbf{p} \cdot \mathbf{x}$. But $p_\mu x^\mu$ is a Lorentz scalar, that is, has the same value in every inertial frame related by a Lorentz transformation to the particle's rest frame.

There is one question we need to ask about de Broglie waves and their super-luminal speed: what about the Fock–Kemmer wavefront discussed in Chapter 17? This wavefront propagates at the speed of light and represents an absolute limit of information transmission. If a particle is accelerated from rest to a definite momentum, then there will always be regions of the universe where no signal whatsoever has propagated, not even de Broglie waves moving at superluminal speeds. *All* disturbances are bounded by the Fock–Kemmer front: that is an absolute mathematical fact, given the hyperbolic differential equation governing the waves.

---

[80] The speed of a given peak of the wave.

Inside that front, it is meaningful to talk about de Broglie waves with super-luminal phase velocities; outside of that front there are no such waves. The Fock–Kemmer front plays the role of a causal correlating agent: as it passes a given chorus member in some frame of reference, that member acquires information about whatever caused the front. Unfortunately, discussions of de Broglie waves that do not take into account Fock–Kemmer fronts are contextually incomplete.

The point is, all discussions in physics are contextual, and we should always take into account how given quantum states of SUOs were created.

The time–energy uncertainty relation

Given a normalized state Ψ, that is, one for which $(\Psi, \Psi) = 1$, and an observable $\hat{A}$, then the expectation value $\langle A \rangle$ of the observable relative to that state is given by $\langle A \rangle \equiv (\Psi, \hat{A}\Psi)$. This is the quantum equivalent of the mean or average value of the classical variable A corresponding to $\hat{A}$, determined by setting up the same state many times in the laboratory and measuring the variable each time.

However, an average is just one particular piece of statistical information about a system, albeit a very useful one. What is often of importance is the spread of the various observations about that expectation value. To obtain this, we define the *mean square deviation* Δ_A^2 of the observable A relative to the state Ψ as

$$\Delta_A^2 \equiv (\Psi, (\hat{A} - \langle A \rangle)^2 \Psi) = \langle A^2 \rangle - \langle A \rangle^2. \qquad (24.13)$$

The square root of Δ_A^2 is known as the *standard deviation* in statistics and as the *uncertainty* in QM and is denoted by Δ_A.

Consider two observables $\hat{A}$, $\hat{B}$ which need not commute, and a given normalized Ψ. It is a standard QM exercise to show that

$$\Delta_A \Delta_B \geqslant \frac{1}{2} | (\Psi, [\hat{A}, \hat{B}], \Psi) |, \qquad (24.14)$$

known as the Kennard–Heisenberg relation [Kennard, 1927]. Here $[\hat{A}, \hat{B}]$ is the commutator $[\hat{A}, \hat{B}] \equiv \hat{A}\hat{B} - \hat{B}\hat{A}$ of the operators $\hat{A}$, $\hat{B}$. Consider now the special case when $\hat{A}$ and $\hat{B}$ represent conjugate dynamical variables and satisfy the canonical commutation relation $[\hat{A}, \hat{B}] = -i\hbar$. Then we deduce $\Delta_A \Delta_B \geqslant \frac{1}{2}\hbar$. In particular, if $\hat{A}$ is the position operator $\hat{x}$ for a single particle and $\hat{p}$ is its conjugate momentum operator, then we arrive at the famous Heisenberg Uncertainty relation [Heisenberg, 1927]

$$\Delta_x \Delta_p \geqslant \frac{1}{2}\hbar. \qquad (24.15)$$

This result concerns the statistics of many repeated measurements, each one done once on one of an ensemble of identically prepared pure states. It says nothing about a one-off act of measurement such as a single particle decay or a single photon impact on a detector. We cannot predict any individual outcome with certainty. This uncertainty is not the result of imprecision in our measuring apparatus; neither does it reflect some unavoidable disturbance to an otherwise well-defined classical value. The uncertainty relation tells us how closely we can ever come to believing in the existence of a classical, absolute reality before any measurement is taken.

We come now to the point. We saw in Chapter 15 that in our extended phase space, the original time parameter t appeared to be conjugate to (minus) the Hamiltonian. Since QM generally quantizes conjugate pairs of classical variables by replacing them with non-commuting operators, we might be tempted to believe that in QM, time should be represented by an operator, $\hat{t}$. We might even believe that there were quanta associated with such an operator, thereby justify the term 'chronon' for the fundamental quantum of time.

Unfortunately, whilst this is an appealing notion still discussed occasionally by theorists [Muga & Egusquiza, 2008], there are three plausible reasons for not believing this.

1. *Pauli's theorem* asserts that the physics of the known universe cannot support a time 'operator' conjugate to the ordinary Hamiltonian $\hat{H}$ [Pauli, 1933]. The argument is straightforward. Suppose there were such a commutation relation of the form $[\hat{t}, \hat{H}] = -i\hbar$. With reference to our discussion of generators in Chapter 14, we interpret $\hat{H}$ as a generator of translations in t (a reasonable concept) and $\hat{t}$ as a generator of translations in *energy*. This latter conclusion would be in conflict with the fact that for normal Hamiltonians, energies are invariably bounded below. If they were not, then the inevitable and irreversible dissipation of energy into the expanding universe would result in the net local disappearance of SUOs with such Hamiltonians, incompatible with the stability of atoms that we observe all around us. Therefore, the proposed commutator seems incompatible with physics as we know it.

2. If there were such a commutator, then we might expect a time–energy uncertainty relation analogous to (24.15), that is, $\Delta_t \Delta_E \geqslant \frac{1}{2}\hbar$. Now such a relationship is admissible if we look at the way experiments are conducted carefully. In (24.14), take the operator $\hat{B}$ to be the Hamiltonian $\hat{H}$. Then we have at initial time $t = 0$, $\Delta_A \Delta_E \geqslant \frac{1}{2}\hbar |(\Psi_0, [\hat{A}, \hat{H}]\Psi_0)|$. Now for time-independent Hamiltonians, we can show that $|(\Psi_0, [\hat{A}, \hat{H}]\Psi_0)| = \left|\frac{d}{dt}(\Psi, \hat{A}\Psi)\right|_{t=0} \equiv \left|\frac{d}{dt}\langle A \rangle\right|_{t=0}$, so assuming this is not zero, we have

$$\frac{\Delta_A}{\left|\dfrac{d}{dt}\langle A \rangle\right|_{t=0}} \Delta_E \geqslant \tfrac{1}{2}\hbar. \tag{24.16}$$

The physical dimensions of the factor involving the operator $\hat{A}$ on the left-hand side of this inequality has the physical dimensions of a time, so if we *define* $\Delta t \equiv \Delta_A / \left| \dfrac{d}{dt} \langle A \rangle \right|_{t=0}$, we appear to have a time–energy uncertainty relation. The interpretation of this requires some care, because 'time' is not an observable and is not being measured: the above inequality is telling us something about the relationship of the initial uncertainty in A to the initial rate of change of expectation value of A. In other words, it is not a statement about any time operator.

3. In Dirac's constraint analysis discussed in Chapter 15, the set of extended phase-space variables includes what looks like the original t parameter and the Hamiltonian as conjugate variables. However, there is a constraint in the system, viz., the extended Hamiltonian $\tilde{H}$ is zero on the surface of constraints. Not only does this lead to the notorious 'problem of time' in quantum cosmology, but it tells us that the extended variables cannot all be observables. The time coordinate in this interpretation is really like a a gauge-dependent function, which means it is not a physical, dynamical variable: but time was never that in the first place.

The relativistic propagator

There is in relativistic quantum theory a most remarkable confrontation between classical SR and QM, concerning proper time. There appears to be an unresolvable conundrum. It goes as follows.

Consider an inertial frame of reference in SR. Consider the calculation of the quantum amplitude $\mathcal{A}(\mathbf{x}, t \,|\, \mathbf{y}, 0)$ for a particle created at event $(\mathbf{y}, 0)$ to propagate to event $(\mathbf{x}, t)$ for $t > 0$. In this discussion, we have to ignore many complicating factors, such as what sort of field we are talking about, and so on, but none of that will alter the essential conundrum. Now according to Feynman, this amplitude can be calculated by a path integral of the form

$$\mathcal{A}(\mathbf{x}, t \,|\, \mathbf{y}, 0) \sim \sum_{paths} \exp \left\{ i \int_0^t L \, dt' \right\}, \qquad (24.17)$$

where the continuous time Lagrangian L at a point in a given path is given by $L = -mc^2 \gamma^{-1}(v)$, where v is the instantaneous speed and $\gamma(v)$ is the Lorentz factor.

Consider now two different paths, Γ_1 and Γ_2, as shown in Figure 24.1. According to SR, the proper time along each of these paths is specific to that path. So the question is: when the particle has been observed at event $(\mathbf{x}, t)$, what proper time interval is registered? If indeed it is possible to find the reading on the particle's internal clock (assuming it had one), would that not then identify to some extent which path it had traversed? Would that not then mean that the external observer had a method of finding out to some extent which paths had *not* been traversed?

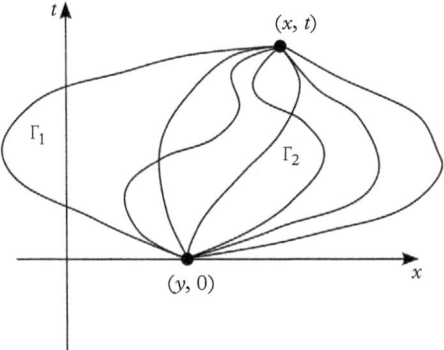

Fig. 24.1 *Paths contributing to the relativistic particle propagator.*

This is a problem that for resolution requires very heavyweight mathematical technology that we cannot discuss in detail here. We refer the reader to [Batlle *et al.*, 1988] for some of the details. The method is to encode the mass–shell constraint $p^\mu p_\mu = m^2 c^2$ that the four-momentum has to satisfy into the path integral using the technology of the so-called BRST charge, involving bosonic Lagrange multipliers and fermionic ghost variables in the path integration. Very briefly, the result of the fermionic ghost integrations depends on the path parameter interval but that contribution gets cancelled precisely by the integrations over the bosonic variables, leaving a result that is independent of proper time.

25

Temporal correlations

Introduction

As physicists continued their exploration of quantum mechanics (QM) into the twentieth century, they encountered phenomena such as quantum interference, quantum randomness, entanglement, and the uncertainty principle: these appeared to undermine the principles of classical mechanics (CM). However, classically minded 'hidden variables' theorists such as Bohm [Bohm, 1952] never accepted the implications of QM and made great efforts to explain away those phenomena in CM terms. The principal basis of their work was a non-contextual interpretation of the Schrödinger wavefunction. For relatively simple systems under observation (SUOs), such as one or two particle systems, they appeared to be successful, so that the CM versus QM debate remained unresolved for decades.

The situation changed significantly in 1964, when J. S. Bell wrote a paper on a two-spin experiment that concluded that CM and QM could give measurably different empirical outcomes in a certain class of experiment that went beyond the relatively simple scenarios hitherto envisaged [Bell, 1964]. In other words, the question could be decided empirically.

Bell's analysis was presented in the form of an equality, now universally referred to as a *Bell inequality*. Bell showed that, in certain experiments, if classical hidden variables were really the explanation for QM, then the data obtained from those experiments would always satisfy a certain inequality, whilst QM could occasionally violate it. Since then, numerous experiments have been done on systems such as photons [Aspect *et al.*, 1982] and the predictions of QM have been validated emphatically.

Bell's analysis involved spatial non-locality, that is, measurements distributed over space. Subsequently, analogous predictions involving temporal non-locality were formulated. We shall discuss one of these, known as the Leggett–Garg (LG) inequality [Leggett & Garg, 1985].

The LG inequality is based on two CM principles: *macrorealism* and *noninvasive measurability*. Macrorealism asserts that a macroscopic (large scale) SUO that can be observed to be in one of two or more macroscopically distinct states will always be in one of those states. Non-invasiveness asserts that the actual state

Images of Time. First Edition. George Jaroszkiewicz.
© George Jaroszkiewicz 2016. Published in 2016 by Oxford University Press.

such an SUO is in can always be determined cost-free, that is, without having any effect on that state or on its subsequent dynamical evolution. These principles are eminently reasonable from a classical perspective: after all, the world around us seems to be just like that.

The LG inequality equality is a relation between temporal correlations, so we need to understand what these are.

Classical bit temporal correlations

Without loss of content we shall restrict the discussion to an SUO consisting of a spatial ensemble[81] of N classical bits. We do not need to worry about what each bit actually is. What matters is that each bit can be found by the observer (us) in only one of two possible states, denoted as *up* or *down*. We consider an experiment where at initial time $t = 0$, N_{up} of these bits are definitely in the *up* state and $N_{down} = N - N_{up}$ are definitely in the *down* state.

According to the above classical principles, the state of each bit is always a definite quantity and can be known to the observer cost free and can be assigned an empirical 'value'. We assign a 'bit value' $Q^i(t) \equiv +1$ to the ith bit if it is in its *up* state at time t and a 'bit value' $Q^i(t) \equiv -1$ if it is in its *down* state at that time. The average bit value $\langle Q(0) \rangle$ over the ensemble at time $t = 0$ is therefore

$$\langle Q(0) \rangle = \frac{1}{N} \sum_{n=1}^{N} Q^i(0) = \frac{N_{up} - N_{down}}{N} = 2f(0) - 1, \qquad (25.1)$$

where $f(0) \equiv N_{up}/N$ is the fraction of the ensemble that is in the *up* state initially. In the limit as N tends to infinity, $f(0)$ becomes the probability of randomly selecting one bit from the ensemble at time $t = 0$ and finding it to be in the *up* state.

The experiment consists of allowing the N bits to evolve over time, monitoring the state of each bit. The discussion at this point is perfectly general but consistent with CM ideology. The same arguments apply whether the bits interact amongst themselves or are effectively isolated.

Consider any given bit. Regardless of its state at initial time $t = 0$, it will be in either state *up* at time t_1 or else in state *down* at that time. If it is in state *up* at that time, then its bit value $Q^i(t_1)$ is $+1$ or else it is in the *down* state and its bit value is $Q^i(t_1) = -1$. Over an extended sequence of observation times $\{t_0, t_1, t_2, \dots\}$ the ith bit will take on a corresponding sequence of bit values $\{Q^i(t_0), Q^i(t_1), Q^i(t_2)\dots\}$. Temporal correlations quantify the relative changes along such a bit value sequence.

---

[81] A collection of identical or near identical objects existing at the same time. In principle, according to Leibniz' principle of identity of indiscernibles, these object cannot be absolutely identical. Spatial position plays the role of distinguishing them.

We define the bit temporal correlation C_{nm}^i to be the product $Q^i(t_n)Q^i(t_m)$ of the bit values $Q^i(t_n)$, $Q^i(t_m)$ at two times t_n and t_m respectively, where $m > n$. The bit states at these two times are said to be perfectly correlated if $C_{nm}^i = +1$ and perfectly anticorrelated if $C_{nm}^i = -1$.

For an SUO consisting of N bits, we define the temporal correlation C_{nm} as the ensemble average

$$C_{nm} \equiv \frac{1}{N} \sum_{i=1}^{N} C_{nm}^i. \tag{25.2}$$

The average of a finite number of real numbers is bounded above and below by the greatest and least numbers in the set respectively, so we readily conclude that $-1 \leqslant C_{nm} \leqslant 1$.

The LG inequality considers an ensemble of bits first evolving from time t_0 to time t_1 and then evolving to time t_2. If we define the LG correlation K by

$$K \equiv C_{01} + C_{12} - C_{02} \tag{25.3}$$

then we can readily prove the LG inequality $-3 \leqslant K \leqslant 1$ as follows. Tracking the ith bit in the ensemble over the time interval $[t_0, t_2]$, we have the following range of possibilities:

Table 25.1 *Calculation of the possible values of the LG correlation for the ith bit.*

$Q^i(t_0)$	-1	-1	-1	-1	1	1	1	1
$Q^i(t_1)$	-1	-1	1	1	-1	-1	1	1
$Q^i(t_2)$	-1	1	-1	1	-1	1	-1	1
K^i	1	1	-3	1	1	-3	1	1

By inspection, the average of the $\{K^i\}$ over any finite number of such bits clearly cannot be outside the interval $[-3, 1]$, which proves the LG inequality. Note that there is absolutely no way that the LG inequality could ever be violated classically, given the above CM paradigm.

Quantum bit temporal correlations

The analogous discussion of the above using QM has profound differences involving how we model reality and time. The QM version of the above classical experiment replaces the two classical bit states of each bit in the ensemble by the two basis vectors of a two-dimensional Hilbert space known as a *quantum bit*, or *qbit*, and then tensors them all together. Therefore we are dealing with a spatial ensemble of N qbits, technically described as a *quantum register* $\mathcal{Q}^N$, a Hilbert space defined as the tensor product of all of the two-dimensional qbit Hilbert spaces. Such a register has dimension 2^N, which means that for the sort of N encountered with macroscopic scale SUOs, we would be faced with an intractable problem. An ensemble of even a hundred qbits would be far too complicated for us to deal with in detail, being a complex vector space of $2^{100} = 1,267,650,600,228,229,401,496,703,205,376$ dimensions. Macroscopic SUOs could easily require a quantum register of 10^{20} qbits,

We will assume for simplicity that the qbits in the quantum register do not interact with each other, but can interact with other elements of their environment. This assumption means that we can meaningfully discuss the evolution of a single bit and then take what amounts to a temporal ensemble average.[82] This is similar to the approach taken in the theory of nuclear magnetic resonance (NMR) [Abragam, 1961]. In NMR, the individual nuclear spins in a sample behave analogously to the qbits we are discussing, each nuclear spin interacting with its local environment only. This consists of externally imposed magnetic fields plus local spin-coupling to neighbouring spins. Provided the number of such neighbours is not excessive, then it can be reasonably assumed that we are not dealing with an intractable quantum register. The irreversible environmental effects acting to dephase the Larmor rotations of individual spins can then be parametrized in terms of various relaxation times [Jaroszkiewicz & Strange, 1985]. These give information about that environment that can then be analysed to give visual images of that environment, the basis of magnetic imaging in medicine [Stehling *et al.*, 1987].

The theory of measurement and observation in QM is still controversial. We shall discuss temporal correlations from the perspective of the so-called *strong* measurement assumption. In a strong measurement of an observable A, a previously prepared quantum state Ψ collapses to one of the eigenstates $|\alpha\rangle$ of the observable and the observer registers a measured value λ_α, the corresponding eigenvalue of that operator. Specifically, according to the von Neumann projection postulate [von Neumann, 1955; Lüders, 1951], the initial state reduces to

$$|\Psi\rangle \rightarrow \frac{\langle\alpha|\Psi\rangle}{|\langle\alpha|\Psi\rangle|}|\alpha\rangle. \tag{25.4}$$

We shall follow the evolution of a typical single qbit in a non-interacting ensemble from initial time t_0 to intermediate time t_1 and then to final time t_2. Without loss of generality, suppose the initial qbit state $|\Psi, t_0\rangle$ is a normalized pure state given by

$$|\Psi, t_0\rangle = \begin{bmatrix} \Psi_{up} \\ \Psi_{down} \end{bmatrix}, \tag{25.5}$$

---

[82] An average over many repetitions over time of the same experiment: this requires only one copy of the SUO, but it is used repeatedly.

where $|\Psi_{up}|^2 + |\Psi_{down}|^2 = 1$. If a strong measurement of the *up–down* observable Q is made just after time t_0, the result is a mixed state given by the density matrix[83]

$$\rho_0 = \begin{bmatrix} |\Psi_{up}|^2 & 0 \\ 0 & |\Psi_{down}|^2 \end{bmatrix}. \tag{25.6}$$

We shall discuss the correlation C_{01} of two measurements of Q, at time t_0 and time t_1. There are several ways to calculate C_{01}. We shall use the density matrix approach in the Heisenberg picture, discussed in Chapter 24. First, we need to specify the quantum evolution operator U_{10} taking states from time t_0 time t_1. The most general unitary operator up to an arbitrary phase is of the form

$$U_{10} = \begin{bmatrix} a_1 & -b_1^* \\ b_1 & a_1^* \end{bmatrix}, \tag{25.7}$$

where $|a_1|^2 + |b_1|^2 = 1$. Then the correlation C_{01} is given by [Wang, 2002]

$$C_{01} = Re\left[Tr\left\{ U_{10}^\dagger Q U_{10} Q \rho_0 \right\} \right] \tag{25.8}$$

where

$$Q \equiv \begin{bmatrix} 1 & 0 \\ 0 & -1 \end{bmatrix}. \tag{25.9}$$

We find

$$C_{01} = |a_1|^2 - |b_1|^2 = 2|a_1|^2 - 1. \tag{25.10}$$

We note this is independent of the initial state of the qbit.

Now consider a further evolution from time t_1 to time t_2 with evolution operator

$$U_{21} = \begin{bmatrix} a_2 & -b_2^* \\ b_2 & a_2^* \end{bmatrix}, \tag{25.11}$$

where $|a_2|^2 + |b_2|^2 = 1$. Then $C_{12} = 2|a_2|^2 - 1$.

According to *QM* principles, evolution from t_0 to t_2 is given by the evolution operator $U_{20} = U_{21}U_{01}$. We find

$$U_{20} = \begin{bmatrix} a_2 & -b_2^* \\ b_2 & a_2^* \end{bmatrix} \begin{bmatrix} a_1 & -b_1^* \\ b_1 & a_1^* \end{bmatrix} = \begin{bmatrix} a_1 a_2 - b_1 b_2^* & -a_1^* b_2^* - b_1^* a_2 \\ a_1 b_2 + b_1 a_2^* & a_1^* a_2^* - b_1^* b_2 \end{bmatrix}, \tag{25.12}$$

from which we deduce

$$C_{02} = 2|a_1 a_2 - b_1 b_2^*|^2 - 1 \tag{25.13}$$

---

[83] This density matrix is the observer's quantum theory *prediction*, at initial time t_0, of the quantum state of the SUO as it *should* be just after time t_0.

The Leggett–Garg correlation

From the above correlations, using definition (25.3) we find

$$K = 2|a_1|^2 + 2|a_2|^2 - 2|a_1 a_2 - b_1 b_2^*|^2 - 1. \tag{25.14}$$

Now the parameters of an experiment are under the control of the experimentalist. Therefore, we shall assume that the parameters a_1, b_1, a_2, and b_2 can be chosen to be whatever we wish, subject to the unitarity constraints $|a_1|^2 + |b_1|^2 = |a_2|^2 + |b_2|^2 = 1$, and the limits of what is possible in the laboratory. Assuming there is no practical barrier, consider the reparametrization

$$a_1 = \cos\theta_1, b_1 = \sin\theta_1 e^{i\phi_1},$$
$$a_2 = \cos\theta_2, b_2 = \sin\theta_2 e^{i\phi_2}, \tag{25.15}$$

where $\theta_1, \theta_2, \phi_1$ and ϕ_2 are real. Now take $\phi_1 = \phi_2$. Plotting K as θ_1 and θ_2 range from $-\pi$ to π each gives the following figure:

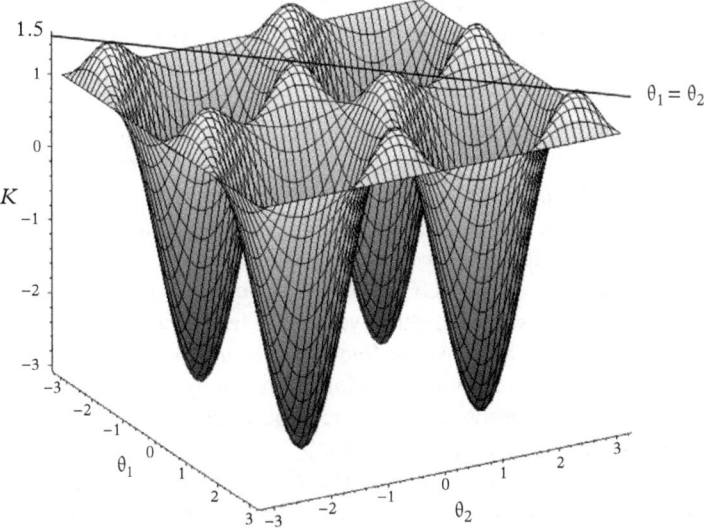

Fig. 25.1 *Values of K greater than 1 demonstrate a quantum mechanical violation of the Leggett–Garg inequality.*

It is clear that there are values of the parameters where K exceeds the classical limit $+1$. In fact, the maximum value of K over the region shown is 1.5 This is an entirely non-classical result. How can we explain it?

Understanding the Leggett–Garg prediction

There is a deep issue being demonstrated in the above analysis, to do with the way humans think about reality and time. It is intimately bound up with the motif running throughout this book, viz., the contextuality of truth.

Let us say right at this point we are *not* going to 'explain' the quantum breakdown of the LG inequality. Merely saying that there is some bizarre quantum superposition effect going on explains nothing. What we can do is to say what is wrong with the classical calculation that set up the LG inequality in the first place. It's not quantum that is wrong, but our preconditioned classical way of thinking about the universe and time.

In fact, we are fortunate that our faulty conditioning works 99.99999 . . . per cent of the time. Indeed, it took very sophisticated technology for quantum phenomena to be detected: they tend to hide themselves very successfully.

So how about the present scenario. Can we see where our classical thinking went wrong?

The answer is given by John Archibold Wheeler's philosophy of observation: *if you haven't actually done it, don't assume anything about it.*

Consider the LG correlation K. It is calculated from three separate correlations. Now classically, these correlations can be regarded as coming from the same experiment. First, a bit evolves from time t_0 to time t_1. *Snap!*—we observe its state at that time, compare it with its initial state, and hence work out the temporal correlation C_{01}, cost free. Then, without further ado, we let it evolve to time t_2 where again, *snap!* we observe its final state and work out not just C_{12} but also C_{02}. We have, after all, obtained all the information required to do those calculations.

But, according to Wheeler, the correlation C_{02} *does not involve any observation at time* t_1. The evolution operator U_{20} is applied on the strict understanding that no attempt is made to extract information between times t_0 and t_2. And that is the essential point. Classically, we have been conditioned to think it must be the case that the 'observed' state of the SUO at time t_1 should play a role in C_{02}. But how can it? We did not observe it when the system ran from t_0 to t_2.

Essentially, there are *three separate 'sub-experiments'* involved in the determination of K; evolution from t_0 to t_1, evolution from t_1 to t_2, and evolution to t_0 to t_2. Classically, we can believe there is just one experiment. Quantum mechanically, we have to recognize context and factor it in.

What appears to be going on with the C_{02} correlation is that our lack of information about the 'actual' classical state of the SUO at time t_1 in the C_{02} sub-experiment allows some sort of quantum interference between the two possible classical states *up* and *down* to occur at that time. This reminds us of the double-slit experiment, where an interference pattern on a screen is observed provided no attempt is made to determine through which slit the particle had gone.

The above is not an explanation, because we cannot account for the mysterious rules of quantum mechanics, such as the Born outcome probability rule, and so much more. But perhaps that is just as well. How could humans ever deceive themselves into thinking they could explain and understand the universe, time, and hence themselves?

26

Time reversal

Introduction

Time reversal experiments are designed to test the Standard Model of particle physics. It is hoped that this research may help answer one of the outstanding questions in science: why is there such a great asymmetry between matter and antimatter, given the temporal symmetry between particles and antiparticles at the fundamental level of particle theory? Did some subtle difference in the laws of physics between particles and antiparticles come into play during the 'Big Bang', at the origin of time?

The basic idea of a time reversal experiment is this. First the observer chorus O prepares an initial state $\Psi(0)$ of a system under observation (SUO) S and observes it evolving from initial time $t_i = 0$ to final time $t_f \equiv T > 0$. Its dynamical behaviour during that interval is assumed described by some laws of physics, Λ. Then O prepares another state $\Psi_T(0)$, possibly of a related but different SUO S_T, such that $\Psi_T(0)$ can be thought of a 'time reversed' version of $\Psi(T)$. O then observes the evolution of $\Psi_T(0)$ over an equivalent interval of time and works out the laws of physics Λ_T that apply to S_T. If O finds that $\Lambda_T = \Lambda$, then O can say that the laws of physics are invariant under this particular time reversal experiment.

There are three points to note about this:

1. The observer O is *not* reversed in time: the relative external context applied to O remains the same, meaning that the direction of time in the wider universe, including Hubble expansion, remains the same.

2. There is a big different between the temporal architecture of a time reversed experiment conducted along classical mechanical (CM) principles and one conducted according to quantum mechanical (QM) principles. In the former, it is meaningful for O to discuss every stage of the SUO's evolution between t_i and t_f, whereas in the latter, O can interact with the SUO at two stages only: state preparation at initial time t_i and outcome detection at final time t_f.[84]

---

[84] State preparation and outcome detection do not take play instantaneously. In this context, t_i and t_f are to be understood as representative of the *stages* of preparation and detection: these take place over

Images of Time. First Edition. George Jaroszkiewicz.
© George Jaroszkiewicz 2016. Published in 2016 by Oxford University Press.

3. Such an experiment involves an *active transformation* of the state Ψ, not a *passive transformation* of $O's$ time coordinate.

Throughout this discussion we shall work with an observer O at rest in an inertial frame $\mathcal{F}$, using standard Cartesian spacetime coordinates $x \equiv (ct, \mathbf{x})$, where $\mathbf{x} = (x, y, z)$ are spatial position coordinates in $O's$ laboratory. We define $Tx \equiv (-ct, \mathbf{x})$.

We need to say how $\Psi_T(0)$ is related to $\Psi(T)$. The standard definition of active time reversal is that time coordinates, velocities, momenta, and angular momenta are all reversed whilst electric charge and mass are left unchanged. We shall discuss a succession of scenarios, working our way up from classical Newtonian mechanics to relativistic quantum field theory.

Classical active time reversal

Our first scenario involves an SUO consisting of a non-relativistic particle of mass m and electric charge q moving through a region of spacetime in which the observer has set up external electric and magnetic fields $\mathbf{E}_0(x)$ and $\mathbf{B}_0(x)$ respectively. We suppose that the observer is at rest in a standard inertial frame and using standard Cartesian coordinates $(t, \mathbf{x})$. According to classical mechanical (CM) principles, the observer can record the particle's trajectory $\mathbf{x}_0(t)$ as a function of time t without affecting the dynamics in any way.

We do not need special relativity to extract some important ideas, so at this point we shall use standard Newtonian CM. For such an SUO, the particle's trajectory satisfies the non-relativistic Lorentz force law

$$m \left(\frac{d^2}{dt^2} \mathbf{x}_0 \right) (t) = q\mathbf{E}_0(t, \mathbf{x}_0(t)) + q \left(\frac{d}{dt} \mathbf{x}_0 \right) (t) \times \mathbf{B}_0(t, \mathbf{x}_0(t)), \qquad (26.1)$$

where for convenience the experiment runs from initial time $t_i = -\frac{1}{2}T$ to final time $t_f = \frac{1}{2}T$.

Suppose now O takes a film of this trajectory and runs it backwards in the projector. As we watch the time reversed film, we denote the position of the time-reversed trajectory by $\mathbf{x}_1(t)$, with $\mathbf{x}_1(t) \equiv \mathbf{x}_0(-t)$. From the rules of basic calculus, this means that

$$\left(\frac{d}{dt} \mathbf{x}_0 \right) (t) = -\left(\frac{d}{dt} \mathbf{x}_1 \right) (-t), \quad \left(\frac{d^2}{dt^2} \mathbf{x}_0 \right) (t) = \left(\frac{d^2}{dt^2} \mathbf{x}_1 \right) (-t), \qquad (26.2)$$

somewhat ill-defined times, a point that in general is marginalized by theorists as being an inessential complicating factor, much like friction in Newtonian mechanics. We do not agree with this policy but the current level of quantum modelling gives little choice in this matter.

where our notation $\left(\dfrac{d}{dt}\mathbf{x}_0\right)(t)$ means the derivative of $\mathbf{x}_0$ with respect to time, evaluated at time t. When these relations are used in (26.1) the force law becomes

$$m\left(\frac{d^2}{dt^2}\mathbf{x}_1\right)(t) = q\mathbf{E}_1(t,\mathbf{x}_1(t)) + q\left(\frac{d}{dt}\mathbf{x}_1\right)(t) \times \mathbf{B}_1(t,\mathbf{x}_1(t)). \tag{26.3}$$

Here

$$\mathbf{E}_1(t,\mathbf{x}) \equiv \mathbf{E}_0(-t,\mathbf{x}), \quad \mathbf{B}_1(t,\mathbf{x}) \equiv -\mathbf{B}_0(-t,\mathbf{x}) \tag{26.4}$$

are the external time-reversed electromagnetic fields we would deduce were acting so as to produce the time-reversed trajectory, *if* we insisted on using the same functional form[85] for the Lorentz force law when applied to both normal and time reversed trajectories.

The value to us here of this analysis is that we take (26.4) as the rule for the time reversal of electromagnetic fields in general. This rule can be explained as arising from the time reversal of the trajectories of the charged particles that would be the sources of the initial electromagnetic fields $\mathbf{E}_0$ and $\mathbf{B}_0$.

Schrödinger wave mechanics

We now extend the discussion to non-relativistic quantum wave mechanics. Consider an SUO S consisting of a spinless particle of mass m and electric charge q in external electromagnetic fields $\mathbf{E}_0$, $\mathbf{B}_0$, where

$$\mathbf{E}_0 \equiv -\nabla\phi_0 - \frac{\partial}{\partial t}\mathbf{A}_0, \quad \mathbf{B}_0 \equiv \nabla \times \mathbf{A}_0. \tag{26.5}$$

Here ϕ_0 is the scalar potential and $\mathbf{A}_0$ is the vector potential generating the initial field configuration.

The appropriate Schrödinger wave equation is

$$\left(i\hbar\frac{\partial}{\partial t} - q\phi_0(t,\mathbf{x})\right)\Psi_0(t,\mathbf{x}) = \frac{1}{2m}\{-i\hbar\nabla - q\mathbf{A}_0(t,\mathbf{x})\}^2\Psi_0(t,\mathbf{x}). \tag{26.6}$$

Now define the wavefunction Ψ_1 for another SUO S_T, such that

$$\Psi_0(t,\mathbf{x}) \equiv \Psi_1^*(-t,\mathbf{x}). \tag{26.7}$$

Then

$$\left(\frac{\partial}{\partial t}\Psi_0\right)(t,\mathbf{x}) = -\left(\frac{\partial}{\partial t}\Psi_1\right)(-t,\mathbf{x}). \tag{26.8}$$

---

[85] Same appearance, different actual values of fields.

The reason for taking the complex conjugate of Ψ_0 is that linear momenta are reversed when we create a time reversed state. To see this, suppose Ψ_0 was an eigenstate of momentum with momentum $\mathbf{p}$. Then by definition,

$$-i\hbar\nabla\Psi_0(t, \mathbf{x}) = \mathbf{p}\Psi_0(t, \mathbf{x}). \qquad (26.9)$$

Given this, it is easy to see that Ψ_1 is also an eigenstate of momentum but with reversed momentum, $-\mathbf{p}$.

Using (26.7) and (26.8) in (26.6) gives

$$\left(-i\hbar\frac{\partial}{\partial t} - q\phi_0(t, \mathbf{x})\right)\Psi_1^*(-t, \mathbf{x}) = \frac{1}{2m}\{-i\hbar\nabla - q\mathbf{A}_0(t, \mathbf{x})\}^2\Psi_1^*(-t, \mathbf{x}) \qquad (26.10)$$

Now taking complex conjugates on both sides and replacing t by $-t$ for convenience, we find

$$\left(i\hbar\frac{\partial}{\partial t} - q\phi_0(-t, \mathbf{x})\right)\Psi_1(t, \mathbf{x}) = \frac{1}{2m}\{-i\hbar\nabla + q\mathbf{A}_0(-t, \mathbf{x})\}^2\Psi_1(t, \mathbf{x}).$$

Defining $\phi_1(t, \mathbf{x}) \equiv \phi_0(-t, \mathbf{x})$, $\mathbf{A}_1(t, \mathbf{x}) \equiv -\mathbf{A}_0(-t, \mathbf{x})$ and

$$\mathbf{E}_1 \equiv -\nabla\phi_1 - \frac{\partial}{\partial t}\mathbf{A}_1, \quad \mathbf{B}_1 \equiv \nabla \times \mathbf{A}_1, \qquad (26.11)$$

we find form invariance of the Schrödinger wave equation and consistency with (26.4).

The time-reversal operator

Our discussion in the previous section of the Schrödinger wave equation involved complex conjugation. This becomes an essential ingredient in the formal theory of time reversal in quantum mechanics [Peres, 1993]. To see this explicitly, consider unitary evolution in quantum theory. Given an initial state vector ψ_1 at time t_1, it evolves to state ψ_2 at time $t_2 > t_1$ according to the dynamical rule

$$\psi_2 = U_{2,1}\psi_1, \qquad (26.12)$$

where $U_{2,1}$ is the unitary evolution operator from time t_1 to time t_2. Next, pick some other initial state ϕ_1. Then it evolves according to the same dynamics as ψ_1, that is, $\phi_2 = U_{2,1}\phi_1$.

The inner product (ϕ_t, ψ_t) is independent of time. To see this, we have

$$(\phi_2, \psi_2) = (U_{2,1}\phi_1, U_{2,1}\psi_1) = (\phi_1, U_{2,1}^\dagger U_{2,1}\psi_1) = (\phi_1, \psi_1), \qquad (26.13)$$

using the unitary property $U_{2,1}^\dagger U_{2,1} = I$ of the evolution operator.

Up to this point, the discussion is perfectly general. We now choose a specific initial state ϕ_1, defined to be the the the active *time-reversed state* of ψ_2, viz., $\phi_1 \equiv \mathcal{T}\psi_2$, where $\mathcal{T}$ is the *time-reversal operator*. This operator reverses velocity, momentum, and angular momentum. Now *if* the dynamics is time reversible we must have $\mathcal{T}U_{2,1}\mathcal{T}^\dagger = U_{1,2} = U_{2,1}^\dagger$. Hence $\phi_1 = U_{2,1}^\dagger \mathcal{T}\psi_1$ and then $\phi_2 = \mathcal{T}\psi_1$. Therefore

$$(\mathcal{T}\psi_1, \psi_2) = (\mathcal{T}\psi_2, \psi_1). \tag{26.14}$$

This means that $\mathcal{T}$ is an antilinear operator, viz. for complex α, $\mathcal{T}\alpha\psi = \alpha^*\mathcal{T}\psi$. Antilinear operators are discussed in the Appendix.

Wigner's theorem states that if an operator $\hat{O}$ satisfies the rule

$$\left| \left(\hat{O}\phi, \hat{O}\psi \right) \right| = |(\phi, \psi)| \tag{26.15}$$

for all state vectors ψ, ϕ, then $\hat{O}$ can be only either unitary or antiunitary. The time-reversal operator $\mathcal{T}$ is generally taken to be an antiunitary operator. An antiunitary operator $\hat{A}$ can be written as the product $\hat{A} = \hat{V}\hat{K}$ of a unitary operator $\hat{V}$ and $\hat{K}$, the *complex conjugation operator* [Gasiorowicz, 1967]. This latter operator satisfies the rules

$$\hat{K}^\dagger = \hat{K}, \quad \hat{K}^2 = 1$$
$$\hat{K}(\alpha\psi + \beta\phi) = \alpha^*\hat{K}\psi + \beta^*\hat{K}\phi \tag{26.16}$$
$$(\hat{K}\psi, \hat{K}\phi) = (\phi, \psi).$$

The Pauli equation

The charged Pauli equation is

$$i\hbar\left(\frac{\partial}{\partial t}\Phi_0\right)(t, \mathbf{x}) = \frac{1}{2m}\{-i\hbar\nabla - q\mathbf{A}_0(t, \mathbf{x})\}^2 \Phi_0(t, \mathbf{x}) \tag{26.17}$$

$$-\frac{q}{m}\mathbf{B}_0(t, \mathbf{x}) \cdot \frac{1}{2}\hbar\boldsymbol{\sigma}\,\Phi_0(t, \mathbf{x}) \tag{26.18}$$

where Φ is a two-component spinor, that is, has the form

$$\Phi(t, \mathbf{x}) = \begin{bmatrix} \Phi_1(t, \mathbf{x}) \\ \Phi_2(t, \mathbf{x}) \end{bmatrix}, \tag{26.19}$$

where Φ_1, Φ_2 are complex valued functions over space-time.

The presence of two components makes the time-reversal discussion slightly more complicated that for the Schrödinger equation. Now we have to take into account the possibility that time reversal mixes up the two Pauli wavefunction components. Accordingly, we define

$$\Phi_0(t, \mathbf{x}) \equiv W^* \Phi_1^*(-t, \mathbf{x}),$$

$$\left(\frac{\partial}{\partial t}\Phi_0\right)(t, \mathbf{x}) = -W^*\left(\frac{\partial}{\partial t}\Phi_1^*\right)(-t, \mathbf{x}), \tag{26.20}$$

for some 2×2 complex matrix W to be determined. Then (26.17) becomes

$$-i\hbar W^*\left(\frac{\partial}{\partial t}\Phi_1^*\right)(-t, \mathbf{x}) = \frac{1}{2m}(-i\hbar\nabla + q\mathbf{A}_1(-t, \mathbf{x}))^2 W^*\Phi_1^*(-t, \mathbf{x})$$

$$+\frac{q}{m}\mathbf{B}_1(-t, \mathbf{x}) \cdot \frac{1}{2}\hbar\boldsymbol{\sigma}\, W^*\Phi_1^*(-t, \mathbf{x}). \tag{26.21}$$

Taking complex conjugates on each side, we find

$$i\hbar W\left(\frac{\partial}{\partial t}\Phi_1\right)(-t, \mathbf{x}) = \frac{1}{2m}(-i\hbar\nabla - q\mathbf{A}_1(-t, \mathbf{x}))^2 W\Phi_1(-t, \mathbf{x})$$

$$+\frac{q}{m}\mathbf{B}_1(-t, \mathbf{x}) \cdot \frac{1}{2}\hbar\boldsymbol{\sigma}^*\, W\Phi_1(-t, \mathbf{x}). \tag{26.22}$$

Multiplying across by W^{-1}, the multiplicative inverse of W, which is assumed to exist, we find

$$i\hbar\frac{\partial}{\partial t}\Phi_1 = \frac{1}{2m}(-i\hbar\nabla - q\mathbf{A}_1)^2\Phi + \frac{q}{m}\mathbf{B}_1 \cdot \frac{1}{2}\hbar W^{-1}\boldsymbol{\sigma}^* W\Phi_1. \tag{26.23}$$

This will give us a good physical interpretation if we find a W matrix such that $W^{-1}\boldsymbol{\sigma}^* W = -\boldsymbol{\sigma}$, because under time reversal, we expect angular momentum to be reversed (from $\mathbf{L} \equiv \mathbf{r} \times \mathbf{p}$).

Now the $\boldsymbol{\sigma} \equiv (\sigma^1, \sigma^2, \sigma^3)$ are the Pauli matrices in standard representation

$$\sigma^1 \equiv \begin{bmatrix} 0 & 1 \\ 1 & 0 \end{bmatrix}, \quad \sigma^2 \equiv \begin{bmatrix} 0 & -i \\ i & 0 \end{bmatrix}, \quad \sigma^3 \equiv \begin{bmatrix} 1 & 0 \\ 0 & -1 \end{bmatrix}. \tag{26.24}$$

If we take $W \equiv \sigma^2$, then we can show that $W^{-1}\sigma^{i*} W = -\sigma^i$. With this choice, the time-reversed equation then becomes

$$i\hbar\frac{\partial}{\partial t}\Phi_1 = \frac{1}{2m}(-i\hbar\nabla - q\mathbf{A}_1)^2\Phi_1 - \frac{q}{m}\mathbf{B}_1 \cdot \frac{1}{2}\boldsymbol{\sigma}\Phi_1, \tag{26.25}$$

that is, this equation is form invariant to time reversal.

To check that the physics of the situation is being modelled correctly, suppose Φ_0 is an eigenstate of spin along the z-axis. That means that $\sigma^3\Phi_0 = \pm\Phi_0$. With the above choice of W, we can readily show that $\sigma^3\Phi_1 = \mp\Phi_1$, which is to be expected: time reversal of a spinning object flips its angular momentum vector.

The Dirac wave equation

Dirac's fundamental equation plays a central role in all modern particle physics. Its properties under time reversal are therefore of critical interest, not least because the equation involves positive and negative energy components, that is, particle and antiparticle attributes. We use the following convenient definitions: $x \equiv (ct, x^i)$, $Tx \equiv (-ct, x^i)$, $\eta x \equiv (ct, -x^i)$ and then the electromagnetic potentials satisfy the relation $A_0^\mu(x) = A_{1\mu}(Tx)$.

The charged Dirac equation is

$$i\hbar\gamma^\mu \left(\partial_\mu \psi_0\right)(x) - q\gamma^\mu \left(A_{0\mu}\psi_0\right)(x) - mc\psi_0(x) = 0. \qquad (26.26)$$

As with the Pauli equation, we assume that under time reversal we have a mix-up of spinor components, of the form

$$\psi_0(x) = \mathsf{T}^*\psi_1^*(Tx), \quad \left(\partial_\mu\psi_0\right)(x) = -\mathsf{T}^* \left(\partial^\mu\psi_1^*\right)(Tx) \qquad (26.27)$$

for some 4×4 matrix T. Hence, under time reversal we find

$$-i\hbar\gamma^\mu\mathsf{T}^* \left(\partial^\mu \psi_1^*\right)(Tx) - q\gamma^\mu\mathsf{T}^* \left(A_1^\mu\psi_1^*\right)(Tx) - mc\mathsf{T}^*\psi_1^*(Tx) = 0. \qquad (26.28)$$

Taking complex conjugates on both sides and multiplying from the left by T^{-1} we arrive at the equation

$$i\hbar\mathsf{T}^{-1}\gamma^{\mu*}\mathsf{T} \left(\partial^\mu \psi_1\right)(Tx) - q\mathsf{T}^{-1}\gamma^{\mu*}\mathsf{T} \left(A_1^\mu\psi_1\right)(Tx) - mc\psi_1(Tx) = 0. \qquad (26.29)$$

If now we find a T such that $\mathsf{T}\gamma^\mu\mathsf{T}^{-1} = \gamma_\mu^*$ then we have achieved our objective and shown form invariance under time reversal. The solution $\mathsf{T} = i\gamma^1\gamma^3$ is given in [Bjorken & Drell, 1965], where the γ^μ matrices are in the standard representation. This is consistent with our Pauli equation analysis above, given that W and T are defined up to arbitrary phases [Bjorken & Drell, 1964].

A particularly interesting result in field theory is that, if $\mathcal{T}$ is the field theoretic time-reversal operator [Bjorken & Drell, 1965], it can be shown that $\mathcal{T}\hat{\psi}(x)\mathcal{T}^{-1} = \mathsf{T}\hat{\psi}(Tx)$: this tells us that a time-reversed electron is still an electron, *not* a positron moving back in time.

TCP theorem

Elementary particle physics has explored the notion that there is a 'Lagrangian for the universe', a key that will unlock all the doors of reality. Over the decades, various symmetries of the currently understood Lagrangian have come and gone. An important development was the overthrow of parity, that is, the observation that the laws of physics are not totally left–right symmetrical.

As more and more details of the Lagrangian of the Standard Model were acquired, one dominant theme emerged. There were various symmetry operations, such as spatial inversion $\mathcal{P}$, charge conjugation $\mathcal{C}$, and time reversal $\mathcal{T}$, that individually might be broken. However, a theorem in quantum field theory asserts that the combination $\mathcal{CPT}$ is a true symmetry of the universal Lagrangian L_U, that is, $(\mathcal{CPT})L_U(\mathcal{CPT})^{-1} = L_U$. Equivalently, $[\mathcal{CPT}, L_U] = 0$. From this it follows that if any two of these operations together do not commute with L_U then the third does not commute either.

Kaons

Kaons are a quartet of unstable elementary particles, some of which have unusual properties involving time. The four species of Kaon are denoted K^+, K^-, K^0, and $\bar{K}^0$. In the standard model, Kaons are described as bound states with the quantum numbers of a quark and an antiquark.

The K^+ and K^- are antiparticles of each other and have equal and opposite electrical charge and the same rest mass. Their lifetime is of the order 1.2×10^{-8} seconds, which is regarded as extremely slow compared to the fastest decay lifetimes observed in particle physics (typically 10^{-24} seconds). The decay of these Kaons has no unusual properties.

The really interesting Kaons are the K^0 and the $\bar{K}^0$. These are antiparticles of each other, having electrical charge zero. There are three unusual temporal properties of these Kaons that can be understood if the K^0 and $\bar{K}^0$ are considered superpositions of two very different components denoted K_S (short) and K_L (long). The K_L has a much longer lifetime than the K_S, living about one thousand times longer. This has the following consequences.

When a beam consisting of K^0 particles was created in some apparatus and carefully monitored in flight, it was observed that the beam would turn into a beam of $\bar{K}^0$ particles and then back into a beam of K^0 particles, and so on. This known as *oscillation*.

When a beam of K^0 particles is directed onto a target, the short-lived component K_S can decay before the beam reaches the target, leaving a pure beam of K_L. But those objects can be considered a superposition of K^0 and $\bar{K}^0$. When they hit the target, these two components react differently with the matter in it, resulting in an emergent beam that now contains some K_S component. This is the phenomenon known as *regeneration*.

Finally, it was observed that some decays of the K_L component violated $\mathcal{CP}$ symmetry: before decay, the K_L is an eigenstate of $\mathcal{CP}$ with eigenvalue -1 but some decays with eigenvalue $+1$ were observed. The spectacular inference from this is that Kaon physics does not respect $\mathcal{T}$ symmetry.

The complete implications of these results are still to be understood and time reversal remains an active area of theoretical and empirical research.

27

Quantized spacetime

Introduction

We pointed out in Chapter 3 the view of the theorist Schwinger that the space-time concept is no more than an idealization abstracted from the context of our apparatus. This was Schwinger's answer to an age old question related to time: *what exactly is space*? Is it, like the concept of manifold time, an objective thing with properties, or is it, like process time, something that is meaningful only in a certain context? Schwinger clearly thought it was contextual.

This question was pondered in Antiquity by philosophers such as Aristotle. He argued that completely empty space or *void* could not exist. Instead, the universe had to be a *plenum*, a space completely filled with some form of matter or material. This is the source of the classical phrase '*Natura abhorret vacuum*' (nature abhors a vacuum).

The debate continued into the Renaissance, with some philosophers such as Descartes attempting to explain phenomena rationally. He attempted to explain gravitation in terms of vortices in the plenum [Descartes, 1644]. It continued into the nineteenth century with the general view that there was an Aether responsible for the transmission of electromagnetic vibrations. In the twentieth century, the aether was apparently abolished with the advent of Special Relativity, but made a sort of comeback with General Relativity. In this chapter we shall discuss some thoughts for and against the two positions, finishing with Hartland Snyder's quantized spacetime model, which in our view holds some promise of things to come.

Mach's relationalism

The physicist Ernst Mach holds a curious position in the Pantheon of Heroes of Physics. He was a scientist who made enduring contributions to physics, yet he did not accept the atomic hypothesis, the assertion that matter is not continuous but comes in integral pieces called atoms. Mach was a *relationalist*, a theorist who does not accept that space and time are absolute. Mach went further and asserted

Images of Time. First Edition. George Jaroszkiewicz.
© George Jaroszkiewicz 2016. Published in 2016 by Oxford University Press.

that inertia was not a property of space per se but a manifestation of how a given particle interacts with the rest of the universe. His ideas stimulated Einstein and others to think about the relationship between space, time, and matter.

Einstein's relationalism

It has been suggested that, under the influence of Mach's ideas concerning the origin of inertia, Einstein abolished the aether. Certainly, a letter exists in which Einstein expressed his appreciation of Mach's ideas, which appear to have influenced Einstein as he was developing general relativity (GR) [Einstein, 1913]. But the situation is not that clear. It has also been said that Mach would not have approved of GR, because after all, the metric is a property associated with space-time that has very definite observable effects (gravitation). At different times in his career Einstein made statements that suggested a degree of fence-sitting. For example, in a popular lecture he gave in 1920, later published in 1922 and translated into English, he said

> Recapitulating, we may say that according to the general theory of relativity space is endowed with physical qualities; in this sense, therefore, there exists an ether. According to the general theory of relativity space without ether is unthinkable; for in such space there not only would be no propagation of light, but also no possibility of existence for standards of space and time (measuring-rods and clocks), nor therefore any space-time intervals in the physical sense. But this ether may not be thought of as endowed with the quality characteristic of ponderable media, as consisting of parts which may be tracked through time. The idea of motion may not be applied to it.
>
> [Einstein, 1922]

Planck, quanta, photons, and existence

The impression is sometimes given that in 1900 Planck postulated the existence of photons to account for the empirical black body spectrum distribution. In fact, he did not. He was interested in regulating (or making finite) the energy held by the atomic oscillators that he imagined were part of the container of the black body radiation field. He did so to account for the empirical data, which clearly showed a rapid fall in energy with increasing electromagnetic frequency [Planck, 1900, 1901]. Our interpretation is that he was discussing the physics of *detectors* of radiation, not of the system under observation (SUO), the radiation field itself.

Not long after, Einstein made the suggestion that it was the radiation field itself that contained discrete lumps of energy that he referred to as *light quanta* [Einstein, 1905a]. The term *photon* was coined many years later by the chemist Lewis [Lewis, 1926]. Clearly, Einstein was attributing some material or objective existence to the radiation field in a classical sense.

The question then arises: in what sense does a radiation field 'exist'. Is there something there even when no one is observing it, or is existence contextual?

Snyder's quantized spacetime

The debate about the existence of the electromagnetic field has an interesting analogue: does empty space exist? In what sense is that a meaningful question?

There are numbers of theorists currently who take the view that there *is* something there, even when no one is looking. The programme of quantum gravity is predicated on the assumption that space and time do have some sort of dynamical structure, some dynamical degrees of freedom that could possibly be observed.

There are two problems with this. First, no one knows quite what the correct degrees of freedom are, and second, no one really knows how to test these ideas. The question arises, is there any sense in speculation?

Our answer is *possibly*. The problem with Wheeler's participatory principle is that it does not actually explain *anything*. It gives us a guide for thinking about our relationship with the universe and how to go about thinking about physics, but it does not actual help us do that physics: it does not help us understand the *information void*.

By this we mean the following. Suppose we agree with Wheeler that the universe must be described empirically. So we set up our preparation devices, create quantum states, and then measure outcome probabilities in our detectors. That is all very fine, but on what basis can we predict those outcome probabilities?

Now it was argued in Chapter 2 that physics should be discussed only in terms of contextually complete generalized propositions. Such a proposition will have a relative internal context and a relative external context. The latter will usually be classical in its description, such as describing the observer as embedded in some four dimensional spacetime with curvature. On the other hand, the former, the internal context, will usually involve quantum theory, and we know that the world of the quantum is bizarre. That is also the world of the information void, the twilight world of ignorance in our theories between preparation devices and detectors.

Now in high-energy physics, the information void has traditionally been modelled by four-dimensional Minkowski spacetime over which quantum fields propagate. But perhaps that is not the best model. It is, after all, a metaphysical assumption to assert that empty spacetime is like Minkowski spacetime. We cannot in principle test whether space is empty without putting something in it to make that test.

Schwinger said that spacetime was a conceptual abstraction of our apparatus. Perhaps we could develop some alternative models for the information void. This is really what we believe Hartland Snyder did in 1947. He wrote two bold papers [Snyder, 1947a,b], setting out a vision of spacetime being associated with a set of operators, satisfying a particular set of commutation relations. It is our view that this may make sense if we view his ideas as a model of the information void.

We shall use Standard International units in the following.

Now one of the principles of SR is that the laws of physics are invariant to standard Poincaré transformations. These are coordinate transformations of the form

$$x^\mu \to x'^\mu = \Lambda^\mu{}_\nu x^\nu + a^\mu, \tag{27.1}$$

where $\left[\Lambda^\mu{}_\nu\right]$ is a Lorentz transformation parameter matrix and the $\{a^\mu\}$ are translation parameters.

Now according to our discussion of canonical transformations in Chapter 14, infinitesimal transformations are represented by the action of generators of infinitesimal transformations, one generator for each parameter of the transformation. There are ten generators associated with the Poincaré transformations: they can be represented by the differential operators [Hamermesh, 1962]

$$X_\mu \mapsto \partial_\mu, \quad X_{\mu\nu} \mapsto x_\nu \partial_\mu - x_\mu \partial_\nu. \tag{27.2}$$

In relativistic quantum wave mechanics, these ten classical generators are replaced by the ten operators:

$$\hat{p}_\mu \mapsto i\hbar X_\mu, \quad \hat{M}_{\mu\nu} \mapsto i\hbar X_{\mu\nu}. \tag{27.3}$$

The $\hat{p}_\mu$ satisfy the commutation rule

$$[\hat{p}_\mu, \hat{p}_\nu] = 0, \tag{27.4}$$

which has the interpretation that $\mathcal{M}^4$ is a flat spacetime. Assuming the spacetime coordinates can be represented by operators, viz., $x^\mu \mapsto \hat{x}^\mu$, then the phase-space coordinate operators $\{\hat{x}^\mu, \hat{p}^\mu : \mu = 0, 1, 2, 3\}$ satisfy the algebra

$$[\hat{x}^\mu, \hat{x}^\nu] = 0, \quad [\hat{p}^\mu, \hat{x}^\nu] = i\hbar\eta^{\mu\nu}, \quad [\hat{p}^\mu, \hat{p}^\nu] = 0, \tag{27.5}$$

where $\hat{p}^\mu \equiv \eta^{\mu\nu}\hat{p}_\nu$. Snyder proposed replacing the algebra (27.5) with

$$[\hat{x}^\mu, \hat{x}^\nu] = \frac{ia^2}{\hbar}\left\{\hat{x}^\mu\hat{p}^\nu - \hat{x}^\nu\hat{p}^\mu\right\},$$

$$[\hat{p}^\mu, \hat{x}^\nu] = i\hbar\eta^{\mu\nu} - i\frac{a^2}{\hbar}\hat{p}^\mu\hat{p}^\nu \tag{27.6}$$

$$[\hat{p}^\mu, \hat{p}^\nu] = 0,$$

where a is a fundamental length parameter [Snyder, 1947a,b]. In the limit $a \to 0$, the algebra (27.6) reduces to (27.5).

Snyder introduced five new dimensionless coordinates θ^{μ}: $\mu = 0, 1, 2, 3$, and θ^4 for a new spacetime, which we denote $\mathcal{S}^5$. The conventional spacetime position operators $\hat{x}^{\mu}$ are represented as vector fields over the tangent bundle $T\mathcal{S}^5$ as follows:

$$\hat{x}^{\mu} \mapsto ia\theta^{\mu}\frac{\partial}{\partial\theta^4} + ia\theta^4\frac{\partial}{\partial\theta_{\mu}}, \quad \hat{p}^{\mu} \mapsto -\frac{\hbar}{a}\frac{\theta^{\mu}}{\theta^4}, \quad \mu = 0, 1, 2, 3. \tag{27.7}$$

Then this representation satisfies the Snyder algebra (27.6).

Expanding out, we find for the coordinate operators

$$\begin{aligned}
\hat{x}^0 \equiv c\hat{t} \equiv ia\left(\theta^0\frac{\partial}{\partial\theta^4} + \theta^4\frac{\partial}{\partial\theta^0}\right) \\
\hat{x}^1 \equiv \hat{x} \equiv ia\left(\theta^1\frac{\partial}{\partial\theta^4} - \theta^4\frac{\partial}{\partial\theta^1}\right) \\
\hat{x}^2 \equiv \hat{y} \equiv ia\left(\theta^2\frac{\partial}{\partial\theta^4} - \theta^4\frac{\partial}{\partial\theta^2}\right) \\
\hat{x}^3 \equiv \hat{z} \equiv ia\left(\theta^3\frac{\partial}{\partial\theta^4} - \theta^4\frac{\partial}{\partial\theta^3}\right).
\end{aligned} \tag{27.8}$$

The line element in $\mathcal{S}^5$ is given by

$$ds_{\mathcal{S}}^2 = (d\theta^0)^2 - (d\theta^1)^2 - (d\theta^2)^2 - (d\theta^3)^2 - (d\theta^4)^2. \tag{27.9}$$

Snyder stated that the $\hat{x}^i$ operators have spectra ma, where m is a positive, negative, or zero integer, whilst $\hat{x}^0$ has a continuous spectrum from $-\infty$ to $+\infty$. To see this, consider the eigenvalue equation

$$ia\left(u\frac{\partial}{\partial v} - v\frac{\partial}{\partial u}\right)\varphi(u, v) = \lambda\varphi(u, v), \tag{27.10}$$

for some real coordinates u, v. Changing coordinates from (u, v) to $[r, \theta]$, where $u = r\cos\theta$, $v = r\sin\theta$, gives

$$\partial_\theta = (\partial_\theta u)\,\partial_u + (\partial_\theta v)\,\partial_v = -r\sin\theta\,\partial_u + r\cos\theta\,\partial_v = u\partial_v - v\partial_u.$$

Then the eigenvalue equation (27.10) becomes $ia\partial_\theta\tilde{\varphi}(r,\theta) = \lambda\tilde{\varphi}(r,\theta)$, where $\tilde{\varphi}(r,\theta) \equiv \varphi(u, v)$. This equation is solved by the ansatz

$$\tilde{\varphi}(r,\theta) = R(r)\,e^{-i\lambda\theta/a}. \tag{27.11}$$

Assuming φ is a single valued function of u and v, then the periodicity of the coordinate θ leads to the quantization condition $\lambda = ma$, where m is some integer.

Comparing the form of the operators $\hat{x}$, $\hat{y}$, $\hat{z}$ in (27.8) to (27.10) leads to the conclusion that the spatial coordinate operators $\hat{x}$, $\hat{y}$, and $\hat{z}$ have discrete eigenvalues, which corresponds to spatial quantization.

On the other hand, consider the eigenvalue equation

$$ia \left(u\frac{\partial}{\partial v} + v\frac{\partial}{\partial u} \right) \varphi\,(u,v) = \lambda\varphi\,(u,v). \tag{27.12}$$

This lead us to the coordinate transformation $u = r\cosh\theta$, $v = r\sinh\theta$, which gives

$$\partial_\theta = (\partial_\theta u)\,\partial_u + (\partial_\theta v)\,\partial_v = r\sinh\theta\,\partial_u + r\cosh\theta\,\partial_v = u\partial_v + v\partial_u. \tag{27.13}$$

Now the eigenvalue equation (27.12) becomes $ia\partial_\theta\tilde{\varphi}\,(r,\theta) = \lambda\tilde{\varphi}\,(r,\theta)$, which has solution $\tilde{\varphi}\,(r,\theta) = R\,(r)\,e^{-i\lambda\theta/a}$. Now, however, there is no condition on λ arising from periodicity and therefore no quantization. By inspection of (27.8), we deduce that $\hat{x}^0$ does not have a discrete spectrum.

The conclusion is that Snyder's spacetime algebra does not amount to a 'quantization' of time, in that there is no chronon in this model. The reason why the spatial coordinate operators have a discrete spectrum whilst the time coordinate does not can be traced to the fact that the Minkowski spacetime metric has a Lorentzian signature.

Snyder's model of spacetime remains of great interest to theorists: we speculate that it has not yet seen its best days.

28

Epilogue

As I was writing this book, I encountered a great variety of images of time. Many images could not be included, either because they were too conjectural, too technical, or there just was no room here for them.

In order to get a handle on this veritable avalanche of ideas, ranging from metaphysical assertion to the scientifically validated theory, I felt obliged to develop a classification scheme to tell me when to leave this or that idea where I found it, or to pick it up and run with it. I do not pretend this scheme is perfect, but contextual completeness really has come to my aid many times. When in doubt, I found I needed to ask only a few simple questions. For example, when I read somewhere that the universe is static and that its wavefunction obeys the Wheeler–de-Witt equation, I know now how to deal with that assertion. I simply ask of the author, in my mind, the direct questions *How do you know?*, *How could you validate that?* *Who wants to know?* and *What are they going to do with this information?* In other words, I throw *Nullius in verba* at them.

We should above all recognize when we are dealing with conjecture and when there is reasonable prospect of evidence. Past generations were perhaps wiser than we are, we with our sophisticated technologies and mass communication. *Nullius in verba* cuts through everything. I think that principle has been forgotten by many otherwise good thinkers. If there is one injunction that I would want the reader to take away above all others it is to remember that motto when reading about time. It helps also to keep in mind that science is not a democracy: popularity does not turn bad ideas into good ones. Some ideas are not even wrong: they have a general proposition classification of zero.

What has delighted me in my digging around the literature on time has been a growing realization that thinkers of the past were easily as good as we are. I include Aristotle in that. I developed a great respect for many past theorists. Perhaps I should have known about them more. I am now much more appreciative of FitzGerald, Larmor, Lorentz, Minkowski, Nordstrøm, Stueckelberg, Snyder, Tangherlini, and many others. They were giants, all of them. We can only benefit by the scope of their work.

Images of Time. First Edition. George Jaroszkiewicz.
© George Jaroszkiewicz 2016. Published in 2016 by Oxford University Press.

Another aspect that has delighted me is that I have found more theoretical avenues for myself to walk through. I can name a few: the chronon as an infinitesimal, Cramer's transactional theory of quantum mechanics, observers in curved spacetime, extension of Snyder's quantized spacetime algebra, and several other ideas that look really exciting. They should occupy me until my own time runs out.

Appendix

Sets

A set is a well-defined collection of objects. A set with no elements in it at all is called the *empty set*, denoted by $\emptyset$. The empty set should not be confused with the number *zero*.

The *Cartesian product* $A \times B$ of two sets A, B is the set of all ordered pairs (a, b), the first element a being from A and the second element b being from B.

A *space* is a set with some additional mathematical structure, such as a *topology or metric*. A *point* is an element of a space.

Groups

A group $(G, *)$ is a set of elements $\{g_1, g_2, \cdots\}$ together with a binary map $*$ from the Cartesian product $G \times G$ back into G, known as *group multiplication*, satisfying the following axioms:

1. (Closure) For each element (g_1, g_2) in $G \times G$, the map $*$ assigns a unique element in G denoted by $g_1 * g_2$.

2. (Associativity) For any three elements g_1, g_2, g_3 in G, we have $(g_1 * g_2) * g_3 = g_1 * (g_2 * g_3)$.

3. (Existence of identity) There is a unique element e in G such that $g * e = e * g = g$ for any g in G.

4. (Existence of inverse) For any element g in G, there exists a unique element in G denoted by g^{-1} such that $g * g^{-1} = g^{-1} * g = e$.

The $*$ symbol in a group product is often dropped, so we write $g_1 g_2$ instead of $g_1 * g_2$.

An *abelian group* is a group for which the ordering in group multiplication is immaterial, that is, for any elements g_1, g_2 of G, we have $g_1 g_2 = g_2 g_1$. A group which is not abelian is called *nonabelian*.

Metric spaces

A *metric space* is a set with a notion of distance between elements of the set. A *metric* (or *distance function* or *distance rule*) d on a set X is a function d from the Cartesian product space $X \times X$ into the real numbers $\mathbb{R}$ satisfying the following conditions for all elements x, y, z in X:

1. (non-negativity) $d(x, y) \geqslant 0$
2. (identity of discernibles) $d(x, y) = 0$ if and only if $x = y$

3. (symmetry) $d(x, y) = d(y, x)$
4. (the triangle inequality) $d(x, z) \leqslant d(x, y) + d(y, z)$

Rings and fields

A *ring* is a set R with two binary operations, + (addition) and · (multiplication), satisfying the following axioms:

1. The pair $(R, +)$ is a commutative (abelian) group, that is, for any elements a and b of R, $a + b$ is an element of R and $a + b = b + a$.

2. The *additive identity* is called *zero* and denoted by 0, that is, for any element a in R, $a + 0 = a$.

3. Multiplication is associative and admits an *identity* element, denoted by 1, such that for any element a, we have $a \cdot 1 = 1 \cdot a = a$.

4. Multiplication is *distributive* (on both sides) over addition, that is, for any elements x, y, and z we have
$$x \cdot (y + z) = (x \cdot y) + (x \cdot z), \quad (x + y) \cdot z = (x \cdot z) + (y \cdot z).$$

If multiplication is commutative, then R is a *commutative ring*.

An element a is an *invertible element* with respect to multiplication if there exists a unique element a^{-1} such that $a \cdot a^{-1} = a^{-1} \cdot a = 1$. If every non-zero element of R is invertible, the R is said to be a division ring.

A field is a commutative division ring. Important fields are $\mathbb{R}$, the *real numbers* and $\mathbb{C}$, the *complex numbers*.

Vector spaces

A vector space $(V, \mathbb{F})$ over a (ground) field $\mathbb{F}$ is a set V of elements $V \equiv \{\mathbf{a}, \mathbf{b}, \dots\}$, known as *vectors*, with the following properties:

1. There is a binary map denoted + from the Cartesian product $V \times V$ into V such that, for any elements $\mathbf{a}, \mathbf{b}$ of V, the object $\mathbf{a} + \mathbf{b}$ is also in V. This is called *addition of vectors*, or just *vector addition*. Vector addition is *commutative*, that is, $\mathbf{a} + \mathbf{b} = \mathbf{b} + \mathbf{a}$. Vector addition is *associative*, that is, $\mathbf{a} + (\mathbf{b} + \mathbf{c}) = (\mathbf{a} + \mathbf{b}) + \mathbf{c}$.

2. There is a unique element in V, known as the *zero vector*, denoted by $\mathbf{0}$, such that for any vector, $\mathbf{a} + \mathbf{0} = \mathbf{a}$.

3. For every vector $\mathbf{a}$, there exists an *additive inverse*, denoted by $-\mathbf{a}$, such that $\mathbf{a} + (-\mathbf{a}) = \mathbf{0}$.
   These properties mean that V is an abelian group under vector addition.

4. For any $\mathbf{a}$ in V and λ in $\mathbb{F}$, then the object $\lambda\mathbf{a}$ is some element in V. This is known as *scalar multiplication*. In this context, the elements of $\mathbb{F}$ are called *scalars*. Scalar multiplication satisfies the property $\lambda(\mu\mathbf{a}) = (\lambda\mu)\mathbf{a}$.

5. Scalar multiplication is *distributive*, that is,

$$\lambda(\mathbf{a} + \mathbf{b}) = (\lambda\mathbf{a}) + (\lambda\mathbf{b}), \quad (\lambda + \mu)\mathbf{a} = (\lambda\mathbf{a}) + (\mu\mathbf{a}). \tag{A.1}$$

If the ground field $\mathbb{F} = \mathbb{R}$ then V is a *real vector space* whilst if $\mathbb{F} = \mathbb{C}$ then V is a *complex vector space*. The space of *three vectors* used to represent position in physical space is a real vector space, whilst the Hilbert space of quantum state vectors is a complex vector space.

An expression of the form $x^1 \mathbf{v}_1 + x^2 \mathbf{v}_2 + \ldots + x^k \mathbf{v}_k$, where $\mathbf{v}_1, \mathbf{v}_2, \ldots, \mathbf{v}_k$ are vectors in V and $x^1, x^2, \ldots, x^k$ scalars in $\mathbb{F}$, is called a *linear combination* of the vectors $\mathbf{v}_1, \mathbf{v}_2, \ldots, \mathbf{v}_k$. If the x^i are not all zero then it is called a *non-trivial* linear combination. Otherwise it is called *trivial*.

A set of vectors $\{\mathbf{v}_1, \mathbf{v}_2, \ldots, \mathbf{v}_k\}$ is *linearly dependent* if there exists a non-trivial linear combination equal to the zero vector $\mathbf{0}$. In other words, $\{\mathbf{v}_1, \mathbf{v}_2, \ldots, \mathbf{v}_k\}$ is linearly dependent if the equation

$$x^1 \mathbf{v}_1 + x^2 \mathbf{v}_2 + \ldots + x^k \mathbf{v}_k = \mathbf{0} \tag{A.2}$$

has a solution for which at least one of the x^i is non-zero.

A set of vectors $\{\mathbf{v}_1, \mathbf{v}_2, \ldots, \mathbf{v}_k\}$ is *linearly independent* if the only solution to equation $(A.2)$ is $x^1 = x^2 = \ldots = x^k = 0$, where 0 is the zero element of the ground field $\mathbb{F}$.

A linearly independent set $\{\mathbf{e}_1, \mathbf{e}_2, \ldots, \mathbf{e}_n\}$ is called a *basis* if every vector $\mathbf{v}$ in V can be expressed in one and only one way as a linear combination of the basis vectors, that is, $\mathbf{v} = x^1 \mathbf{e}_1 + x^2 \mathbf{e}_2 + \ldots + x^n \mathbf{e}_n$, where the components x^i are unique.

All bases of a finite-dimensional vector space have the same number of elements. The *dimension* dim V of a finite-dimensional vector space V is the number of elements of any basis for that vector space.

Hilbert space

A *Hilbert space* $\mathcal{H}$ is a complex vector space with two additional properties:

1. There is a scalar product (ψ, ϕ) mapping any two elements ψ, ϕ of $\mathcal{H}$ into the complex numbers such that for any complex numbers z, w,

$$
\begin{aligned}
(\psi, \phi) &= (\phi, \psi)^* \\
(\psi, z\phi + w\chi) &= z(\psi, \phi) + w(\psi, \chi) \\
\|\psi\|^2 &\equiv (\psi, \psi) \geq 0 \\
(\psi, \psi) &= 0 \text{ if and only if } \psi = 0.
\end{aligned}
\tag{A.3}
$$

2. $\mathcal{H}$ is *complete*, viz. for every *Cauchy sequence*[86] $\{\psi_n\}$ of vectors in $\mathcal{H}$ there exists a unique limit vector ψ in $\mathcal{H}$ such that

$$\lim_{n \to \infty} \|\psi_n - \psi\| = 0, \tag{A.4}$$

that is the sequence $\{\psi_n\}$ converges *strongly* to ψ.

The scalar product (ψ, ϕ) allows us to define a non-negative 'distance' $d(\phi, \psi) \equiv |(\phi, \psi)|$ between any two elements ϕ, ψ of the Hilbert space, and therefore a Hilbert space is also a metric space. We define $\|\psi\| \equiv \sqrt{(\psi, \psi)}$ to be the 'length' or norm of the vector ψ.

---

[86] A sequence of vectors that are getting closer and closer to each other in a well-defined sense as we move further and further along the sequence.

If we allowed some elements to have a negative squared norm, that is, $(\psi, \psi) < 0$, then the space would no longer be a metric space but an example of a pseudo-Hilbert space. Many theorems that can be proved for Hilbert spaces are not valid in pseudo-Hilbert spaces. For example, the Schwarz inequality $\|\phi\|\|\psi\| \geq |(\phi, \psi)|$ does not hold for pseudo-Hilbert spaces.

The non-negativity requirement of the square of the norm in (A.3) is crucial to the probability interpretation of quantum mechanics (QM). In QM, Hilbert spaces are used to represent states of systems under observation (SUOs). According to the generally accepted Born rule [1926], the square modulus of the inner product $|(\phi, \psi)|^2$ is interpreted as proportional to the probability of a prepared state ψ of an SUO being detected as state ϕ. Because probabilities are required normally to be non-negative, the space of quantum states must have a definite metric. There are examples, however, where negative squared norms and negative probabilities have been discussed in physics, such as in the Gupta–Bleuler approach to electromagnetism [Gupta, 1950; Bleuler, 1950] and quantum computation [Feynman, 1982].

For functions ψ, ϕ we define the scalar product of (ψ, ϕ) to be the overlap integral

$$(\psi, \phi) \equiv \int \psi^* (x) \, \phi (x) \, dx. \qquad (A.5)$$

This integral must exist (be finite) to make mathematical sense and hence have value in any physical application. Consequently, only square integrable functions are admitted in a Hilbert space whose inner product is defined in this way. There are problems with vectors of infinite norm.

A Hilbert space is *separable* if there exists a *countable* basis set, viz. there exists a set of vectors $\{u_1, u_2, \dots\}$ such that for any vector ψ in $\mathcal{H}$, we can write $\psi = \sum_{i=1}^{\infty} \psi_i u_i$, where the coefficients ψ_i are complex numbers. Otherwise $\mathcal{H}$ is *non-separable*.

Separability does not preclude the possibility of representing vectors in such a Hilbert space with basis sets with continuous indices.

Given an operator $\hat{O}$ on a Hilbert space, the *adjoint operator* $\hat{O}^\dagger$ satisfies the rule $(\hat{O}\Phi, \Psi) = (\Phi, \hat{O}^\dagger \Psi)$ for any vectors Φ, Ψ in the Hilbert space.

Observables

An *observable* is a quantum operator that corresponds to a classically observable quantity, such as energy and linear momentum. For this to be the case, an observable $\hat{O}$ should have real eigenvalues, that is, if Ψ is an eigenstate of an observable $\hat{O}$ then we have $\hat{O}\Psi = \lambda\Psi$, where λ is a real number. Such eigenvalues correspond to measured quantities in the laboratory. In other words, they represent real physics.

An operator is *self-adjoint* or *Hermitian* if for any states Φ, Ψ in the Hilbert space, we have $(\Phi, \hat{O}\Psi) = (\Psi, \hat{O}\Phi)^*$, that is, $\hat{O}^\dagger = \hat{O}$. The eigenvalues of Hermitian operators are real, so observables are generally represented by Hermitian operators.

Antilinear and antiunitary operators

In the following, $\mathcal{H}$ is a complex Hilbert space with inner product (ϕ, ψ) between elements ϕ and ψ of $\mathcal{H}$.

A operator A mapping elements of $\mathcal{H}$ back into $\mathcal{H}$ is *antilinear* if for any vectors ϕ, ψ in $\mathcal{H}$ and any complex numbers α, β we have

$$A(\alpha\phi + \beta\psi) = \alpha^* A\phi + \beta^* A\psi. \tag{A.6}$$

If A is an antilinear operator over $\mathcal{H}$ then it is also *antiunitary* if

$$(A\phi, A\psi) = (\psi, \phi). \tag{A.7}$$

The *adjoint* $A^\dagger$ of an antiunitary operator A satisfies the rule

$$(A^\dagger\phi, \psi) = (\phi, A\psi)^* = (A\psi, \phi) \tag{A.8}$$

for any vectors ϕ, ψ in $\mathcal{H}$. In addition, $A^\dagger A = AA^\dagger = I_\mathcal{H}$, the identity operator over $\mathcal{H}$.

The adjoint of an antiunitary operator is also an antiunitary operator.

The product of any two antilinear operators is linear.

The product of any two antiunitary operators is unitary.

Affine spaces

Let V be a vector space of dimension n over base field $\mathbb{F}$ and let $\mathcal{A}$ be a non-empty set with the following properties: given any point P in $\mathcal{A}$ and any vector $\mathbf{a}$ in V, there is a map from the Cartesian product $\mathcal{A} \times V$ into $\mathcal{A}$, such that $(P, \mathbf{a})$ is taken to a point in $\mathcal{A}$ denoted by symbolically by $P + \mathbf{a}$, subject to the conditions

1. $(P + \mathbf{a}) + \mathbf{b} = P + (\mathbf{a} + \mathbf{b})$, for any vectors $\mathbf{a}, \mathbf{b}$.
2. For any Q in $\mathcal{A}$, there exists a unique vector $\mathbf{a}$ in V such that $Q = P + \mathbf{a}$.

For any ordered pair (P, Q) of elements of $\mathcal{A}$, it is often convenient to write $Q = P + \mathbf{PQ}$, where $\mathbf{PQ}$ is an element of the vector space V called the *vector from P to Q*.

Using (1) and (2) we can prove that for any P in $\mathcal{A}$, $P + \mathbf{0} = P$, where $\mathbf{0}$ is the zero vector in V.

An important property of an affine $\mathcal{A}$ space is that it can be described in terms of coordinates called *affine coordinates*, set up as follows. First, choose a special point O in $\mathcal{A}$ known as the *origin of coordinates*. This choice is quite arbitrary, but once chosen is fixed. Next, choose a basis $B(V) \equiv \{\mathbf{e}_i : i = 1, 2, \ldots, n\}$ for V. Then from property (2) we know that for any point P in $\mathcal{A}$ we can always write $P = O + \sum_{i=1}^{n} x_i \mathbf{e}_i$, where $\mathbf{x} \equiv (x_1, x_2, \ldots, x_n)$ is an ordered set of elements of $\mathbb{F}$ known as the affine coordinates of P relative to the origin O and basis $B(V)$.

A significant point about affine coordinates is that they are *global*, that is, a single choice of origin and a single choice of basis for V can be used to give affine coordinates for *every* point of $\mathcal{A}$. This is not necessarily true of *manifolds* in general. When the ground field is $\mathbb{R}$, then we may use affine coordinates to identify an affine space $\mathcal{A}$ of dimension n with $\mathbb{R}^n$. The relevance to us is that Euclidean space and Minkowski spacetime can be modelled as affine spaces. In particular, inertial frames are examples of affine coordinate frames.

Manifolds

Manifolds are used extensively by mathematical physicists to model space and time as well as many other concepts such as Lie groups. To discuss manifolds in sufficient detail we first need to set up a framework based on $\mathbb{R}$, the set of real numbers discussed in Chapter 7. We define $\mathbb{R}^n$ to be the Cartesian product

$$\underbrace{\mathbb{R} \times \mathbb{R} \times \ldots \times \mathbb{R},}_{n \text{ copies}}$$

of n copies of $\mathbb{R}$. Each point of $\mathbb{R}^n$ is uniquely labelled by specifying its Cartesian coordinates $(x_1, x_2, \ldots, x_n)$, x_i real, $1 \leqslant i \leqslant n$.

Manifolds come in a variety of types and it has been found necessary to set up a rather elaborate procedure for describing them using coordinates. An n-dimensional *manifold* is a set whose points may be labelled by coordinates in *open* subsets known as *patches* or *charts*, of $\mathbb{R}^n$. Some manifolds can be covered by just one patch, but others may require two or more patches to be completely covered. In such cases, a set of covering patches is called an atlas. It is possible for a single point in a manifold to have coordinates in two or more overlapping patches.

Signature

Suppose $M \equiv [M_{ij}]$ is an $n \times n$ real symmetric matrix. A standard theorem in linear algebra tells us that the eigenvalue equation $M\mathbf{v} = \lambda\mathbf{v}$ has n real eigenvalues $\lambda_1, \lambda_2, \ldots, \lambda_n$, some or all of which may be repeated [Tropper, 1969]. Here $\mathbf{v}$ is a column vector, a vertical array of n real components. If p is the number of positive eigenvalues, q the number of negative eigenvalues, and r the number of eigenvalues equal to zero, we define the *signature* of M to be (p, q, r). Then $p + q + r = n$. Without loss of generality, we will assume $p \leqslant q$.

If $\mathbf{x}$ is a column vector with n components, we can use M to define an *inner product* $(\mathbf{x}, \mathbf{x})$ on the vector space V of such vectors as

$$(\mathbf{x}, \mathbf{x}) \equiv \mathbf{x}^T M \mathbf{x} = \sum_{i,j=1}^{n} M_{ij} x_j x_j, \tag{A.9}$$

where $\mathbf{x}^T \equiv [x_1, x_2, \ldots, x_n]$ is the *transpose* of $\mathbf{x}$. (A.9) is also known as a *quadratic form*. In general, a quadratic form contains not only terms proportional to $(x_i)^2$ for $i = 1, 2, \ldots, n$, but also cross terms proportional to $x_1 x_2$, $x_1 x_3$, and so on.

Consider now a transformation of the components $[x_i]$ to another set $[X_i]$ by a real, invertible linear transformation of the form $\mathbf{X} \equiv P^{-1}\mathbf{x}$ where P is some non-singular real $n \times n$ matrix. Then we can write $\mathbf{x} = P\mathbf{X}$, and hence $(\mathbf{x}, \mathbf{x}) = \mathbf{X}^T P^T M P \mathbf{X}$.

It is always possible to find a transformation matrix P such that the matrix $D \equiv P^T M P$ is diagonal, with its entries given by the signs of the eigenvalues of the original matrix, arranged so that

$$(\mathbf{x}, \mathbf{x}) = \underbrace{X_1^2 + X_2^2 + \ldots + X_p^2}_{p \text{ terms}} \underbrace{- X_{p+1}^2 - X_{p+1}^2 - \ldots - X_{p+q}^2}_{q \text{ terms}}. \tag{A.10}$$

Our convention is that the first p terms are positive and the next q terms are negative, with the overall sign of M chosen so that $p \leqslant q$. If M has $r \equiv n{-}p{-}q > 0$ eigenvalues equal to zero, then there are no terms involving $X_{p+q+1}^2, \ldots, X_n^2$ in (A.10). Such a transformation can be interpreted as a change of basis in the vector space V spanned by the column vectors $\{\mathbf{x}\}$.

The *signature* (p, q, r) of a real quadratic form is the number of positive, negative and zero eigenvalues of the real symmetric matrix in (A.9). *Sylvester's law of inertia* states that these numbers do not depend on the choice of basis for the vector space V. Signature is an important property of metric tensors because relative to a convenient basis, a metric tensor at a spacetime point is given by a real symmetric matrix of components $[g_{\mu\nu}]$.

Metric tensor component matrices are typically *non-degenerate*, which means $r = 0$. The physical interpretation of a non-degenerate metric is then characterized by the values of p and q. Since $p + q = n$, an equivalent characterization is the value of $q - p$, also referred to as the signature.

A *Riemannian* metric is one for which $p = r = 0$, $q = n$, which is locally equivalent to an n-dimensional Euclidean space. A *Lorentzian* metric is one such that $p = 1$, $q = n{-}1$, $r = 0$.

Lorentzian signature metrics are fundamental to General Relativity (GR), and therefore to our discussion of time, because the signature $(1, 3, 0)$ is that of GR spacetime metric tensors. Such spacetimes can be discussed in terms of observers with a single time parameter identified with the value $p = 1$ in the signature.

Variational derivation of Einstein's field equations from the Hilbert action

Given a Lorentzian signature metric tensor g over a four-dimensional spacetime manifold, we define the action $A[g]$ based on that given by Hilbert in 1915:

$$A[g] \equiv \frac{1}{2\kappa c} \int d^4x \sqrt{-|g|} \, \{2\kappa \mathcal{L}_M - 2\Lambda + R\}, \quad \kappa \equiv \frac{8\pi G}{c^4}. \quad \text{(A.11)}$$

Here $|g|$ is the determinant of the matrix of components of the metric tensor, relative to the coordinate patch used, $\mathcal{L}_M$ is the energy–momentum Lagrange density, Λ is the cosmological constant, and R is the Ricci scalar field, constructed from the Riemann curvature tensor, itself constructed from the metric tensor [Misner & Wheeler, 1973].

The idea here is to monitor the changes in the action integral when the metric tensor components $g_{\alpha\beta}$ are changed by infinitesimal amounts, viz., $g_{\alpha\beta} \rightarrow g_{\alpha\beta} + \delta g_{\alpha\beta}$. Standard manipulations then give the change $\delta A[g]$ in the action integral to lowest order in the $\delta g_{\alpha\beta}$ to be proportional to

$$\int d^4x \sqrt{-|g|} \, \delta g^{\alpha\beta} \left\{ 4\kappa \frac{\partial \mathcal{L}_M}{\partial g^{\alpha\beta}} - g_{\alpha\beta}(2\kappa \mathcal{L}_M - R + 2\Lambda) + 2R_{\alpha\beta} \right\} + ST, \quad \text{(A.12)}$$

where $R_{\alpha\beta}$ are the components of the Ricci tensor (also derived from the curvature tensor) and ST denotes surface terms on the boundary of the spacetime region.

At this point we postulate Hilbert's action principle that asserts that $\delta A[g] = 0$ for infinitesimal variations away from the true equations of motion. Ignoring the surface terms, which will not contribute inside the region concerned, this principle leads to

$$\underbrace{R_{\alpha\beta} - \tfrac{1}{2}Rg_{\alpha\beta}}_{G_{\alpha\beta}} + \Lambda g_{\alpha\beta} = -\kappa \underbrace{\left\{ 2\frac{\partial \mathcal{L}_M}{\partial g^{\alpha\beta}} - g_{\alpha\beta}\mathcal{L}_M \right\}}_{T_{\alpha\beta}}, \tag{A.13}$$

which is recognized as the Einstein field equations of general relativity.

Doppler shifts

Doppler shifts are fundamental to cosmology. We give a brief sketch of how to derive the special relativistic Doppler shift formula as follows. Suppose we have two standard inertial frames of reference $\mathcal{F}$ and $\mathcal{F}'$, in relative motion, with Lorentz transformation

$$t' = \gamma \left(t - \frac{\mathbf{v} \cdot \mathbf{x}}{c^2} \right), \quad \mathbf{x}' = \mathbf{x} + \left\{ \frac{(\gamma - 1)}{v^2}\mathbf{x} \cdot \mathbf{v} - \gamma t \right\} \mathbf{v}. \tag{A.14}$$

This is a generalization of the standard Lorentz transformation (17.3), where we leave the direction of the relative velocity arbitrary.

Consider a Lorentz scalar field φ representing a wave process with frequency ν and wavelength λ relative to $\mathcal{F}$ and frequency ν' and wavelength λ' relative to $\mathcal{F}'$, given by

$$\varphi(x) = \exp \left\{ i2\pi t\nu - i2\pi \frac{\mathbf{x} \cdot \mathbf{n}}{\lambda} \right\} = \exp \left\{ i2\pi t'\nu' - i2\pi \frac{\mathbf{x}' \cdot \mathbf{n}'}{\lambda'} \right\}, \tag{A.15}$$

where $\mathbf{n}$ and $\mathbf{n}'$ are unit vectors. The speed of the wave is c, the speed of light, and so we must have $\nu\lambda = \nu'\lambda' = c$. The wave propagates in direction $\mathbf{n}$ as seen in $\mathcal{F}$ and direction $\mathbf{n}'$ as seen in $\mathcal{F}'$.

Applying the transformation rule (A.14) to the last term on the right in (A.15) and comparing coefficients of t and $\mathbf{x}$ separately we deduce various relations from which we can derive not only the formulae for longitudinal and transverse Doppler shifts, but the relativistic aberration formula. For example, comparing coefficients of t gives

$$\nu = \gamma \left\{ 1 + \frac{\mathbf{v} \cdot \mathbf{n}'}{c} \right\} \nu'. \tag{A.16}$$

The important point about result (A.16) is the appearance of the Lorentz factor $\gamma \equiv 1/\sqrt{1 - v^2/c^2}$, reflecting time dilation, in addition to the other factor, which is seen in the non-relativistic formula for the Doppler shift.

Bibliography

Abragam, A. 1961. *The Principles of Nuclear Magnetism*. The Clarendon Press, Oxford.

Abraham, R., & Marsden, J. E. 2008. *Foundations of Mechanics*. Second edn. AMS Celsea Publishing.

Allmendinger, F., Heil, W., Karpuk, S., Kilian, W., Scharth, A., Schmidt, U., Schnabel, A., Sobolev, Yu., & Tullney, K. 2014. New Limit on Lorentz and CPT Violating Neutron Spin Interactions Using a Free Precession 3He-129Xe Co-magnetometer. *Phys. Rev. Lett.*, **112**, 110801.

Antchev G., et al,. 2012. Luminosity-independent measurements of total, elastic and inelastic cross-sections at sqrt(s) = 7 TeV. *Preprint Report number CERN-PH-EP-2012-353*, 1–7.

Arfken, G. 1985. *Mathematical Methods for Physicists*. Third edn. Academic Press Inc.

Aristotle. 1930. *Physica [The Physics]*. The Clarendon Press, Oxford.

Asimov, I. 1950. *Pebble in the Sky*. First edn. Empire Series. Doubleday.

Asimov, I. 1955. *The End of Eternity*, Doubleday, New York.

Aspect, A., Grangier, P., & Roger, G. 1982. Experimental Realization of Einstein-Podolsky-Rosen-Bohm Gedankenexperiment: A New Violation of Bell's Inequalities. *Phys. Rev. Lett.*, **49**, 91–94.

Augustine. 398. *Confessions*. Translated by E. B. Pusey (1838), J. H. Parker, London.

Bailly, F., Longo, G., & Montévil, M. 2011. A 2-Dimensional Geometry for Biological Time. *Progress in Biophysics and Molecular Biology*, **106**(3), 474–84.

Barbour, Julian. 1999. *The End of Time*. Phoenix.

Barrow, John D. 1992. *Pi in the Sky*. Clarendon Press, Oxford.

Batlle, C., Gomis, J., & Roca, J. 1988. The Propagator of a Free Relativistic Particle in a Generic Gauge $d\lambda/d\tau = f(\lambda)$. *Class. Quantum Grav.*, 5, 1663–7.

Bekenstein, J. D. 1973. Generalised Second Law of Thermodynamics in Black-Hole Physics. *Phys. Rev. D*, 3292–300.

Belavkin, V. 2003. Quantum Trajectories, State Diffusion and Time Asymmetric Eventum Mechanics. *Int. J. Theor. Phys.*, 42(10), 2461–85.

Bell, J. S. 1964. On the Einstein-Podolsky-Rosen paradox. *Physics*, **1**, 195–200.

Bender C. M., Milton K. A., Sharp D. H., Simmons L. M. Jr, & Strong, R. 1985. Discrete-time Quantum Mechanics. *Phys. Rev. D*, 32(6), 1476–85.

Benza, V. & Caldirola, P. 1981. de Sitter Microuniverse Associated to the Electron. *Il Nuovo Cimento*, **62A**(3), 175–85.

Berkeley, G. 1721. *De Motu or Sive de Motus Principio & Natura et de Causa Communicationis Motuum* [The Principle and Nature of Motion and the Cause of the Communication of Motions]. Translated by A. A. Luce, Published in: Michael R. Ayers: George Berkeley: *Philosophical Works*, Everyman, London 1993, pp. 253–76.

Bishop, E. 1977. Book Review of Elementary Calculus, by H. J. Keisler. *Bull. Am. Math. Soc.*, **83**, 205–08.

Bjorken, J. D. & Drell, S. D. 1964. *Relativistic Quantum Mechanics*. McGraw-Hill.

Bjorken, J. D. & Drell, S. D. 1965. *Relativistic Quantum Fields*. McGraw-Hill Inc.

Bleuler, K. 1950. Eine neue Methode zur Behandlung der longitudinalen und skalaren Photonen. *Helv. Phys. Acta*, **23**(V), 567–86.

Bohm, D. 1952. A Suggested Interpretation of the Quantum Theory in Terms of 'Hidden Variables', I and II. *Phys. Rev.*, **85**, 166–93.

Bohr, N. 1913. On the Constitution of Atoms and Molecules. *Philos. Mag.*, **26**(1), 1–24.

Bohr, N. 1935. Can Quantum-Mechanical Description of Physical Reality be Considered Complete? *Phys. Rev.*, **48**, 696–702.

Bombelli L., Lee J., Meyer D., & Sorkin, R. 1987. Space-Time as a Causal Set. *Phys. Rev. Lett.*, **59**(5), 521–4.

Born, M. 1926. Zur Quantenmechanik der Stossvorgange. [The Quantum Mechanics of the Impact Process] *Zeitschrift fur Physik*, **38**, 803–27.

Brenner, Andrew. 2010. Aquinas on Eternity, Tense, and Temporal Becoming. *Florida Philosophical Review*, **10**(1), 16–24

Britannica, Encyclopaedia. 2000. *Time*. CD Rom edn. Britannica.co.uk.

Brown, K. S., Marean, C W., Jacobs, Z., Schoville, B. J., Oestmo, S., Fisher, E. C., Bernatchez, J., Karkanas, P., & Matthews, T. 2012. An Early and Enduring Advanced Technology Originating 71,000 years ago in South Africa. *Nature*, **491**, 590–3.

Brown, L. M. 2005. *Feynman's Thesis, A New Approach to Quantum Theory*. World Scientific.

Buccheri, R. 2000. Studies on the structure of time. *In:* Buccheri, R., & Di Gesù, C. (eds), *Studies on the Structure of Time: From Physics to Psycho(patho)logy*. Kluwer Academic/Plenum Publishers.

Cadzow, J. A. 1970. Discrete calculus of variations. *Int. J. Control*, **11**(3), 393–407.

Caldirola, P. 1978. The Chronon in the Quantum Theory of the Electron and the Existence of Heavy Leptons. *Il Nuovo Cimento*, **45**(4), 549–79.

Caldirola, P. 1979. On a Relativistic Model of the Electron. *Il Nuovo Cimento*, **49**(4), 497–511.

Calvao, M. O., Soares, I. D., & Tiomno, J. 1990. Geodesics in Gödel-type space-times. *Gen. Relat. and Gravit.*, **22**(6), 683–705.

Cantor, Georg. 1869. Beiträge zur Begründung der transfiniten Mengenlehre [Contributions to the founding of the theory of transfinite numbers]. *Mathematische Annalen*, 481–512.

Carter, Rita. 1998. *Mapping the Mind*. Weidenfeld and Nicolson, London.

Chang, T. 1979. Maxwell's Equations in Anisotropic Space. *Phys. Lett.*, **70A**(1), 1–2.

Chew, Geoffrey F. 1966. *The Analytic S Matrix*. W. A. Benjamin, Inc., New York, Amsterdam.

Chou, C. W., Hume, D. B., Rosenband, T., & Wineland, D. J. 2010. Optical Clocks and Relativity. *Science*, **329**(5999), 1630–33.

Christenson, J. H., Cronin, J. W., Fitch, V. L., & Turlay, R. 1964. Evidence for the 2π decay of the K_2^0 meson. *Phys. Rev. Lett.*, **13**(4), 138–40.

Clayton, P. 2013. On the plurality of complexity-producing mechanisms. In Lineweaver, C. H., Davies, P. C. W., & Ruse, M. (eds), *Complexity and the Arrow of Time*. Cambridge University Press.

Colosi, D., & Rovelli, C. 2009. What is a Particle? *Classical Quan. Grav.*, **26**, 025002.

Cummings, R. K. 1922. *The Girl in the Golden Atom*. Project Gutenberg (online).

de Broglie, L. 1924. Recherches sur la Théorie des Quanta [Researches on Quantum Theory]. Ph.D. thesis, Faculty of Sciences at Paris University.

de Chardin, Pierre. 1947. *The Phenomenon of Man*. Harper (1958).

de Laplace, Pierre Simon. 1812. *Essai Philosophique sur les Probabilites [A Philosophical Essay on Probabilities]*. Courcier. English translation from sixth edition by F. W. Truscott and F. L. Emory (1902).

De Magalhães, J P., Costa, J., & Church, G. M. 2007. An Analysis of the Relationship Between Metabolism, Developmental Schedules, and Longevity Using Phylogenetic Independent Contrasts. *J. Geront. A-biol.*, **62A**(2), 149–160.

Descartes, R. 1644. *Les Princpes de la Philosophie*. F-G Levrault, Paris.

Descartes, R. 2006. *Discours de la méthode*. Project Gutenberg Ebook.

Deutsch, D. 1997. *The Fabric of Reality*. The Penguin Press.

Deutsch, D. 2001. The Structure of the Multiverse. *arXiv:quant-ph/0104033*, 1–21.

Deutsch, D. 2004. Qubit Field Theory. *arXiv:quant-ph/0401024*, 1–23.

DeWitt, Bryce S. 1965. *Dynamical Theory of Groups and Fields*. Documents on Modern Physics. Blackie and Son Limited.

DeWitt, B.S. 1967. Quantum Theory of Gravity I. The Canonical Theory. *Phys. Rev.*, **160**, 1113–48.

Dingle, H. 1967. The Case Against Special Relativity. *Nature*, **216**, 119–22.

Dirac, P. A. M. 1925. The Fundamental Equations of Quantum Mechanics. *Proc. Roy. Soc. Lond. A*, **109**, 642–653.

Dirac, P. A. M. 1928. The Quantum Theory of the Electron. *Proc. Roy. Soc. A*, **117**(778), 610–24.

Dirac, P. A. M. 1933. The Lagrangian in Quantum Mechanics. *Physikalische Zeitschrift der Sowjetunion*, **3**(1), 64–72.

Dirac, P. A. M. 1938. Classical Theory of Radiating Electrons. *Proc. Roy. Soc. A*, **167**, 148–69.

Dirac, P. A. M. 1958. *The Principles of Quantum Mechanics*. Clarendon Press.

Dirac, P. A. M. 1964. *Lectures on Quantum Mechanics*. Belfer Graduate School of Science Monograph Series no. 2, Yeshiva University, New York.

Donoghue, J. F. 1994. General relativity as an effective field theory: The leading quantum corrections. *Phys. Rev. D*, **50**, 2874–3888.

Dragovich, B., & Rakić, Z. 2010. Path Integrals for Quadratic Lagrangians on p-Adic and Adelic Spaces. *p-Adic Numbers, Ultrametric Analysis and Applications*, **2**(4), 322–340.

Drever, R. W. P. 1961. A Search for Anisotropy of Inertial Mass Using a Free Precession Technique. *Phil. Mag.*, **6**(65), 683–87.

Droste, J. 1917. The Field of a Single Centre in Einstein's Theory of Gravitation, and the Motion of a Particle in that Field. *Proc. Roy. Netherlands Acad. Arts and Science*, **19**(1), 197–215.

Dunne, J. W. 1934. *An Experiment with Time*. Faber, London.

Eddington, A. S. 1929. *The Nature of the Physical World*. The Gifford Lectures 1927. Cambridge University Press.

Einstein, A. 1905a. Über einen die Erzeugung und Verwandlung des Lichtes betreffenden heuristischen Gesichtspunkt [Concerning an Heuristic Point of View Toward the Emission and Transformation of Light]. *Annalen der Physik*, **17**, 132–48. Translation into English *American Journal of Physics*, v. 33, n. 5, May 1965.

Einstein, A. 1905b. Zur Electrodynamik Bewgter Körper. *Annalen der Physik*, 17, 891–921. *On the Electrodynamics of Moving Bodies*, translation in *The Principle of Relativity*, Dover Publications, Inc.

Einstein, A. 1913. *Letter to Ernst Mach*. Reprinted with commentary *in Gravitation* by C. Misner, K. Thorne, and J. Wheeler, W. H. Freeman, San Francisco, 1973.

Einstein, A. 1915a. Die Feldgleichungen der Gravitation. *Sitz. Preus. Akad. Wiss. Berlin*, 844–47.

Einstein, A. 1915b. *Explanation of the Perihelion Motion of Mercury from General Relativity Theory*. Königlich Preussische Akademie der Wissenschaften, Berlin.

Einstein, A. 1920. Ether and the Theory of Relativity. An address delivered May 5, 1920 at the University of Leyden.

Einstein, A. 1922. *Sidelights on Relativity: I: Ether and Relativity, II: Geometry and Experience*. Methuen and Co. Ltd.

Einstein, A., Podolsky, B., & Rosen, N. 1935. Can Quantum-Mechanical Description of Physical Reality be Considered Complete? *Phys. Rev.*, **47**, 777–80.

Farias, R. H. A., & Recami, E. 2010. Introduction of a Quantum of Time ("chronon") and its Consequences for Quantum Mechanics. *Adv. Imag. Elec. Phys.*, **163**, 33–115.

Farmelo, G. 2009. *The Strangest Man: The Hidden Life of Paul Dirac, Quantum Genius*. Faber and Faber Ltd.

Feynman, R. P. 1948. Space-Time Approach to Non-Relativistic Quantum Mechanics. *Rev. Mod. Phys.*, **20**(2), 367–87.

Feynman, R. P. 1949. The Theory of Positrons. *Phys. Rev.*, **76**(6), 749–59.

Feynman, R. P. 1982. Simulating Physics with Computers. *Int. Journal. Theor. Phys.*, **21**(6/7), 467–88.

Feynman, R. P., & Hibbs, A. R. 1965. *Quantum Mechanics and Path Integrals*. McGraw-Hill, New York.

Fink, H., & Leschke, H. 2000. Is the Universe a Quantum System? *Found. Phys. Lett.*, **13**(4), 345–56.

Finkelstein, D. 1968. The space-time code. *International Centre for Theoretical Physics Preprints*, **IC/68/19**, 1–24.

FitzGerald, G. F. 1889. The Ether and the Earth's Atmosphere. *Science*, **13**, 390.

Fock, V. 1964. *The Theory of Space, Time and Gravitation*. Pergamon Press. Translated by N. Kemmer.

Foster, R. G., & Roenneberg, T. 2008. Human Responses to the Geophysical Daily, Annual and Lunar Cycles. *Curr. Biol.*, **18**(17), R784–R794.

Freeman, K., 1983. *Ancilla to PreSocratic Philosophers: A Complete Translation of the Fragments in Diels, Fragmente der Vorsokratiker*. Harvard University Press.

French, S., & Krause, D. 2002. Identity, Individuality and Modern Physics. *http://www.academia.edu/2678138/*.

Gagnon, P. 2012. Is the moon full? Just ask the LHC operators. *http://www.quantumdiaries.org/2012/06/07/*.

Galileo, G. 1632. *Dialogue on the Two Chief World Systems*. Translated by S. Drake, University of California Press, 1997.

Gasiorowicz, S. 1967. *Elementary Particle Physics*. John Wiley and Sons.

Ghez, A. M., Salim, S., Weinberg, N. N., Lu, J. R., Do, T., Dunn, J. K., Matthews, K., Morris, M., Yelda, S., Becklin, E. E., Kremenek, T., Milosavljevic, M., & Naiman, J. 2008. Measuring Distance and Properties of the Milky Way's Central Supermassive Black Hole with Stellar Orbits. *Astrophysics Journal*, **689**(2), 1044–62.

Glauber, R. J. 1963a. Photon Correlations. *Phys. Rev. Lett.*, **10**(3), 84–86.

Glauber, R. J. 1963b. The Quantum Theory of Optical Coherence. *Phys. Rev.*, **130**(6), 2529–39.

Gödel, K. 1949. An Example of a New Type of Cosmological Solutions of Einstein's Field Equations of Gravity. *Rev. Mod. Phys.*, **21**(3), 447–50.

Goleman, D. 1996. The Experience of Change: Interview with H. H. the Dalai Lama. *Tricycle*, 24(4).

Goldstein, H., Poole, C., & Safko, J. 2002. *Classical Mechanics*. Third edn. Addison-Wesley.

Gould, S. J. 1987. *Time's Arrow, Time's Cycle: Myth and Metaphor in the Discovery of Geological Time*. Harvard University Press.

Gregory, A. 2013 *Ancient Greek Cosmogony*. Bristol Classical Press.

Grøn, Ø. 2004. *Relativity in Rotating Frames*. Springer.

Grøn, O. & Johannesen, S. 2010. A spacetime with closed timelike geodesics everywhere. *Nuovo Cimento B*, **125**(10), 1215–21.

Gross, R. S. & Chao, B. F. 2006. The Rotational and Gravitational Signature of the December 26, 2004 Sumatran Earthquake. *Surv. Geophys.*, **27**(6), 615–32.

Gupta, S. 1950. Theory of Longitudinal Photons in Quantum Electrodynamics. *Proc. Phys. Soc. A*, **63A**(7).

Hafele, J. C. & Keating, R. E. 1972a. Around-the-World Atomic Clocks: Observed Relativistic Time Gains. *Science*, **177**, 168–70.

Hafele, J. C. & Keating, R. E. 1972b. Around-the-World Atomic Clocks: Predicted Relativistic Time Gains. *Science*, **177**, 166–8.

Halligan, Peter, & Oakley, David. 2000. Greatest Myth of All. New Scientist, 18 November, 34–39.

Hamermesh, M. 1962. *Group Theory and its Applications to Physical Problems*. Addison Wesley Publishing Company, Inc.

Hamilton, W. R. 1837. Theory of Conjugate Functions, or Algebraic Couples; with a Preliminary and Elementary Essay on Algebra as the Science of Pure Time. *Trans. Roy. Irish Acad.*, **17**, 293–422.

Harding, K. 1993. Causality Then and Now: Al-Ghazali and Quantum Theory. *American Journal of Social Sciences*, **10**(2), 165–77.

Hartle, J. B., & Hawking, S. W. 1983. Wave Function of the Universe. *Phys. Rev.*, **D28**, 2960–75.

Hawking, S. W. 1976. Black Holes and Thermodynamics. *Phys. Rev.*, 101–97.

Hawking, S. W., & Ellis, G. F. R. 1973. *The Large Scale Structure of Space and Time*. Cambridge Monographs on Mathematical Physics. Cambridge University Press.

Hawking, S. W. & Turok, N. 1998. Open Inflation without False Vacua. *Phys. Lett. B*, **425**, 25–32.

Hecht S., Schlaer S., & Pirenne, M.H. 1942. Energy, Quanta and Vision. *J. Gen. Physiol.* 25(6), 819–840.

Heisenberg, W. 1927. Uber den Anschaulichen Inhalt der Quanten Theoretischen Kinematik und Mechanik. *Zeits. Physik*, **43**, 172–198. Reprinted [The physical content of quantum kinematics and mechanics in *Quantum Theory of Measurement*, ed. by J. A. Wheeler and W. H. Zurek, Princeton University Press, N. J. (1983)].

Heisenberg, W. 1930. *The Physical Principles of the Quantum Theory*. Dover Edition, 1949 edn. University of Chicago Press.

Heisenberg, W. 1952. Questions of Principle in Modern Physics. *In Philosophic Problems in Nuclear Science*. Faber and Faber, London.

Hewish, A., Bell, S. J., Pilkington, J. D. H., Scott, P. F., & Collins, R. A. 1968. Observation of a Rapidly Pulsating Radio Source. *Nature*, **217**, 709–13.

Hilbert, D. 1915. Die Grundlahen der Physik. *Konigl. Gesell. d. Wiss. Göttingen, Nachr. Math.-Phys.*, **3**, 395–407.

Hobson, M. P., Efstathiou, G., & Lasenby, A. N. 2006. *General Relativity*. Cambridge University Press.

Hodges, A. 2014. *Alan Turing: The Enigma*. Vintage Books, London.

Howson, A. G. 1972. *A Handbook of Terms Used in Algebra and Analysis*. CUP.

Hubble, E. 1929. A Relation Between Distance and Radial Velocity Among Extra-Galactic Nebulae. *Proc. Nat. Acad. Sci. USA*, **15**, 168–73.

Hughes, V. W., Robinson, H. G., & Beltran-Lopez, V. 1960. Upper Limit for the Anisotropy of Inertial Mass from Nuclear Resonance Experiments. *Phys. Rev. Lett.*, **4**(7), 342–44.

Hume, D. 1739. *A Treatise of Human Nature: Being An Attempt to Introduce the Experimental Method of Reasoning Into Moral Subjects*. Clarendon Press, Oxford 1896.

Hume, D. 1748. *An Enquiry Concerning Human Understanding*. Clarendon Press, Oxford, 2000.

Ikeda, M. & Maeda, S. 1978. On symmetries in a discrete model of mechanical systems. *Math. Japonica*, **23**(2), 231–44.

International des Poids et Mesures, Bureau. 2006. *The International System of Units (SI)*. Eighth edn. Organisation Intergouvernementale de la Convention du Mètre.

Itano, W. M., Heinzen, D. J., Bollinger, J. J., & Wineland, D. J. 1990. Quantum Zeno Effect. *Phys. Rev. A*, **41**(5), 2295–300.

Ives, H. E. & Stilwell, G. R. 1938. An Experimental Study of the Rate of a Moving Atomic Clock. *J. Optical Soc. America*, **28**(7), 215–19.

Jaroszkiewicz, G. 2008a. Quantized Detector Networks: A Review of Recent Developments. *Int. J. Mod. Phys. B*, **22**(3), 123–88.

Jaroszkiewicz, G. 2008b. Quantized Detector Networks, Particle Decays and the Quantum Zeno Effect. *J. Phys. A: Math. Theor.*, **41**(9), 095301.

Jaroszkiewicz, G. 2010. Towards a Dynamical Theory of Observation. *Proc. Roy. Soc. A*, **466**(2124), 3715–39.

Jaroszkiewicz, G. 2014. *Principles of Discrete Time Mechanics*. Cambridge University Press.

Jaroszkiewicz, G. A. & Strange, J. H. 1985. Motion on Inequivalent Lattice Sites: NMR Theory and Application to LaF_3. *J. Phy. C*, **18**(11), 2331–49.

Jeans, J. H. 1905. On the Laws of Radiation. *Proc. Roy. Soc. A*, **76**(513), 545–52.

Joseph, R. 2011. Evolution of Paleolithic Cosmology and Spiritual Consciousness, and the Temporal and Frontal Lobes. *Journal of Cosmology*. **14**, 4400–40.

Kastner, R. E. 2013. *The Transactional Interpretation of Quantum Mechanics*. Cambridge University Press.

Kemmer, N. 1971. Advanced Electromagnetic Theory. *Unpublished Edinburgh undergraduate lecture notes taken by G. Jaroszkiewicz*, 1–55.

Kennard, E. H. 1927. Zur Quantenmechanik einfacher Bewegungstypen. *Zeits. f. Physik*, **44**(4-5), 326–52.

Kleiber, M. 1932. Body Size and Metabolism. *Hligardia*, **6**(11), 315–53.

Kochen, S. & Specker, E. 1967. The Problem of Hidden Variables in Quantum Mechanics. *J. Math. Mech.*, **17**, 59–87.

Korzybski, Alfred. 1994. *Science and Sanity: An Introduction to Non-Aristotelian Systems and General Semantics*. Fifth edn. Institute of General Semantics.

Kraus, K. 1983. *States, Effects, and Operations*. Lecture Notes in Physics (190). Springer-Verlag, Berlin, Heidelberg, New York, Tokyo.

Kronz, F. M. 1997. Theory and Experience of Time: Philosophical Aspects. *In*: Atmanspacher, H. & Ruhnau, E. (eds) *Time, Temporality, Now*. Springer.

Lampa, A. 1924. Wie erscheint nach der Relativitätstheorie ein bewegter Stab einem ruhenden Beobachter? *Z. Phys.*, **27**(1), 138–48.

Larmor, J. J. 1897. On a Dynamical Theory of the Electric and Luminiferous Medium, Part 3, Relations with Material Media. *Phil. Trans. Roy. Soc.*, **190**, 205–300.

Lee, T. D. 1983. Can time be a discrete dynamical variable? *Phys. Lett.*, **122B**(3,4), 217–20.

Lee, T. D. & Yang, C. N. 1956. Question of Parity Conservation in Weak Interactions. *Phys. Rev.*, **104**, 254–58.

Leech. 1965. *Classical Mechanics*. Methuen and Co. Ltd.

Leggett, A. J. & Garg, A. 1985. Quantum Mechanics Versus Macroscopic Realism: Is the Flux There When Nobody Looks? *Phys. Rev. Lett.*, **54**(9), 857–60.

Lehmann, H., Symanzik, K., & Zimmermann, W. 1955. Zur Formulierung Quantisierter Feldtheorien. *Il Nuovo Cimento*, **1**(1), 205–25.

Leibniz, G. W. 1714. *The Monadology*. Oxford University Press.

Leibniz, G. W. 1717. *Samuel Clarke, A Collection of Papers, Which passed between the late Learned Mr. Leibnitz, and Dr. Clarke, In the Years 1715 and 1716* (London: 1717). The Newton Project (online, 2006).

Lemaître, G. 1927. Un Univers homogène de masse constante et de rayon croissant rendant compte de la vitesse radiale des nébuleuses extra-galactiques. *Annales de la Société Scientifique de Bruxelles*, **A47**, 49–59.

Lewis, G. N. 1926. The Conservation of Photons. *Nature*, **118**, 874–5.

Lüders, G. 1951. Über die Zustandsänderung durch den Meßprozeß. [Regarding the state-change due to the measurement process] *Ann. Physik*, **8**(6), 322–8.

Ludwig, G. 1983a. *Foundations of Quantum Mechanics I*. Springer, New York, Heidelberg, Berlin.

Ludwig, G. 1983b. *Foundations of Quantum Mechanics II*. Springer, New York, Heidelberg, Berlin.

Mach, E. 1866. Bemerkungen über Die Entwicklung der Raumvorstellungen. *Zeitschrift fur Philosophie und philosophische Kritik*, **49**, 227–32.

Maeda, S. 1980. Canonical Structure and Symmetries for Discrete Systems. *Math. Japonica*, **25**(4), 405–420.

Maeda, S. 1981. Extension of discrete Noether theorem. *Math. Japonica*, **26**(1), 85–90.

Maimionides, M. 1190. *The Guide for the Perplexed*. George Routledge, London, translated by M Freidlander (1904).

McAllister, R. W. & Hofstadter, Robert. 1956. Elastic Scattering of 188 MeV Electrons from Proton and the Alpha Particle. *Phys. Rev.*, **102**, 851–6.

McPherron, S. P., Alemseged, Z., Marean, C. W., Wynn, J. G., Reed, D., Geraads, D., Bobe, R., & Béarat, H. A. 2010. Evidence for stone-tool-assisted consumption of animal tissues before 3.39 million years ago at Dikika, Ethiopia. *Nature*, **466**, 857–60.

McTaggart, J. E. 1908. The Unreality of Time. *Mind*, **17**, 456–73.

Meschini, D. 2006. Planck-Scale Physics: Facts and Belief. *Found. Science*, 1233–1821.

Meschini, D. 2008. A Metageometric Enquiry Concerning Time, Space and Quantum Physics. Ph.D. thesis, Dept. Phys., Univ. Jyväskylä.

Michelson, A. A. & Morley, E. W. 1887. On the Relative Motion of the Earth and the Luminiferous Ether. *Am. J. Sci.*, **34**, 333–45.

Mill, J. S. 1882. *A System of Logic, Ratiocinative and Inductive*, Book III. Eighth edn, Harper & Brothers.

Minkel, J. R. 2002. If the Universe were a Computer. *Phys. Rev. Focus*, **9**, 27.

Minkowski, H. 1908. Space and Time. A translation of an Address delivered at the 80th Assembly of German Natural Scientists and Physicians, at Cologne, 21 September, 75–91. Reprinted in *The Principle of Relativity*, Dover Publications, Inc, a collection of original memoirs on the special and general theory of relativity, by H. A. Lorentz, A. Einstein, H. Minkowski, and H. Weyl.

Misner C.W., Thorne K.S., & Wheeler, J.A. 1973. *Gravitation*. W H Freeman and Company, NY. 21st Printing, 1998.

Misra, B. & Sudarshan, E. C. G. 1977. The Zeno's Paradox in Quantum Theory. *J. Math. Phy.*, **18**(4), 756–63.

Mlodinow, L. & Brun, T. A. 2014. On the Relation between the Psychological and Thermodynamic Arrow of Time. *Phys. Rev. E*, **89**, 052102.

Mott, N. 1929. The Wave Mechanics of Alpha-Ray Tracks. *Proc. Roy. Soc.*, **A126**, 79–84.

Muga J. G., Sala Mayato R., & Egusquiza, Í. L. (eds). 2008. *Time in Quantum Mechanics*. Second edn. Springer.

Nahin, P. J. 1999. *Time Machines: Time Travel in Physics, Metaphysics, and Science Fiction*. Springer-Verlag, New York, Inc.

Newton, I. 1687. *The Principia [Philosophiae Naturalis Principia Mathematica]*. New translation by I. B. Cohen and Anne Whitman, University of California Press (1999).

Nordström, G. 1913. Zur Theorie der Gravitation vom Standpunkt des Relativitätsprinzips. *Ann. d. Phys.(Leipzig)*, **42**, 533–54.

Novikov, I. D. 1998. *The River of Time*. Cambridge University Press.

Ohtomo, Y., Kakegawa, T, Ishida, A., Nagase, T., & Rosing, M. T. 2013. Evidence for biogenic graphite in early Archaean Isua metasedimentary rocks. *Nat. Geosci.*, **7**, 25–28.

Otte, M. 2009. The Paleolithic-Mesolithic Transition. *In*: Camps, M., & Chauhan, P. (eds), *Sourcebook of Paleolithic Transitions*. Springer Science, New York.

Page, D. N. 1995. Information Loss in Black Holes and/or Conscious Beings? *Heat Kernel Techniques and Quantum Gravity*. Texas A&M University Department of Mathematics, College Station, Texas.

Pásztor, E. 2011. Prehistoric Astronomers? Ancient Knowledge Created by Modern Myth. *J. Cosmology*, **14**.

Pauli, W. 1933. Die allgemeinen Prinzipien der Wellenmechanik. *Handbuch der Physik*, **24**. Part 1, 88–272.

Peierls, R. E. 1952. The commutation laws of relativistic field theory. *Proc. Roy. Soc. (London)*, **A214**, 143–57.

Penrose, R. 1990. *The Emperor's New Mind*. Oxford University Press.

Penzias, A. A. & Wilson, R. W. 1967. Isotropy of Cosmic Background Radiation at 4080 Megahertz. *Science*, **156**, 1100–01.

Peres, A. 1993. *Quantum Theory: Concepts and Methods*. Kluwer Academic Publishers.

Peres, A. 2000a. Classical Interventions in Quantum Systems. I. The Measuring Process. *Phys. Rev. A*, **61**, 022116 1–9.

Peres, A. 2000b. Classical Interventions in Quantum Systems. II. Relativistic Invariance. *Phys. Rev. A*, **61**, 022117 1–8.

Petigura, E. A., Howard, A. W., & Marcy, G. W. 2013. Prevalence of Earth-size Planets Orbiting Sun-like stars. *PNAS*, **110**(48), 19175–6.

Petkov, V. 2012. Introduction The not-fully-appreciated Minkowski, *In*: *Space and Time: Minkowski's Papers on Relativity*. Minkowski Institute Press. pages 1–37.

Pinney, E. 1958. *Ordinary Difference-Differential Equations*. University of California Press, Berkeley 4, California.

Planck, M. 1900. On an Improvement of Wein's Equation for the Spectrum. *Verhandl. Dtsch. Phys. Ges.*, **2**, 202–4.

Planck, M. 1901. On the Law of Distribution of Energy in the Normal Spectrum. *Annalen der Physik*, **309**(3), 553–63.

Preskill, J. 1992. Do Black Holes Destroy Information? In: Kalara, S & Nanopoulos, D. V. (eds) *International Symposium on Black Holes, Membranes, Wormholes, and Superstrings.* World Scientific, p116.

Price, H. 1997. *Time's Arrow.* Oxford University Press.

Ramsey, A. M. 1925. The speed of the Roman Imperial Post. *JRS*, **15**, 60–74.

Rayleigh, J. W. S. 1900. Remarks Upon the Law of Complete Radiation. *Phil. Mag.*, **49**, 539–40.

Reichenbach, H. 1958. *The Philosophy of Space and Time.* New York Dover Publications.

Rembieliński, J. 1980. The Relativistic Ether Hypothesis. *Phys. Lett.*, **78A**(1), 33–36.

Ridout, D. P. & Sorkin, R. D. 2000. A Classical Sequential Growth Dynamics for Causal Sets. *Phys. Rev.*, **D61**, 024002.

Rindler, W. 1969. *Essential Relativity.* Van Nostrand Reinhold Company.

Robinson, A. 1966. *Non-standard Analysis.* North-Holland Publishing Co., Amsterdam.

Rossi, B. & Hall, D. B. 1941. Variation of the Rate of Decay of Mesotrons with Momentum. *Phys. Rev.*, **59**(3), 223–8.

Russell, B. 1967. *A History of Western Philosophy.* Simon and Schuster.

Salpeter, E. E. & Bethe, H. A. 1951. A Relativistic Equation for Bound-State Problems. *Phys. Rev.*, **84**(6), 1232–42.

Scarani, V., Tittel, W., Zbinden, H., & Gisin, N. 2000. The Speed of Quantum Information and the Preferred Frame: Analysis of Experimental Data. *Phys. Lett.*, **A276**, 1–7.

Schrödinger, E. 1926a. Der stetige Übergang von der Mikrozur Makromechanik. *Die Naturwissenschaften*, **14**(Jahrg. Heft 28), 664–6.

Schrödinger, E. 1926b. Quantisierung als Eigenwertproblem (Dritte Mitteilung). *Ann. Phys.*, **80**, 437–90.

Schrödinger, E. 1926c. Quantisierung als Eigenwertproblem (Erste Mitteilung). *Ann. Phys.*, **79**, 361–76.

Schrödinger, E. 1926d. Quantisierung als Eigenwertproblem (Vierte Mitteilung). *Ann. Phys.*, **81**, 109–39.

Schrödinger, E. 1926e. Quantisierung als Eigenwertproblem (Zweite Mitteilung). *Ann. Phys.*, **79**, 489–527.

Schrödinger, E. 1926f. Über das Verhältnis der Heisenberg-Born-Jordanschen Quantenmechanik zu der meinen. *Ann. Phys.*, **79**, 734–56.

Schrödinger, E. 1926. An Undulatory Theory of the Mechanics of Atoms and Molecules. *Phys. Rev.*, **28**(6), 1049–70.

Schutz, B. 1980. *Geometrical Methods of Mathematical Physics.* Cambridge University Press.

Schwarzschild, K. 1916. Über das Gravitationsfeld eines Massenpunktes nach der Einsteinschen Theorie. *Sitzungsberichte der Königlich Preussischen Akademie der Wissenschaften*, **7**, 189–96.

Schwinger, J. 1958. Spin, Statistics and the TCP theorem. *Proc. N. A. S.*, **44**, 223–8.

Schwinger, J. 1969. *Particles and Sources.* Gordon and Breach.

Smith, J. D. H. 2002. Time in Biology and Physics. *In:* R. Buccheri, M. Saniga, & Stuckey, W. M. (eds), *The Nature of Time: Geometry, Physics and Perception.* Nato Science Series. II. Mathematics, Physics and Chemistry, Vol. 95. Kluwer Academic Publishers.

Snyder, H. S. 1947a. The Electromagnetic Field in Quantized Space-Time. *Phys. Rev.*, **72**(1), 68–71.

Snyder, H. S. 1947b. Quantized Space-Time. *Phys. Rev.*, **71**(1), 38–41.

Som, M. M. & Raychaudhuri, A. K. 1968. Cylindrically symmetric charged dust distributions in rigid rotation in general relativity. *Proc. Roy. Soc. London A*, **304**(1476), 81–86.

Stoppard, T. 1993. *Arcadia*. Faber and Faber, London, Boston.

Stehling, M., Chapman, B., Glover, P., Ordidge, R. J., Mansfield, P., Dutka, D., Howseman, A., Coxon, R., Turner, R., Jaroszkiewicz, E. M., Morris, G. K., Worthington, B.S., & Coupland, R. E. 1987. Real-time NMR Imaging of Coronary Vessels. *The Lancet*, **330**(8565), 964–5.

Stern, S. 2007. *Time and Processes in Ancient Judaism*. Littman.

Stuckey, M. 1999. Pregeometry and the Trans-Temporal Object. *In*: R. Buccheri, V. Di Gesu and M. Saniga (eds) *Studies of the Structure of Time: from Physics to Psycho(Patho)Logy*, 121–128. Kluwer Dordrecht.

Stueckelberg, E. C. G. 1941. Remarque À Propos de la Création de Paires de Particules En Théorie de la Relativité. *Helvetica Physica Acta*, **14**, 588–94.

Sudarshan, E. C. G. & Mukunda, N. 1983. *Classical Dynamics: A Modern Perspective*. Robert E Kreiger Publishing Company, Malabar, Florida.

Svensson, B. E. Y. 2013. Pedagogical review of quantum measurement theory with an emphasis on weak measurement. *Quanta*, **2**(Issue 1), 18–49.

't Hooft, G. & Vandoren, S. 2014. *Time in Powers of Ten*. World Scientific.

Tangherlini, F. R. 1961. An Introduction to the General Theory of Relativity. *Nuovo Cimento*, **20**(Supplement 1), 1–86.

Tangherlini, F. R. 1958. The Velocity of Light in Uniformly Moving Frame. Ph.D. thesis, Standford University.

Tapia, V. 1988. Second Order Field Theory And Nonstandard Lagrangians. *Nuovo Cimento*, **B101**, 183–96.

Tegmark, M. 1997. On the Dimensionality of Spacetime. *Class. Quantum Grav.*, **14**, L69–L75.

Tegmark, M. 2014. *Our Mathematical Universe: My Quest for the Ultimate Nature of Reality*. Alfred A. Knopf, New York.

Thorpe, S., Fize, D., & Marlot, C. 1996. Speed of processing in the human visual system. *Nature*, **381**(6582), 520–22.

Tifft, W. G. 1996. Three-dimensional quantized time in cosmology. *Astrophysics and Space Science*, **244**, 187–210.

Tolman, R. C. 1914. The principle of similitude. *Phys. Rev.*, **4**, 244–55.

Tropper, A. M. 1969. *Linear Algebra*. Thomas Nelson and Sons Ltd.

Ulansey, D. 1991. *Origins of the Mithraic Mysteries*. Oxford University Press.

Unger, R. M. & Smolin, L. 2015. *The Singular Universe and the Reality of Time*. Cambridge University Press.

Unruh, W. G. 1976. Notes on Black-Hole Evaporation. *Phys. Rev. D*, **14**(4), 870–92.

Ussher, J. 1658. *The Annals of the World*. Bennie Blount Ministries International (online).

v. Eötvös, R. 1890. Über die Anziehung der Erde auf verschiedene Substanzen. [On the Attraction of the Earth on Various Substances] *Mathematische und Naturwissenschaftliche Berichte aus Ungarn*, **8**, 65–8.

van Stockum, W. J. 1937. The Gravitational Field of a Distribution of Particles Rotating about an Axis of Symmetry. *Proc. Roy. Soc. Edinburgh*, **57**, 135–54.

von Neumann, J. 1955. *The Mathematical Foundations of Quantum Mechanics*. Princeton University Press.

Wang, M. S. 2002. Multitime measurements in quantum mechanics. *Phys. Rev. A*, **65**, 022103-1-6.

Wei, W., Xie, Z., Cooper, L. N., Seidel, G. M., & Maris, H. J. 2015. Study of Exotic Ions in Superfluid Helium and the Possible Fission of the Electron Wave Function. *J. Low Temp. Phys.*, **178**(1–2), 78–117.

Weinberg, J. R. 1942. The Fifth Letter of Nicholas of Autrecourt to Bernard of Arezzo. *J. Hist. Ideas*, **3**(2), 220–227.

Weiss, P. 1936. On the Quantization of a Theory Arising from a Variational Principle for Multiple Integrals with Application to Born's Electrodynamics. *Proc. Roy. Soc. Lond. A*, **156**, 192–220.

Wells, H. G. 1895. *The Time Machine*. William Heinemann.

Wheeler, J. A. 1979. From the Big Bang to the Big Crunch. *Cosmic Search Magazine*, **1**(4). Interview with J. A. Wheeler.

Wheeler, J. A. & Feynman, R. P. 1945. Interaction with the absorber as the mechanism of radiation. *Rev. Mod. Phys.*, **17**(2,3), 157–81.

Whitehead, A. N. 1929. *Process and Reality*. Cambridge University Press.

Whitrow, G. J. 1980. *The Natural Philosophy of Time*. Second edn. Clarendon Press, Oxford.

WMAP. 2013. Wilkinson Microwave Anisotropy Probe Empirical Results. *http://lambda. gsfc.nasa.gov/product/map/current/parameters.cfm*.

Woit, P. 2006. *Not Even Wrong: The Failure of String Theory and the Search for Unity in Physical Law*. Basic Books.

Wolfe, T. 1940. *You Can't Go Home Again*. Harper and Brow, New York and London.

Wu, C. S., Ambler, E., Hayward, R. W., Hoppes, D. D., & Hudson, R. P. 1957. Experimental Test of Parity Conservation in Beta Decay. *Phys. Rev.*, **105**(4), 1413–15.

Yamauchi, E. M. 1974. Easter, Myth, Hallucination, or History? I and II. *Christianity Today*, **18**(12, 13), 4–7, 12–16.

Zimecki, M. 2006. The lunar cycle: effects on human and animal behavior and physiology. *Postepy Higieny i Medycyny Doswiadczalnej*, **60**, 1–7.

Index